Ammonia Combustion Applications for Energy Systems

Other related titles:

You may also like

- PBPO0890 | Basile | Hydrogen Production, Separation and Purification for Energy | 2017
- PBPO1010 | Ting | Methane and Hydrogen for Energy Storage | 2016
- PBPO205 | Spazzafumo | Hydrogen Energy Systems | (in production 2025)

We also publish a wide range of books on the following topics:
Computing and Networks
Control, Robotics and Sensors
Electrical Regulations
Electromagnetics and Radar
Energy Engineering
Healthcare Technologies
History and Management of Technology
IET Codes and Guidance
Materials, Circuits and Devices
Model Forms
Nanomaterials and Nanotechnologies
Optics, Photonics and Lasers
Production, Design and Manufacturing
Security
Telecommunications
Transportation

All books are available in print via https://shop.theiet.org or as eBooks via our Digital Library https://digital-library.theiet.org.

IET ENERGY ENGINEERING SERIES 224

Ammonia Combustion Applications for Energy Systems

Edited by
Agustin Valera-Medina, Yuyang Li, Hao Shi,
Dongsheng Dong, Mara de Joannon and Daria Bellotti

The Institution of Engineering and Technology

About the IET

This book is published by the Institution of Engineering and Technology (The IET).

We inspire, inform and influence the global engineering community to engineer a better world. As a diverse home across engineering and technology, we share knowledge that helps make better sense of the world, to accelerate innovation and solve the global challenges that matter.

The IET is a not-for-profit organisation. The surplus we make from our books is used to support activities and products for the engineering community and promote the positive role of science, engineering and technology in the world. This includes education resources and outreach, scholarships and awards, events and courses, publications, professional development and mentoring, and advocacy to governments.

To discover more about the IET please visit https://www.theiet.org/.

About IET books

The IET publishes books across many engineering and technology disciplines. Our authors and editors offer fresh perspectives from universities and industry. Within our subject areas, we have several book series steered by editorial boards made up of leading subject experts.

We peer review each book at the proposal stage to ensure the quality and relevance of our publications.

Get involved

If you are interested in becoming an author, editor, series advisor, or peer reviewer please visit https://www.theiet.org/publishing/publishing-with-iet-books/ or contact author_support@theiet.org.

Discovering our electronic content

All of our books are available online via the IET's Digital Library. Our Digital Library is the home of technical documents, eBooks, conference publications, real-life case studies and journal articles. To find out more, please visit https://digital-library.theiet.org.

In collaboration with the United Nations and the International Publishers Association, the IET is a Signatory member of the Sustainable Development Goal (SDG) Publishers Compact. The Compact aims to accelerate progress to achieve the SDGs by 2030. Signatories aspire to develop sustainable practices and act as champions of the SDGs during the Decade of Action (2020–30), publishing books and journals that will help inform, develop, and inspire action in that direction.

In line with our sustainable goals, our UK printing partner has FSC accreditation, which is reducing our environmental impact to the planet. We use a print-on-demand model to further reduce our carbon footprint.

Published by The Institution of Engineering and Technology, London, United Kingdom

The Institution of Engineering and Technology (the "**Publisher**") is registered as a Charity in England & Wales (no. 211014) and Scotland (no. SC038698).

First published 2024

The Institution of Engineering and Technology
Futures Place,
Kings Way, Stevenage,
Herts, SG1 2UA
United Kingdom

www.theiet.org

British Library Cataloguing in Publication Data
A catalogue record for this product is available from the British Library

ISBN 978-1-83953-659-5 (hardback)
ISBN 978-1-83953-660-1 (PDF)

Typeset in India by MPS Limited

Cover images:

Flame 1/Agustin Valera-Medina

Evaporator/Agustin Valera-Medina

Burner/Agustin Valera-Medina

Collage by Agustin Valera-Medina

Contents

About the editors

Agustin Valera-Medina is a professor at Cardiff University, UK. Prior positions include researcher posts at the National Council of Science and Technology, and the Materials Science Institute of the National University of Mexico. His research focus is on combustion dynamics and thermofluids, in particular for energy applications. His awards include the Innovation Award and Sustainable Awards of the Southwales Energy Institute.

Yuyang Li is a professor at the Institute of Aerospace Propulsion at Shanghai Jiao Tong University, China. Earlier positions include a post at the State Key Laboratory of Fire Science. His research concentrates on reaction kinetics and flames for energy transformation in power machinery. He was awarded, among others, two Energy and Fuels Rising Star from Energy and Fuels, and a Research Excellence Award from the Combustion Institute.

Hao Shi is a research associate at Darmstadt TU, Germany. His research is centered on optical and laser diagnostics for combustion characteristics of fuels, including particle image velocimetry, laser-induced fluorescence, and tunable diode laser absorption spectroscopy, to investigate the spray, flame, and its interaction with the turbulent flow field, the wall, and the composition of the mixture and exhaust.

Dongsheng Dong is an assistant professor at Wuhan University of Technology, China. He currently works at Dalian University of Technology. His research focus includes pollutants formation, ammonia dual-fuel combustion, effects of hydrogen, multi-objective optimization of a methanol-diesel dual-fuel engine, and chemical reaction network analysis.

Mara de Joannon is a research director at the Istituto di Scienze e Tecnologie per l'Energia e la Mobilità Sostenibile of the Consiglio Nazionale delle Ricerche, Italy. Her scientific interests include the study of advanced combustion processes and the use of new energy carriers. She is a member of Board of Director (2020–2026) and a fellow of the Combustion Institute, and an editorial board member of ISI Journals *Combustion and Flame* and *Proceedings of the Combustion Institute*.

Daria Bellotti is a researcher with the Thermochemical Power Group (TPG) at the University of Genoa, Italy. She focuses on electric power, optimization and control problems, fuel cell systems, and innovative methods for the storage of energy from renewable sources.

Chapter 1
Introduction. Ammonia as a fuel

M.O. Vigueras-Zuniga[1], M. De Joannon[2], H. Shi[3] and A. Valera-Medina[4]

Ammonia has raised its profile to become a potential fuel for the near and long future in applications such as marine engines, boilers, furnaces and even large gas turbines. Although the chemical suffers some disadvantages compared with currently used fossil fuels in terms of flame speed and ignition delay time, it is also known that ammonia can be fully combusted under the right conditions, whilst also increasing its reactivity when either doped with other agents or reacted at high temperatures. These features make ammonia a good candidate not only to distribute hydrogen but also to use it directly as a fuel, hence reducing conversion inefficiencies and maximising the energy embedded in such a chemical. However, it is critical to raise awareness in the issues carried out when ammonia is employed for such purposes, thus providing potential users of a well-informed assessment of the possible use of ammonia as an energy vector. Along those lines, this chapter briefly addresses current case studies, national projects and international agendas that attempt to decode the best methods to use ammonia as a combustion fuel, aiming at providing evidence of ongoing efforts to maximise energy efficiency, stranded resources recovery and an entire mitigation of unfriendly emissions, all important parameters for the progression towards a net zero economy supported by ammonia combustion.

1.1 Introduction

The promise of a hydrogen economy, where no carbon dioxide (CO_2) would be produced, dates from the nineteenth century, a period when creative minds such as Jules Verne's would write,

[1]Engineering Faculty, Universidad Veracruzana, Mexico
[2]Instituto di Ricerche sulla Combustione, IRC-CNR, Italy
[3]Reactive Flows and Diagnostics, Technische Universitat Darmstadt, Germany
[4]Centre of Excellence on Ammonia Technologies, Cardiff University, UK

> And what will they burn instead of coal? asked Pencroft. 'Water', replied Harding. 'Water as fuel for steamers and engines! Water to heat water!' 'But water decomposed into its primitive elements'. Replied Cyrus Harding. I believe that water will one day be employed as fuel, that hydrogen and oxygen which constitute it will furnish an inexhaustible source of heat and light. Water will be the coal of the future [1]

However, nobody would be thinking along the context that another chemical, ammonia (NH_3), could take also a role in delivering energy for a great variety of applications. Actually, ammonia's history dates back to the eighteenth century, when notable chemists such as Joseph Black (Scotland), Peter Woulfe (Ireland), Carl Wilhelm Scheele (Sweden/Germany) and Joseph Priestley (England) isolated the molecule. It would be in 1785 when Claude Louis Berthollet would determine ammonia's elemental composition [2], hence opening the doors to a revolution with similar magnitude to the industrial revolution.

Around the beginning of the nineteenth century, a better understanding of nitrogen-based composites was taking place. As decades passed on, scientists recognised that nitrogen bounded in molecules such as ammonia was the needed component for crop growing processes. Along with potassium, phosphorus, water and sunlight, nitrogen is essential for plants. In earlier times, plants would consume the nitrogen, grow and eventually die, returning a large amount of this nitrogen to the soil. However, as the agricultural revolution occurred, less nitrogen would return to the soil, hence making imperative for farmers to find new methods to prevent crop yields declining. Manure and compost (rich in nitrogen), or crop rotation with support from legumes (with nitrogen replenishing bacteria) made the trick. But as nitrogen was increased, so was the yields obtained from the fields.

Knowing that ammonia contained nitrogen, the chemical started to be used as a fertilising compound. From the recovery of birds' excrement in Peruvian islands to guano from Chilean caves where large bat populations would coexist [3], the recovery of ammonia became a commercial venture. However, the increase in industrialisation and the rapid growth of large populated areas across Europe and America made critical the development of chemical processes to obtain ammonia. Since the 1860s, ammonia started to be obtained from the destructive distillation of coal. Ammonia liquor was recovered by sulfuric gas absorption that was employed after scrubbing coal gas with water. Ammonium sulphate, the resultant from the process, was then used as fertiliser [4]. It is estimated that approximately 140,000 tons/year of this chemical were produced around Europe by the end of the nineteenth century.

It was in 1889 that Ludwig Mond, a German–British chemist, discovered, during his search of a process to manufacture ammonium sulphate, that coal combustion could produce ammonia when the reaction takes place with air and steam. 'Mond' gas was produced at Brunner from 1902, which was followed by implementation at various other sites in Britain, Argentina, Spain and the US by 1903, with the estimated production of ~220,000 tons/year by that year [4,5]. Figure 1.1 shows a Mond Gas facility used in those years.

Figure 1.1 The Former South Staffordshire Mond Gas Company works [7]. Courtesy of the National Grid Archive, UK.

Fundamental catalysis programs were started for the development of both methods and materials to produce a great variety of chemicals, including ammonia, using less capital-intensive processes whilst increasing efficiency in the production of these compounds. German chemist Fritz Haber proposed the production of ammonia using hydrogen (H_2) and nitrogen (N_2). However, the yield was initially too low to be commercially viable. Haber, with his assistant Robert Le Rossignol, eventually developed high-pressure devices and catalysts needed to demonstrate the process at a laboratory scale, with a patented method to recycle unreacted gases. BASF acquired the patent to develop a commercial process. Various scientists, including Carl Bosch, Alwin Mittasch and other chemists, would test more than 2,500 catalysts and various conditions (including these catalysts at high temperature and pressure) to produce ammonia at commercial scales with relatively high conversion efficiencies. Great success was achieved in 1910, leading to one of the most important chemical breakthroughs in recent history worth a Nobel prize [6].

Ammonia producing plants convert fossil fuels such as propane, methane and coal into gaseous hydrocarbons through steam reforming processes (i.e. these fuels react with steam). The hydrogen produced is then combined with nitrogen extracted from the air via nitrogen separation, and then combined through the Haber–Bosch process. The process is carried out at 400–500 °C and a pressure of 10–30 MPa applied over an iron catalyst. This process is energy-intensive and yields low ammonia per pass, with a low conversion efficiency of 25%, which acts as a limitation. The unreacted H_2 and N_2 are recycled to the process to achieve increased conversion to ammonia [8].

This process has led to an increase in population in more than 40% of what otherwise would have been a natural growth. An estimated 7.8 billion people were

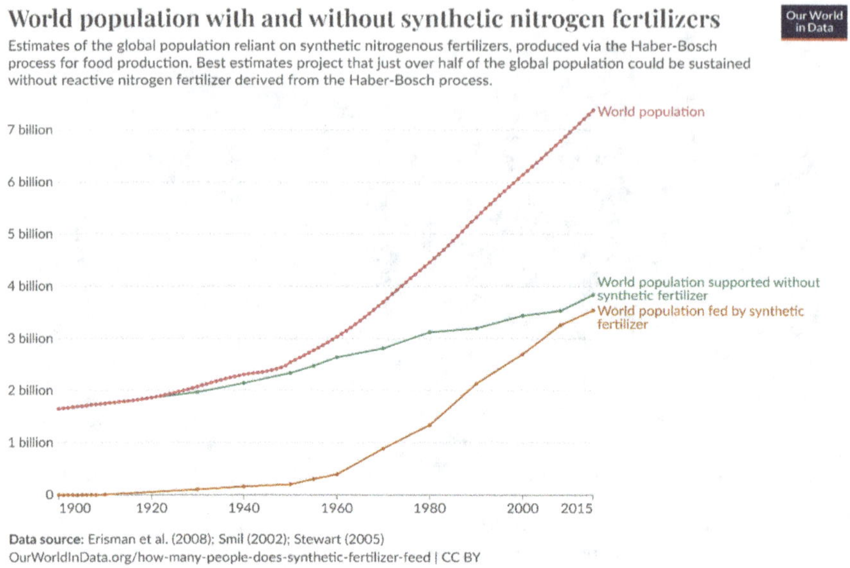

Figure 1.2 Population growth with and without synthetic nitrogen fertilisers. Reprinted, with permission, from [9].

estimated by the middle of the last decade, Figure 1.2, out of which ~3.5 billion were supported by synthetic nitrogen fertilisers, being ammonia the most important of all. However, nothing comes for free. This raise in chemical production has come with a critical payback in the form of carbon emissions, which are regularly dumped into the atmosphere. It is calculated that the global production of ammonia, the second most commercialised chemical, leads to the production of ~3% of all human-made carbon emissions. These quantities make the process extremely unfriendly towards the environment and a major promoter of climate change.

Being such an important chemical, and with the need to reduce its carbon intensity, further developments have been conceived to produce ammonia in a more environmentally friendly manner. From concepts that evaluate the sequestration and capture of carbon, through the use of renewable energy, to the production of ammonia from nuclear energy, several ideas have been presented for the production of blue, green or pink ammonia, respectively (Figure 1.3). These three types of ammonia have the potential to largely decarbonise the production of ammonia, whilst also making use of sources such as fossil fuels without carbon emissions (blue), recovery of stranded renewable sources (green) or maximisation of large energetic processes (pink).

Relevant to the production of green ammonia, it is expected that such an ammonia will be the leading source of nitrogen in the coming decades, hence increasing the recovery of renewables to augment the sustainability of ammonia production. Along those lines, many large ammonia producers are migrating to the use of electrolysis to produce their hydrogen and then covert it into ammonia via the Haber–Bosch process. This type of ammonia method falls into what has been

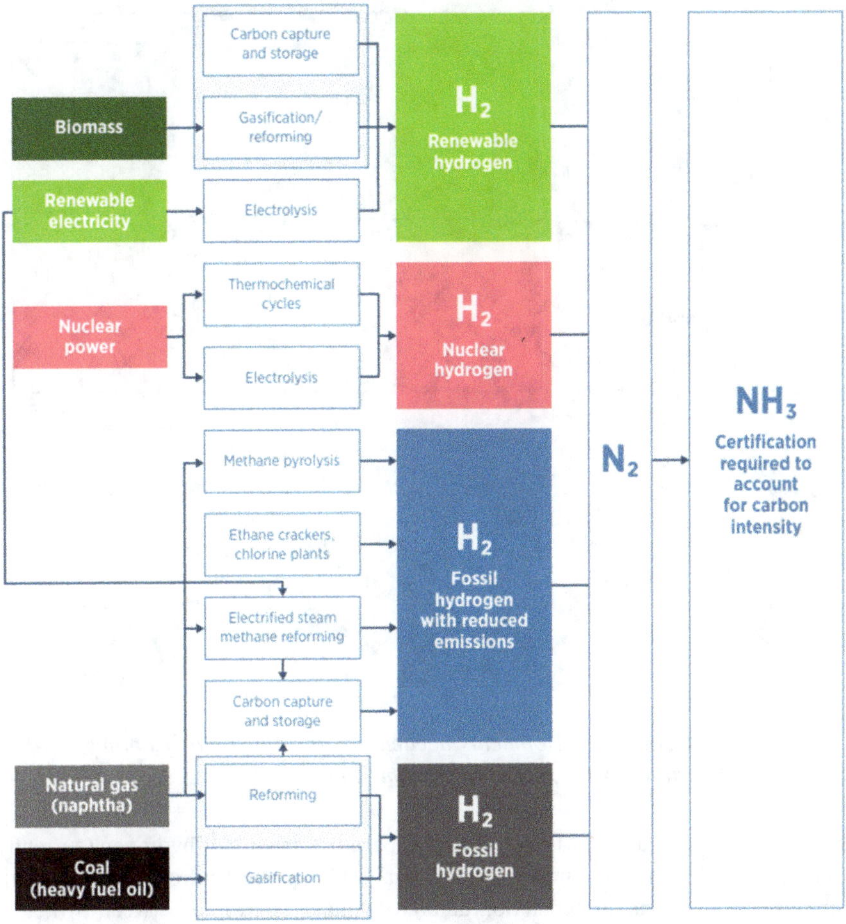

Figure 1.3 Different types of hydrogen and ammonia based on their source and carbon intensity. Reprinted, with permission, from [10].

categorised as 'Generation II', with 'Generation I' being the largely industrialised Haber–Bosch process that feeds from natural gas and steam methane reforming [11]. However, the technological progress is not stopping there. A new 'Generation III' is also appearing in the literature and some spin-out ventures. Generation III ammonia is the product of using water, air and renewable energy. All combined will deliver ammonia without the need of passing through the Haber–Bosch process (Figure 1.4).

A few companies such as NitroVolt [12] and Jupiter Ionics [13] are currently working on promising concepts with the potential of delivering this technology. Other institutes in Asia and the US are already developing their own technologies to deliver a solution capable of converting air and water into ammonia, thus providing limitless resources to the fertilising and refrigeration industries.

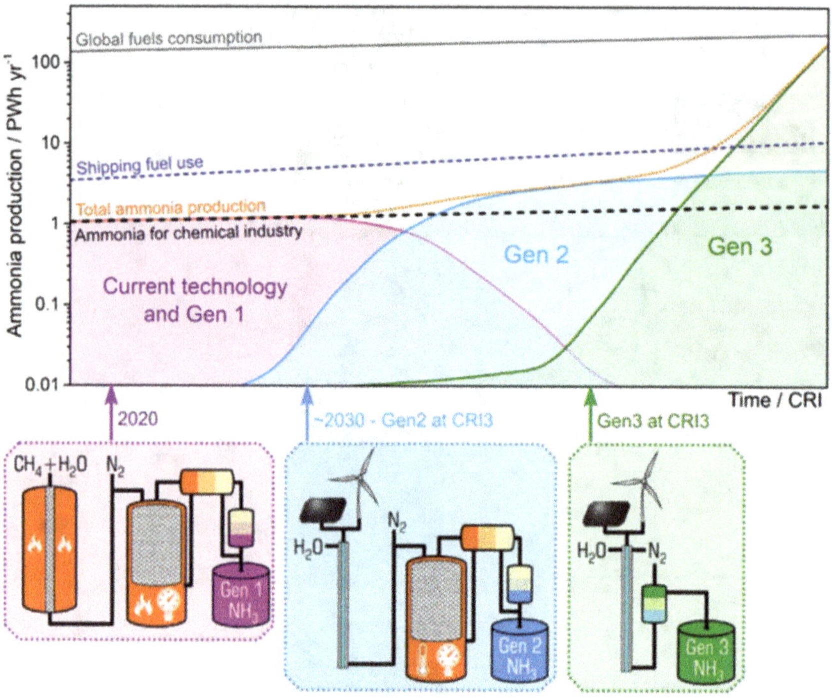

Figure 1.4 Generation developments needed for an 'Ammonia Economy'. Reprinted, with permission, from [11].

The principle of operation of these systems is based on the use of hydrogen (currently produced via electrolysis) and electrochemical cells that use nitrogen from the air (currently obtained via air separation units). A voltage is created between two electrodes. This allows molecules such as Li+ to surpass the energy threshold to break the nitrogen bonding, hence producing molecules such as Li_3N. Similarly, the hydrogen is split into its radical form, H+, and molecules such as ylide carry the hydrogen, which is then mixed with the Li_3N to produce ammonia whilst regenerating Li+. The concept, innovative and with large potential (as it still relies on external H_2 and N_2 producing units, which can potentially be coupled into fuel cell units to integrate the electrolytic step of hydrogen production), can be the first to allow the production of decentralised ammonia where renewable energies and water resources are available (Figure 1.5). Although the concept requires a lot of energy, steps towards the creation of new integrated systems and improvements in the electrolytes might soon deliver a technology capable of producing ammonia anywhere in the world.

Other processes for the production of ammonia currently in trend include the production of the chemical via nuclear power. However, different to conventional large power stations, various companies and institutions are working to develop small reactors of the size of a 40 ft container. These units operate with ultra-high purity molten salts which have the benefit of being highly efficient with a high safety

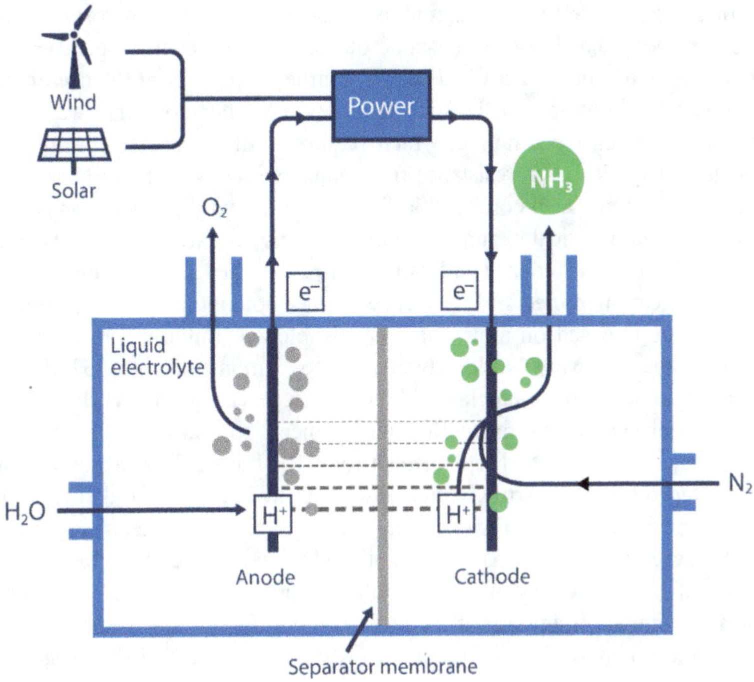

Figure 1.5 Production of ammonia using H_2 and N_2 without employing the H–B process. Reprinted, with permission, from [13].

record. According to developers of such a technology, leaks or air entrainment would lead to the cooling of the molten salts, stopping the reaction and avoiding a nuclear crisis. The concept, still to be demonstrated at large scale, is promising for the future of ammonia production, one of its critical paths towards commercialisation [14].

With such a promising technological and market scenario, current developments for the production of green ammonia are ongoing. It is estimated that the capacity of low-emission ammonia worldwide will reach ~293 million tons in the near future, with the conversion of facilities currently running, new developments and the announcement of new sites operating with advanced green technologies [15].

1.2 Ammonia for power

It is at this point where it can be said that ammonia can be produced from renewable energies or nuclear power generation at competitive costs, hence enabling the use of the molecule not only for fertilising or refrigeration applications (since ammonia is a fantastic refrigerant) but also for the storage of hydrogen and the use of the molecule for direct fuelling applications.

It is on the latter in which this book will focus. Ammonia has been conceived in the last 15 years as a potential fuel that can support the net zero agenda to decarbonise large-scale industries and a diversity of sectors across the world. However, ammonia has been in the scope of the scientific energy agenda since the middle of the 19th century. Sir Goldsworthy Gurney – pioneer of the use of oxygen and hydrogen for lighting and transportation devices – develop what was probably the first ammonia gas locomotive, which required little change from an ordinary steam engine [16]. However, resistance from others with vested interests opposed the new invention and he faced competition from steam railways, which was backed by large financial and political capital. Similarly, there were attempts to replace horse-drawn streetcars in New Orleans with ammonia-powered engines, although these lost out to electrification projects [17]. Hence it was not until the 20th century that transport solutions based on ammonia were developed. Initial advancements were produced by the company Nork, which tried to demonstrate in 1933 the use of ammonia for some of their vehicles [18]. However, it was during World War II that the lack of fuel availability made Belgian engineers to conceive a process where ammonia could be used as fuel replacement for ~100 buses. These accomplishments were followed by the US Army, which saw in ammonia a potential fuel for jeeps, helicopters and air transport. However, the features of ammonia (i.e. its low reactivity and lower energy density compared to fossil fuels) led to the end of these programs, which would be retaken by some institutions across the following decades but without major impact in the use of the molecule [19].

The development of the Viking engine (XLR-10) for the XLR-99 program gave life back to the use of ammonia as a fuel source. Although the program reached record velocities, complexities of using liquid oxyfuel-ammonia combustion and the sudden disinterest from the technical team working on the program would lead to an end of such a venture [20]. It was during the beginning of the 21st century that ammonia took a refreshed position partially thanks to the efforts of associations such as the Ammonia Energy Association, partially due to the requirements of a less carbon intensive molecule for the production of energy via renewable energy recovery. Ammonia would retake a place in the energy agenda by the 2010s, period when the Cross-Ministerial Strategic Innovation Program of Japan (led by Tokyo Gas) would re-instate their confidence in ammonia as a hydrogen molecule, demonstrating in a unique program the potential of using the molecule in power and heat production for medium to large industrial processes [21]. Nowadays, ammonia is considered a potential fuel for the future. In 2018, the International Energy Agency gave its support to ammonia as a fuel, appearing in presentations and white papers that would shift the trajectory of ammonia as an energy vector [22].

Developments and applications of ammonia can be found in a great variety of areas. From small systems used for cracking the molecule to produce hydrogen (small Bunsen burners that support the delivery of H_2 for fuel cells [23]) to large power plants currently running at GWs of power, with ammonia delivering MWs within combustion systems previously employed for the production of energy via coal combustion [24]. Relative to this book, the authors have focused on small- to medium-scale units capable of delivering power between 1 kW>P>1 MW.

Small burners ~3–45 kW have been tested and demonstrated for their utilisation in heat production in the ceramics industry in China [24]. Recent work performed at Foshan Xianhu Laboratory (Guangdong, China) has shown that ammonia is a promising substitute for natural gas in the industrial production of ceramic tiles [25]. Similarly, units operating at >30 kW have been employed for ammonia co-firing demonstration projects in the area of steel making for waste gas recuperation [26], micro gas turbines [27] or coal combustion [28]. In terms of internal combustion engines (ICE), fundamental units from small power ratios to medium power up to 1 MW are under commissioning at laboratory and industrial scale, respectively [29,30]. Spark ignition seems to be the preferred option when ammonia is employed as the feeding fuel, although compression ignition (CI), homogeneous charge compression ignition (HCCI) and stratified charge ignition (SCI) with other variants have also been attempted to mitigate emissions whilst increase power ratios. It is important to raise that ICEs can be used not only in automotive applications but also in marine propulsion (usually in two-stroke configurations), backup power production and heavy load transportation (i.e. trains). Thus, ammonia offers for these systems a unique advantage in terms of transportation and storage that other molecules such as hydrogen do not possess. Another important point to consider is that many of these applications also employ doping agents to enhance the properties of ammonia combustion (which are addressed in the next pages). The use of these chemical enhancers also contributes to the production of unwanted emissions, making a real case for R&D considerations and further support. A vast number of universities are currently working on the development of units capable of working with stable ammonia contents, whilst mitigating slip ammonia and unwanted NO_x emissions, i.e. the major emissions produced from ammonia combustion systems.

Another important area for practical implementation of ammonia combustion relates to large power generation and heat production in medium- to large-scale furnaces and boilers. The concept has been demonstrated for the production of steam, cement, glass, ceramic tiles and steel, all of them benefitting from the use of ammonia for the reduction of CO_2 and also NO_x when coal is replaced by such a fuel. Ammonia combustion in industrial furnaces has been conducted over the years by the company Duiker Combustion Engineers [19]. The company has based their technology in the development of systems capable of burning ammonia from oil and gas processes by following a stoichiometry-controlled oxidation strategy. Similarly, Numata *et al.* [26] also studied the use of ammonia in degreasing steel sheets using an impinging jet burner. Due to the challenges of the use of ammonia for these applications, an improved burner needs to be designed specifically for the use of ammonia-natural gas co-firing.

At much larger scales, the Mizushima Power Station was retrofitted to enable the use of ammonia for power generation in their 156 MW unit. Although 1 MW was produced using ammonia, the concept is now applied to replace 20% coal by ammonia at Hekinan demonstration project [31], with the intention of reaching 50% in the near future and 100% replacement by the end of the decade. South Korea and China are also working towards technologies capable of using ammonia in large power stations. Major Korean boilermakers and chemical companies as well as research institutes and universities are engaged in a project led by KEPCO and the Ministry of Industry, Trade and Energy (MOTIE) for the demonstration of hydrogen/ammonia as energy vectors. The 5-year demonstration project aims at 20% ammonia co-firing in various commercial-scale power plants (Shin Boryeong and Samcheok) [24]. China's vision for decarbonisation is also exploring ammonia as an alternative fuel. In January 2022, the Chinese government took a significant step towards promoting the development of new energy storage, releasing the '14th Five-Year Plan for New Energy Storage Development Implementation Plan' jointly with the National Development and Reform Commission (NDRC) and National Energy Administration (NEA). Notably, ammonia was included as an energy storage medium for the first time [24]. Recently, a 300 MW coal power plant in Anhui Province was co-fired with up to 35% ammonia at the Tongling Power Plant (Figure 1.7) [32]. The project was a collaboration between the Institute of Energy, the Hefei Comprehensive National Science Center and the An Hui Wenergy Company Ltd. An 8.3 MW pure ammonia burner was developed and tested in the coal boilers. The power plant operated at a thermal load of 100–300 MW with 10–35% ammonia additions, resulting in a maximum ammonia consumption rate of 21 tons per hour.

Developments of medium scale units are also happening worldwide. An example is the work from Cardiff University, FloGas and ERM, sponsored by the Department of Energy Security and Net Zero (DESNZ) in the UK [33]. The project

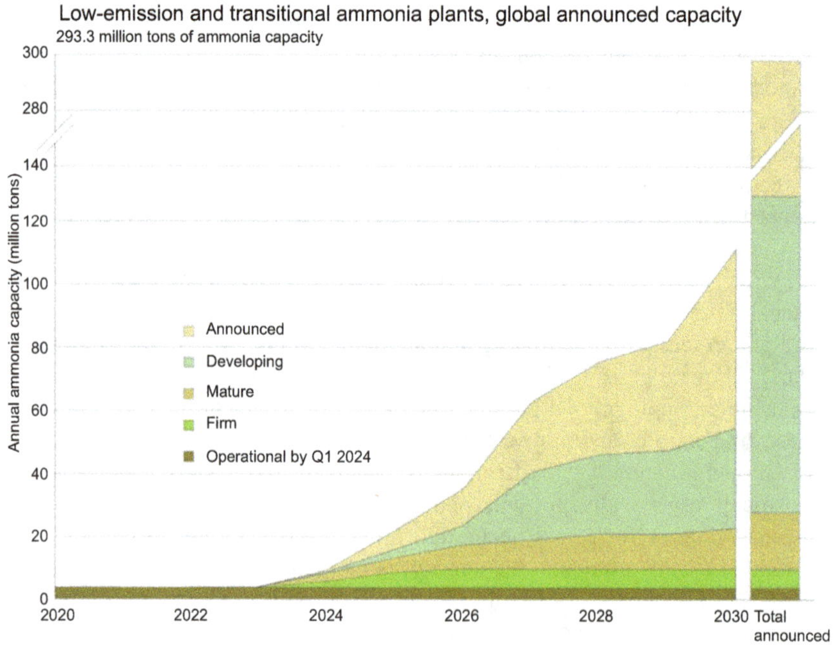

Figure 1.6 Low emission and transitional ammonia plants. Courtesy of the Ammonia Energy Association.

Figure 1.7 Tongling Power Plant, Anhui Province, China. Reprinted, with permission, from [32].

proposes to replace LPG by ammonia in rural locations. Ammonia-fuelled boilers remove the large energy requirement for cracking, reducing costs and carbon emissions compared to ammonia used to deliver hydrogen. With the phasing out of fossil fuels (including propane), the approach will enable the decentralisation of heat and power generation in small communities and medium farming and industrial sites.

Finally, gas turbines manufacturers are seriously working on the development of technologies capable of using NH_3. From the initial demonstration of the AIST 150 kW power unit to the onset of the IHI 2 MW gas turbine fuelled with ammonia, works have continued to develop a unit capable of running up to 40 MW power using only this chemical [34]. Mitsubishi has announced the potential commercialisation of such a system by the end of the decade, whilst companies such as GE, Siemens and Solar are also working towards the use of ammonia in their gas turbine systems. The concept can be supported by a great variety of studies, including those by Oxford University where the delivery of ammonia would be more suitable than hydrogen as a consequence of the reduced energy density of H_2. As an example, the amount of hydrogen for a small ~80 MW unit would require the storage of >3 salt caverns, whilst operating a larger ~870 MW system would need a ~2,500 km H_2 line of 1.4 m in diameter [35]. These quantities can be provided by approximately two large ammonia concrete-made conventional tanks of 50,000 tons each. These facts have led to the potential purchase and commissioning of two 40 MW power units in Japan and Ireland. As for the case in Ireland, the company Centrica is working towards these developments with the intension of having a pure ammonia unit at the Whitegate gas-fired power plant close to Cork, southwest Ireland, by the end of the decade (Figure 1.8) [36].

Interestingly, and going back to the Viking engine, the aerospace sector has regained interest in using ammonia for propulsion applications. A consortium led by the University of Central Florida brings together efforts from Georgia Tech,

Figure 1.8 Whitegate Combined Cycle Gas Turbine (CCGT) power station in Cork, Ireland. Reprinted, with permission, from [36].

Boeing and NASA to demonstrate the use of the chemical for aerospace. The program attempts to demonstrate the versatility of ammonia at large power rates, high-pressure ratios and aeronautical conditions, hence progressing on the use of NH_3 as a fuel for such a sector [37].

Along those lines, it is clear that ammonia has the potential to become a critical player in the decarbonisation process of many sectors and industries, ranging from small scale to large power generation. Although it is recognised that ammonia will not be the final solution to many areas (i.e. it is unlikely it will compete with automotive fuel cells for urban utilisation), it has been acknowledged that there are particular niche areas where ammonia is competitive and a feasible solution for the replacement of fossil fuels and energy storage. Therefore, it is likely that we will see ammonia-based energy solutions progressing at an accelerated pace over the following years. If that is the case, one might question – how good is actually ammonia compared to other fuels?

1.3 Ammonia characteristics

As discussed, ammonia still faces a long way before being entirely recognised as a fuel for power applications. Although technical barriers are overcome by continuous, high-quality research combined with advanced innovation, i.e. as the ones presented in previous sections, one of the main reasons of this trend is the fierce competition between ammonia and other fuels. Ammonia should not be regarded as a competitor to the hydrogen economy, but as an enabler. Therefore, ammonia can find its niche applications amongst some other fuels that are currently under evaluation. In order to recognise these applications, it is necessary to acknowledge some specific characteristics of ammonia to allow comparison with other hydrogen sources employed in power systems. Out of the vast range of alternatives, methanol and pure hydrogen were selected for a detailed analysis (Table 1.1). Methane is also mentioned as baseline. Furthermore, when it comes

Table 1.1 Key properties of ammonia and common fuels [38,39]

	Ammonia	Hydrogen	Methane	Propane	Methanol	Ethanol	Gasoline
Lower heating value (MJ/kg)	18.8	120.1	50	46.4	19.7	26.8	44.5
Maximum laminar burning velocity (m/s)	0.07	2.91	0.37	0.43	0.50	0.47	0.47
Flammability limit (Equivalence ratio)	0.63–1.4	0.10–7.1	0.50–1.7	0.51–2.5	0.55–2.9	0.66–2.4	0.7–4
Auto-ignition temperature (°C)	651	571	537	450	470	365	230
Minimum ignition energy (mJ)	8	0.011	0.28	0.25	0.14	0.28	0.8
Density (g/L, 25 °C, 1 atm)	0.703	0.082	0.657	493	787	789	740

to engines, ammonia tends to be compared also to diesel and gasoline. It is emphasised that compared to backup power units, which can operate using gaseous ammonia, automotive units might end up using liquid ammonia. It is believed that diesel or gasoline will have to be combined with ammonia as the technology progresses, whilst eventually ammonia/hydrogen or ammonia with other non-carbonaceous doping agents will become a substitute for fossil fuels in various applications. Further discussion and more details about current trends are presented in the following chapters.

It is clear that pure ammonia shows both low specific energy and laminar burning velocities combined with high auto-ignition temperatures and elevated ignition energies, making it more difficult to burn in its pure form. As a consequence, blends with hydrogen, which can be obtained relatively easy by cracking the ammonia molecule, have been mostly attempted in many practical applications. Moreover, the use of ammonia with gases such as methane, i.e. that shares similar density, viscosity and heat capacity, makes its implementation in co-firing applications relatively easy. Thus, ammonia has considerable potential to compete with other hydrogen-carrier fuels. However, before ammonia can be implemented in power devices, further understanding of its combustion behaviour is required, mainly by further research that not only demonstrates the concept but also details the chemistry of burning ammonia for enhancement and control of such a process.

1.4 Conclusions

Therefore, this book addresses the current applications where ammonia is being used as a combustion fuel. The chapters include recent advancements and ongoing research conducted to understand the use of the molecule and mitigate unwanted emissions whilst reducing the energy and efficiency costs of such an approach. From the versatility of using ammonia as a replacement fuel in fossil-based systems, through the complex methods to evaluate ammonia flames, to the intricate development of engines, turbines, boilers and furnaces, this book will explore how ammonia can be properly burned to achieve its deserved status within the net zero fuels of the future.

References

[1] Verne J. *The Mysterious Island* (translation by W.H.G. Kingston) Hetzel; 1874.

[2] American Chemical Society. *Ammonia.* 2021. Available: https://www.acs.org/molecule-of-the-week/archive/a/ammonia.html (accessed 21 January 2023).

[3] Johnston P. The Smithsonian and the 19th century guano trade: this poop is crap. *Smithson National Museum of American History*. 2017. Available: https://americanhistory.si.edu/explore/stories/smithsonian-and-19th-century-guano-trade-poop-crap (accessed 12 March 2024).

[4] Valera-Medina A, and Roldan A. Ammonia from steelworks. In Inamuddin, Boddula, R., and Asiri, A. (eds.), *Sustainable Ammonia Production.* Cham: Springer International Publishing; 2020, pp. 69–80.

[5] Wood RDD. *Mond Gas*. Philadelphia Co; 1903. Available: https://archive.org/stream/mondgas00woodrich#page/n5/mode/2up (accessed 19 June 2019).

[6] Pattabathula V, and Richardson J. Introduction to ammonia production. *Am Inst Chem Eng*. 2016. Available: https://www.aiche.org/resources/publications/cep/2016/september/introduction-ammonia-production.

[7] Thomas R. Gasworks profile D: produced gas plants. 2014. Available: https://claire.co.uk/component/phocadownload/category/27-wall-documents?download=433:gasworks-profile-d-producer-gas-plants.

[8] Berwal P, Kumar S, and Khandelwal B. A comprehensive review on synthesis, chemical kinetics, and practical application of ammonia as future fuel for combustion. *J Energy Inst*. 2021;99:273–98. https://doi.org/10.1016/j.joei.2021.10.001.

[9] Ritchie H. How many people does synthetic fertilizer feed? *OurWorldInDataOrg*. 2017. Available: https://ourworldindata.org/how-many-people-does-synthetic-fertilizer-feed (accessed 13 March 2024).

[10] IRENA and AEA. *Ammonia Pathways*. 2022. Available: https://www.irena.org/-/media/Files/IRENA/Agency/Publication/2022/May/IRENA_Innovation_Outlook_Ammonia_2022.pdf (accessed 02 April 2024).

[11] MacFarlane DR, Cherepanov PV, Choi J., *et al.* A roadmap to the ammonia economy. *Joule*. 2020;4(6)1186–205. https://doi.org/10.1016/j.joule.2020.04.004.

[12] Andersen SZ, and Saccoccio M. *NitroVolt*. 2024. Available: https://www.nitrovolt.com/ (accessed 24 May 2024).

[13] MacFarlane DR. *Jupiter Ionics*. 2024. Available: https://jupiterionics.com/ (accessed 25 May 2024).

[14] Copenhagen Atomics. Clean ammonia powered by thorium. 2023;40. Available: https://ammoniaenergy.org/wp-content/uploads/2023/09/Project-Features-speaker-slides-Sept-2023.pdf.

[15] Low-Emission Ammonia Data (LEAD). Ammonia Energy Association. 2024. Available: https://ammoniaenergy.org/lead/ (accessed 12 November 2024).

[16] Fullmer JZ. *The Life and Times of Sir Goldsworthy Gurney: Gentleman Scientist and Inventor, 1793–1875* by Dale H. Porter. *Isis*. 1999;90(1)137–8. https://doi.org/10.1086/384287.

[17] Orleans City, Experience New. *New Orleans Streetcars*. 2015. [Online]. Available: http://www.experienceeworleans.com/new-orleans-streetcars.html (accessed 24 October 2016).

[18] Mounaïm-Rousselle C, Bréquigny P, Valera Medina A, Boulet E, Emberso D, and Løvås T. Ammonia as fuel for transportation to mitigate zero carbon impact. In Kalghatgi, G., Agarwal, A.K., Leach, F., and Senecal, K. (eds.), *Engines and Fuels for Future Transport*. Singapore: Springer; 2022, pp. 257–279.

[19] Valera-Medina A, and Banares-Alcantara R. *Techno-Economic Challenges of Green Ammonia as an Energy Vector*. 1st edn. New York: Academic Press; 2020. https://doi.org/10.1016/c2019-0-01417-3.

[20] Seaman RW, and Huson G. The choice of NH3 to fuel the X-15 Rocket plane. *NH3 Assoc.* 2013. Available: https://nh3fuel.files.wordpress.com/2013/01/2011-seaman-huson.pdf (accessed 3 October 2019).

[21] Kobayashi H, Hayakawa A, Somarathne KDKA, and Okafor EC. Science and technology of ammonia combustion. *Proc Combust Inst.* 2019;37:109–33. https://doi.org/10.1016/j.proci.2018.09.029.

[22] Birol F. Hydrogen: accelerating & expanding deployment. *IEA.* 2018:7. Available: https://www.nedo.go.jp/content/100885441.pdf (accessed 10 September 2019).

[23] AFC Energy. 2023. Available: https://www.afcenergy.com/ (accessed 25 May 2024).

[24] Valera-Medina A, Vigueras-Zuniga MO, Shi H., *et al.* Ammonia combustion in furnaces: a review. *Int J Hydrogen Energy.* 2024;49:1597–618. https://doi.org/10.1016/j.ijhydene.2023.10.241.

[25] Eco China The world's first zero-carbon ammonia fuel fired green tile was baked in Foshan. 2022. Available: https://www.eco.gov.cn/news_info/60897.html (accessed 14 November 2023).

[26] Numata M, Matsuda T, Hagiwara Y, and Yamamoto Y. Development of impinging jet burner using ammonia fuel for degreasing steel sheets. *J Combust Soc Japan.* 2019;61:326–330.

[27] Okafor EC, Kurata O, Yamashita H., *et al.* Liquid ammonia spray combustion in two-stage micro gas turbine combustors at 0.25 MPa; relevance of combustion enhancement to flame stability and NO_x control. *Appl Energy Combust Sci.* 2021;7:100038. https://doi.org/10.1016/j.jaecs.2021.100038.

[28] Kimoto M, Yamamoto A, Ozawa Y, and Hara S. Ammonia co-firing technology applicable to pulverized coal fired boilers. *J Combust Soc Japan.* 2019;61:299–303.

[29] Lhuillier C, Brequigny P, Contino F., *et al.* Ammonia/hydrogen mixtures in an SI-engine: engine performance and analysis of a proposed fuel system. *Fuel.* 2019;90:854–64. https://doi.org/10.1016/j.pecs.2018.07.001.

[30] Zitouni S-E, Mashruk S, Mukundakumar N., *et al.* Ammonia blended fuels - energy solutions for a green future. *Gas Turbines a Carbon-Neutral Soc 10th Int Gas Turbine Conf,* 2021, p. IGTC21-62.

[31] Atchison J. JERA targets 50% ammonia-coal co-firing by 2030. Ashburn, VA: Ammonia Energy Association; 2022. Available: https://www.ammoniaenergy.org/articles/jera-targets-50-ammonia-coal-co-firing-by-2030/ (accessed 2 January 2023).

[32] Anhui Energy Group. A breakthrough in ammonia energy clean combustion power generation (in Chinese). Anhui State Enterp Netw; 2023. Available: http://xhs.anhuinews.com/ahgq1/snyjt/202304/t20230412_6790419.html (10 November 2023).

[33] ERM. *Amburn.* 2023. Available: https://amburn.co.uk/about/ammonia-as-a-low-carbon-fuel/ (10 February 2024).

[34] Alnaeli M, Alnajideen M, Navaratne R., *et al.* High-temperature materials for complex components in ammonia/hydrogen gas turbines: a critical review. *Energies.* 2023;16:6973. https://doi.org/10.3390/en16196973.

[35] Cesaro Z, and Banares R. Hydrogen requirements for two different sized gas turbines. *UK-India Ammon Conf,* 2020.

[36] Centrica. Centrica energy, Bord Gáis energy and Mitsubishi Power announce development of Europe's first ammonia fired power generation facility. 2023. Available: https://www.centrica.com/media-centre/news/2023/centrica-energy-bord-gais-energy-and-mitsubishi-power-announce-development-of-europes-first-ammonia-fired-power-generation-facility/ (accessed 14 May 2024).

[37] Brinkmann P. NASA, Boeing, UCF to study zero-carbon ammonia jet fuel. UPI 2022. Available: https://www.upi.com/Science_News/2022/01/28/NASA-Boeing-zero-carbon-ammonia-jet-fuel/5781643303801/ (10 August 2023).

[38] Chai WS, Bao Y, Jin P, Tang G, and Zhou L. A review on ammonia, ammonia-hydrogen and ammonia-methane fuels. *Renew Sustain Energy Rev.* 2021;147:111254. https://doi.org/10.1016/j.rser.2021.111254.

[39] Li J, Huang H, Kobayashi N, Wang C, and Yuan H. Numerical study on laminar burning velocity and ignition delay time of ammonia flame with hydrogen addition. *Energy.* 2017;126:796–809. https://doi.org/10.1016/j.energy.2017.03.085.

Chapter 2
Ammonia global distribution

C. Anfosso[1], S. Crosa[1] and D. Bellotti[1]

2.1 Introduction

Ammonia production is currently mostly centralized relying on Haber–Bosch (H–B) process [1,2], with China dominating the market [3]. During 2021–2023, the global production of ammonia was estimated at ~150 million metric tons, with China accounting for ~30% of the overall production (Figure 2.1); Russia, the USA and India also play an important role, with each producing more than 10 million metric tons per year (Mt/y). The European Union (EU) and the Middle East account for a further 8–10% each [3,4]. The major manufacturers in the market include CF Industries Holdings Inc. with nine plants with a capacity greater than 350,000 tons, owning also the largest ammonia production complex in the world (Donaldsonville (Louisiana) – 4,335,000 tons) [5], Yara with a capacity to produce 7.7 million tons of ammonia in 2022 with the target of reaching 8.6 million tons by 2025 [6], and Nutrien Ltd with 7 million tons of gross ammonia capacity [7]. This is followed by OCI-Nitrogen and Qatar Fertilizer Company with expected production of up to 6.4 and 3.8 million tons of ammonia per year, respectively [8,9]. Figure 2.2 reports the market share of ammonia [10] where it is underlined that, nowadays, 80% of ammonia produced is used in the agricultural sector for fertilizing purposes. Other applications are in the textile sector (fiber production, i.e. nylon, polyamide) or explosives (10% and 5%, respectively), pharmaceuticals sector, and refrigeration sector which cover together less than 5%. Ammonia can be used directly as fertilizer, or it can be used to produce derivatives such as ammonium nitrate and sulfate, ammonium hydrogen phosphate, and urea to avoid the environmental and health risks associated with pure ammonia toxicity [11].

In this scenario, since ammonia is one of the chemicals that dominate the market with several applications in different end-user sectors, to reach all users, the ammonia storage and distribution network in the world is solid and consolidated [12].

[1]Thermochemical Power Group (TPG), Department of Mechanical, Energetic, Management and Transport Engineering, University of Genoa, Italy

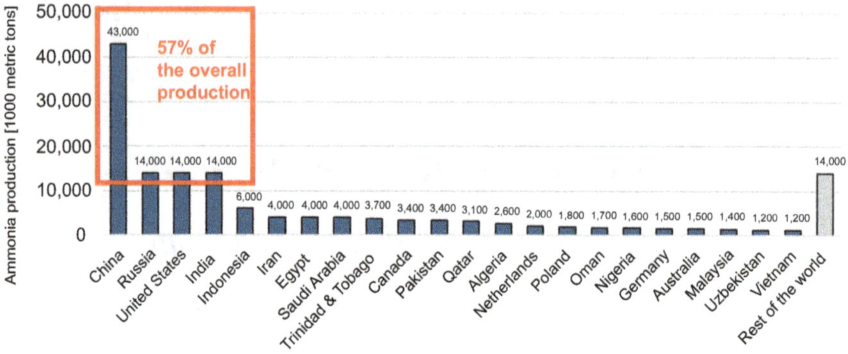

Figure 2.1 Ammonia production worldwide in 2023 by country (based on data from [3])

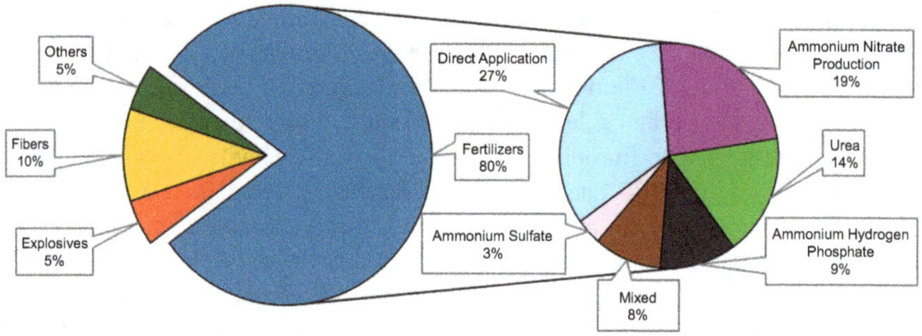

Figure 2.2 Ammonia market share (based on data from [10])

2.2 Ammonia storage

Several ways to store ammonia have been studied and experimented over the last few decades. The best option depends on the amount of ammonia and its intended final use. For large-scale applications, liquid ammonia is the primary solution for both storage and transportation. Conversely, for small-scale facilities, ammonia can be stored in solid materials such as zeolites and then transported [13,14]. Another possibility, mainly used to reduce the risk associated with ammonia toxicity, is to store it in chemical compounds (e.g. urea or ammonium nitrate) that can be easily and safely managed.

2.2.1 Liquid ammonia storage

Nowadays, liquid ammonia is the main solution for storage and transport procedures [13]. NH_3 can be liquefied in two ways: the first involves increasing the pressure for room temperature storage, while the second involves reducing the temperature for atmospheric pressure storage.

Three liquefied ammonia storage conditions are typically applied depending on the amount stored [15]:

- Pressurized storage at moderate pressure (16–25 barg) and ambient temperature for capacities up to 1,500 tons.
- Semi-refrigerated storage in insulated reduced-pressure spherical vessels at 0 °C for capacities up to 2,500 tons.
- Low-temperature/fully refrigerated storage in vertical tanks at 1.1–1.3 bar and −33 °C for capacities up to 50,000 tons.

In the case of pressurized storage, the containers could have a cylindrical shape or a spherical one according to the desired storage capacity. The cylindrical one, usually horizontal, is used for up to 150 tons, whereas the spherical container is used to store 250–1,500 tons [16]. Typically, cylindrical pressure vessels are designed to operate at 25 barg; instead, the spherical vessels are designed to operate only for about 16 barg. This is performed in order to avoid wall thicknesses above 30 mm. In hot weather, non-refrigerated pressurized storage containers are usually protected from overheating by an outer insulation [17]. This kind of system is especially suitable for storing small quantities of ammonia as intermediate storage between the ammonia plant and the final user, to balance the NH_3 production variations, for loading and unloading vessels carrying pressurized ammonia, and to reduce fuel fluctuation at the entrance and at the exit from pipeline systems [17].

Semi-refrigerated systems allow the storage of liquid ammonia at low pressure in an insulated container. In this tank, ammonia is stored at a moderately low temperature of around 0 °C with a pressure lower than 4 barg: in this way, it is possible to adopt much lighter steel compared to the one required for the pressurized tank. Semi-refrigerated storage vessels have usually a spherical shape and may have capacities of up to 2,500 tons [17]. The refrigeration system of these vessels implies compression of the ammonia vapor and subsequent liquefaction by the water cooling system before the liquid is returned to the tank. Therefore, the system is similar in principle to fully refrigerated storage, but much less sophisticated and relatively inexpensive [18].

The fully refrigerated tanks are the common way to store liquid ammonia, especially in large quantities. This kind of system is also known as atmospheric ammonia storage tanks. These containers have a cylindrical shape and operate at atmospheric pressure and −33 °C (ammonia boiling temperature) with a capacity of up to 50,000 tons [15,18]. These systems can be realized with single- or double-wall insulation. In single-wall tanks, the insulation consists mainly of panels of rock wool, foam glass, Styrofoam, and polyurethane foam, whereas the double-wall ones are insulated with dry gases such as nitrogen or air which flow in the interstitial space. However, tanks never have perfect insulation and so, the stored liquid ammonia can absorb heat because of the temperature difference between the tank and the surrounding. In this case, an NH_3 fraction can evaporate and create boil-off gases at the thermodynamics equilibrium with the stored liquid. To contain this risk, refrigeration is provided by a refrigeration unit, normally coupled with a two-stage ammonia compressor. During the procedure, the temperature is kept down by slow vaporization, and the ammonia vapor is continuously compressed back to a liquid [19].

2.2.2 Solid-state ammonia storage

The use of ammonia pressurized storage is gaining more attention due to the increase in small-scale ammonia plants, where the space and the possibility of installing large atmospheric tanks is limited. However, a non-negligible disadvantage related to conventional storage methods is the high ammonia vapor pressure (7–8 barg) at ambient temperature [20], which, in addition to its toxicity [11], raises safety concerns when storing liquid ammonia for end-user applications (NH_3 concentration of 300 ppm is the IDLH value-Immediately Dangerous to Life or Health [21]). Therefore, the use of a solid material such as zeolites as ammonia storage media has become a promising storage method for practical applications [21]. In this way, it is possible to increase the safety of the system, as NH_3 can only be desorbed when heated or depressurized, preventing leakage.

Among the materials today under investigation, metal halides (MX_m) seems a promising solution [22]. MX_m absorb NH_3 at ambient temperature, forming metal ammines $M(NH_3)_nX_m$, which are solid salts [23]. The chemical reaction where M is a metal (i.e., Li, Na, Mg, Ca, Mn, Co, Ni, Cu, Zn, Sr, Ba, Ag) and X is a halogen (i.e., Cl, Br, I) is reported in (2.1) [24]:

$$MX_m + NH_3 \leftrightarrow M(NH_3)_nX_m \tag{2.1}$$

In general, the adsorption heat increases with the metal electronegativity since NH_3 is adsorbed on halides directing the N atom to M one. This means that Li, Na, and Ca-based materials have a lower absolute value compared to Mg and Mn [25]. Referring to the storage capacity, the best performances have been reported for LiCl and $MgCl_2$ showing a gravimetric density of 62 wt% and 52 wt%, respectively (-34 and -56 kJ mol^{-1} as the adsorption heat) [24,26].

However, despite the good opportunity of metal halides to be an ammonia storage media, the absorption capacity is low (1–5% of the equilibrium capacity) and, moreover, MX_m are unstable under absorption conditions and after each loading/unloading cycle the breakthrough of the subsequent cycle reduces, making their use in commercial applications not feasible [22].

Table 2.1 summarizes the characteristics of the four reported types of storage methods.

Table 2.1 Ammonia tank storage methods [19]

Type	TRL	Typical pressure [Barg]	Temp [°C]	Capacity [ton of ammonia]
Pressure storage	9	16–25	20–25	< 270 or > 1,500
Semi-refrigerated storage	9	3–5	0	450–2,700
Low-temperature storage	9	1.1–1.2	−33	4,500–50,000
Solid-state storage	3–4	1–30	25–250	–

2.3 Ammonia distribution

Once storage has been secured, distribution of the ammonia is fundamental. As it is one of the largest chemical markets in the world, NH_3 is transported around the globe using different types of methods. Pipelines, ships, vessels of varying dimensions, trains, and trucks are employed according to the location and the amount of ammonia to be transported. However, all these methods have advantages and disadvantages that must be analyzed to fully understand the most common ways in which ammonia is distributed, reducing complexity and, in the meanwhile, ensuring the storage system's integrity.

Ammonia distribution infrastructures require different capacities including both large-scale transportation from producers to terminals and retailer-level transport. Figure 2.3 reports in 2019 the major NH_3 trade routes [27]. Millions of NH_3 tons are transported every year to the USA, EU, and Asia. The reported countries stand for 96% of exports and 95% of imports in international trade and the lines stand for trade size between 0.1 Mt and 2 Mt/y. The five largest exporters of anhydrous ammonia in the world held together almost 70% of the entire export value in 2019. Saudi Arabia leads with around 22%, followed by Russia (18%), Trinidad and Tobago (15.5%), and Indonesia and Canada which represent less than 10%. EU exports of ammonia account for 8% [28]. Moreover, Figure 2.3 underlines the importance of maritime ammonia trade; in fact, NH_3 terminals already exist in 200 harbors worldwide [4]. Pipelines and shipping are the preferential procedures for long distances followed by the rail system; truck transportation is suitable just for less than 150–200 km due to safety issues [11].

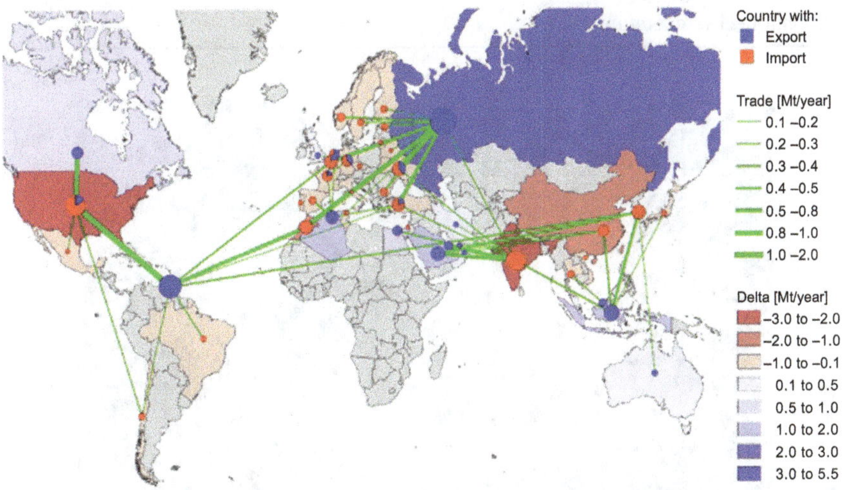

Figure 2.3 Global ammonia trade flows and balances larger than 0.1 Mt/y in 2019. Reprinted, with permission, from [27].

2.3.1 Transport by ship

Sea transport is the most suitable solution for long distances. In this scenario, around 17.5 Mt of ammonia is transported annually by ship [29], making it more convenient than pipelines for routes longer than 400 km [30].

Ammonia can be transported by three different ship types, depending on how the cargo is stored: fully refrigerated (FR), semi-refrigerated (SR), and pressurized (P). The choice of vessel depends on the amount of NH_3 cargo and the sailing distance. The most commonly used storage conditions in the maritime sector, according to the type of ammonia vessel, are reported in Table 2.2 [31,32].

Anhydrous ammonia is today typically transported in gas carriers designed for LPG as NH_3 shares similar properties to propane (i.e. saturated vapor pressure). In Figure 2.4, the tanker "AXIS RIVER" owned by Kawasaki Heavy Industries is reported; it is an 86,700 m^3 LPG and ammonia carrier powered by LPG [33].

In general, for maritime transport, the ship must be designed according to the requirements of the 2014 International Code for the Construction and Equipment of

Table 2.2 Ammonia tank storage condition [31,32]

	P	**SR**	**FR**
Temperature (°C)	25	−10	−40
Pressure (bar)	18	3–5	1.013
Density (kg/m^3)	$6.76 \cdot 10^2$	$6.10 \cdot 10^2$	$6.45 \cdot 10^2$
Δh_{vap} (kJ/kg)	$11.2 \cdot 10^2$	$12.1 \cdot 10^2$	$13.9 \cdot 10^2$
c_p (kJ/kgK)	2.08	2.05	2.02
U (W/m^{2}K)	0.32	0.32	0.32
Tank heat capacity (kJ/K)	25.2	25.2	25.2
Heel mass (% of the total mass)	5%	5%	5%
Ammonia heel temperature (°C)	17	−8	−38

Figure 2.4 Kawasaki Heavy Industries tanker for LPG and ammonia. Reprinted, with permission, from [33].

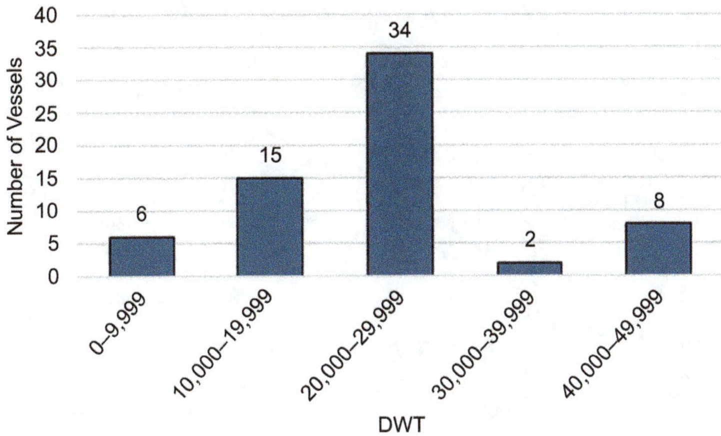

Figure 2.5 Global ammonia shipping fleet in 2021 in terms of Dead Weight Tonnage (DWT) (based on data from [34])

Ships Carrying Liquefied Gases in Bulk (IGC Code) transportation [34]. Ammonia has been carried as cargo aboard vessels for decades. These ships can carry upward of about 80,000 m^3 of ammonia at a time. When storing and transporting large volumes of ammonia, it is commonly stored as a refrigerated liquid aboard ships to keep the pressure low. These ammonia tanker vessels are thought to be likely early adopters of ammonia as a fuel given their familiarity with safe handling, and the opportunity to use marginal amounts of their cargo as a fuel (as LNG carriers commonly do) [35]. However, for smaller amounts of NH_3 to be transported, semi-refrigerated or fully pressurized methods are preferred as they eliminate the need for re-liquefication equipment and simplify the bunkering process [32]. Storing ammonia in pressurized form is also a less energy-intensive process. Figure 2.5 shows the number of existing vessels in 2021 according to their cargo capacity [34].

Usually, ships within the range of 20,000–50,000 m^3 are used for ammonia transportation, whereas larger ships are mostly employed for LPG transportation. In Russia, however, ships larger than 50,000 m^3 are also used to transport ammonia [36]. The ships are typically in the range of 15,000–85,000 m^3, with three common commercial sizes for LPG/ammonia trades of 30,000 m^3, 52,000 m^3, and 80,000 m^3 [37].

The cost of transport will depend on the financial arrangements of exporters and importers. The charter rates of energy carriers are highly volatile based on market conditions and energy demands (i.e. the COVID-19 pandemic caused the charter rate of LNG ships to fall from 80,000€ to 27,000€ per day). Even under less extreme economic conditions, the charter cost can be highly variable [38]. To establish the transport price of ammonia by ship, it is possible to compare the transport prices available in the literature as a function of the distance (km) to apply a linear regression (Figure 2.6). This study has been carried out by Salmon supposing that the ammonia itself is used as a fuel and therefore it costs around 450€

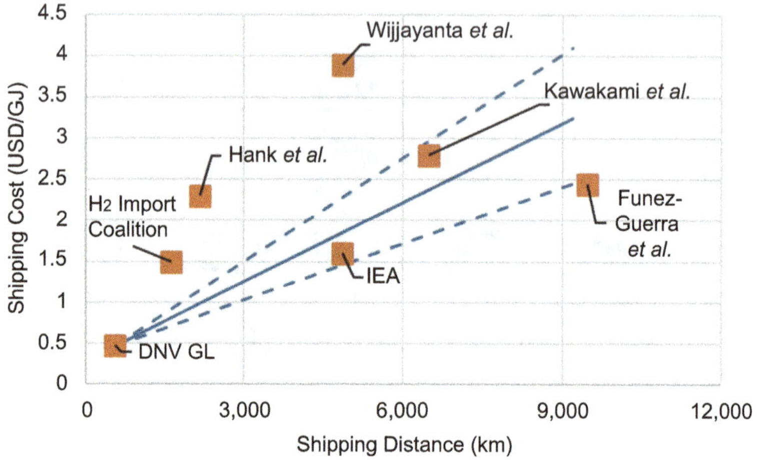

Figure 2.6 Cost of ammonia transport by ship. Reprinted, with permission, from [39].

(500 USD/tonNH_3) [39]; dashed lines in Figure 2.6 represent, instead, the linear regression in case of using a fuel price of 300 USD/tonNH_3 and 700 USD/tonNH_3.

2.3.2 *Transport by pipeline*

Pipeline use has several advantages, such as higher safety, a reduced carbon footprint, and good efficiency due to minimal leaks. Moreover, it is suitable for transporting large volumes over long distances, even though the cost is 4.5 times higher than that of shipping over a distance of 2,700 km [15].

Liquid ammonia is usually transported at −23 °C to 17 °C under a pressure of 2–20 bar, but in some cases, a pressure of 60 bar can be reached, allowing for higher working temperatures [15]. The diameter of the pipe usually ranges between 10 and 25 cm, although they are often oversized to account for possible sudden volume changes in the case of cryogenic ammonia transport [40]. A safety measure adopted in areas where the leakage risk is high is the use of double-walled pipelines filled with N_2. Additionally, water tanks are placed along the pipeline to remove excess ammonia from the system and ensure a pressure balance within the tubes. A small amount of water, at least 0.2 wt%, must be mixed with the liquid to prevent stress corrosion cracking, which can cause significant damage even under an oxygen concentration as low as 0.5 ppm [25]. Ammonia evaporation causes temperature changes, adding to the stress and chemical corrosive properties. For these reasons, steel is preferred over copper, zinc, and aluminum, which are frequently used in industrial process lines [40]. The same considerations apply to storage tanks. At the pipeline end, there is usually a refrigerated terminal that reduces the temperature to the conditions required for NH_3 storage.

However, today, there are still insufficient infrastructures, which are mainly located in the USA and Russia. The pipeline network in the USA has a diameter of 10–25 cm and it is constructed of plain carbon steel. It has a total length of approximately 3,200 km with regularly spaced pump stations. It goes from Louisiana to the Corn Belt and retail ammonia outlets are present almost in every State. (i.e. Iowa has 800 outlets); however, these exits tend to be concentrated primarily in the rural Midwest [41].

The ammonia pipeline in Russia originates in Togliatti but crosses also Ukraine ending in Odessa. It became operational in 1979 and nowadays pumps 2.5 million tons of ammonia annually. It has a total length of about 2,400 and the system is an automatically controlled system with several pumping stations coupled with an independent digital system for recording the operation parameters of the system [42,43].

Figures 2.7 and 2.8 report the ammonia network pipeline in the USA and Russia, respectively [41,42].

In the EU, the infrastructure differs significantly; there are only a few pipelines with relatively short routes (no longer than 100 km) but with a high overall annual capacity. Table 2.3 reports the main infrastructure all over the world.

2.3.3 Transport by rail and truck

Rail transport is widely used worldwide for both passengers and goods and does not require specific connections to pipelines or port terminals. For this reason, it is quite advantageous for NH_3 transport among land vehicles. Ammonia wagons are classified

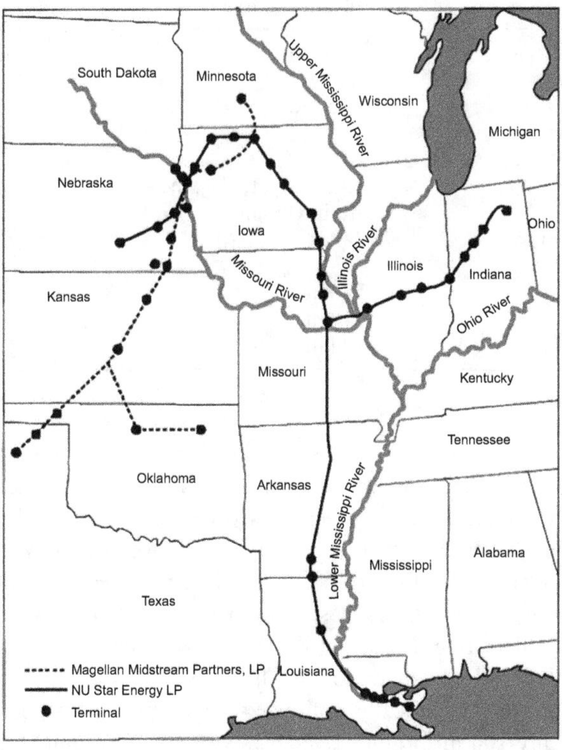

Figure 2.7 Ammonia pipeline network in the USA. Reprinted, with permission, from [41].

Figure 2.8 Ammonia pipeline infrastructure in Russia

Table 2.3 Ammonia pipeline examples in the world [15,40,41,42,44]

	Distance	Annual capacity
USA: from Louisiana to Corn Belt region	3,200 km	1.5 million tons
USA: from Texas to Minnesota (under dismission)	1,750 km	0.9 million tons
Russia: from Togliatti to Odessa	2,417 km (1,021 km are in Ukraine)	2.5 million tons
Italy: from Ferrara to Ravenna	74 km	0.3 million tons
Belgium	12 km	1.1 million tons
Germany	24 km	0.9 million tons
Spain	10 km	0.3 million tons

as tanks for liquefied gases and gases dissolved under pressure. NH_3 is typically stored at 15–16 bar, filling 57–85% of the volume to account for potential fuel expansion due to temperature variations. Utilizing technology similar to that used for LPG, these railway tank cars have a volume of 50–110 m^3, with a total capacity ranging from 25 tons to 90 tons, and are made from carbon steel with a thickness of about 1.75 cm [15].

In Europe, rail transport is the most commonly used method, with approximately 1.5 Mt of NH_3 transported annually [40]. VTG is currently the leading European company handling rail transportation, with a fleet of over 90,000 rail wagons, including four models specifically optimized for ammonia transport, each with a capacity of roughly 100 m^3. Figure 2.9 shows an example of ammonia rail tank car (left side) and an ammonia train (right side) [44].

Figure 2.9 Ammonia rail tank car (left side) and ammonia train (right side). Reprinted, with permission, from [44].

Figure 2.10 Cargo tanks owned by the Indian company BNH gas tanks. Reprinted, with permission, from [45].

Regarding truck transport, cargo tanks are commonly used for routes shorter than 150 km, as it is the most expensive transport method and poses safety issues. Therefore, truck routes are generally limited to short distances. In the USA, trailer trucks transport liquid ammonia at around 20 bar, holding 27 tons, while in Europe, a lower pressure of 10 bar is authorized for transporting 100 m^3 of liquid ammonia and 30 m^3 of aqueous ammonia. Less commonly, chassis-mounted bobtail tractors are used for smaller quantities, with a maximum capacity of 45 m^3 at 22 bar [15]. Regulations require that the tanks be equipped with unloading pumps and various accessories, including tires, mudguards, king pins (2 or 3.5 inches), safety fittings, valves, pipelines, excess flow valves, and internal valves [45]. Figure 2.10 shows a transportation truck for anhydrous ammonia [45].

2.4 Ammonia traditional production process

Traditional ammonia production systems (Figure 2.11) [28,46] synthesize NH_3 from steam-methane-reformed hydrogen and nitrogen, at about 730 K and 200 bar [46]. Natural gas (NG) is preheated before entering the desulphurization unit (DeS), in which H_2S and other sulfur-based compounds are removed from the gas stream to avoid poisoning of Ni-based catalyst used in the primary reformer (I REF). After this treatment, NG is partially converted in the primary reformer, with steam injection. The resulting gas is composed of the unreacted CH_4 (~34% [46]), H_2, CO, and CO_2. Then, compressed air is added to the gas stream in the secondary reformer (II REF) to allow methane complete conversion. To avoid carbon traces in the NH_3 synthesis reactor, CO is converted to CO_2 through high-temperature and low-temperature water gas shift reactions (HT and LT WGS), and the carbon dioxide is then removed using a solvent (CO_2 removal). Any remaining carbon traces are then removed by methanation. The resulting syngas (primarily H_2 and N_2) is then compressed and undergoes the ammonia synthesis process [46].

The traditional ammonia production is based on five existing patents, listed in Table 2.4 [46].

The first one (U.S.-1202995) is the ammonia production process patented by H–B [47], in which a recycle of the un-reacted N_2 and H_2 gases inside the ammonia reactor was introduced, separated by ammonia condensation, to increase the total ammonia globally synthesized (Figure 2.12) [52]. The H–B process requires reactant gas temperatures between 620 and 770 K and pressures between 150 and 300 bar coupled with the use of an iron catalyst inside the reactor [28].

In [48], Wright *et al.* introduce a compact ammonia synthesis system based on a pair of ammonia catalytic converters with a heat exchanger in between. In this

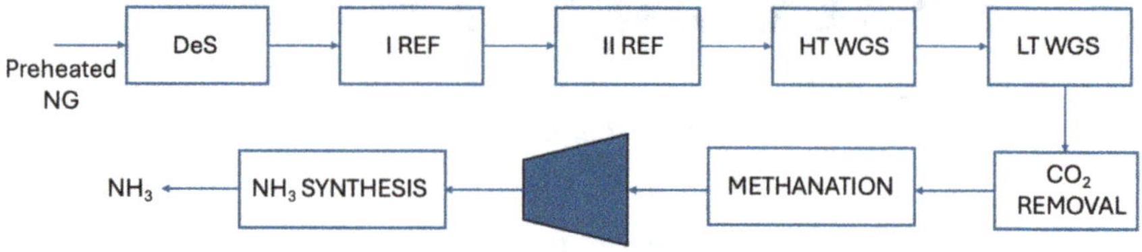

Figure 2.11 Traditional ammonia production systems (based on data from [28,46])

Table 2.4 Ammonia production patents [46]

Invention title	Patent number	Year
Production of ammonia	U.S.-1202995	1916 [47]
Ammonia synthesis system	U.S.-3721532	1973 [48]
Process for the ammonia production	U.S.-4107277	1976 [49]
Low-energy ammonia synthesis process	U.S.-4148866	1978 [50]
Preparation of ammonia synthesis	U.S.-4479925	1983 [51]

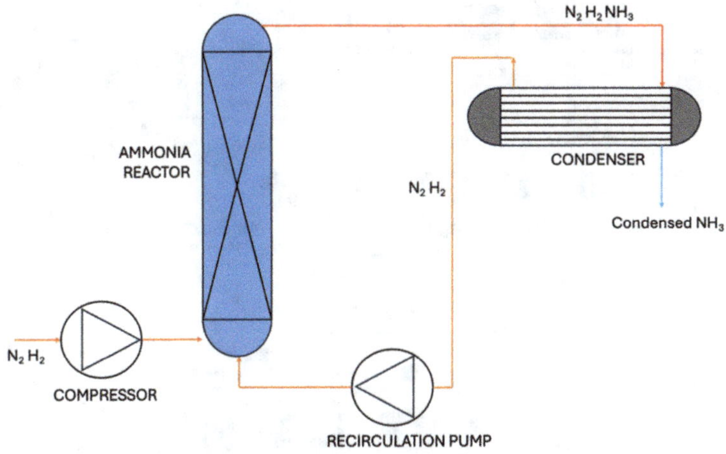

Figure 2.12 Haber–Bosch ammonia synthesis process

way, the fed gases are heated by the partially converted gas stream by the first converter. In the patent number U.S.-4107277, an innovative ammonia synthesis system based on H_2 production by a high-pressure electrolyzer is presented [49]: in that way, the use of the fed gas compressor is avoided, while the resulting oxygen as a by-product is used for refrigeration in the plant. This was the basis for NH_3 production without greenhouse gas (GHG) emissions. In 1978, Becker [50] patented an ammonia synthesis process at low pressure (below 100 atm), to be compliant with the pressures used in synthesis gas production from methane steam reforming. In that process, ammonia is recovered by a water absorption system exploiting heat from the WGS unit. The use of autothermal reformed synthesis gas in ammonia synthesis, instead, is investigated by Shires *et al.* [51].

Nowadays, traditional production plants synthesize ammonia starting from natural gas steam reforming using technologies derived from the original H–B process. The first important improvement to the original plant configuration was the Kellogg Brown and Root (KBR) one [53]. KBR introduces a centrifugal compressor for the synthesis of gas, an enhanced waste process heat recovery, and a steam generation with the waste heat from the primary reformer to be used in a steam turbine. Nowadays, the technology licensor of KBR, together with the one of Haldor Topsøe and ThyssenKrupp Industrial Solutions (TKIS), is governing the ammonia production plant market (Figure 2.13) [53]. Haldor Topsøe technology uses a side-fired reformer for syngas production, a proprietary iron catalyst, a radial-flow ammonia synthesis reactor, and a bayonet-tube waste heat boiler. TKIS, instead, introduces its own secondary reformer design, waste-heat boiler, and radial-flow reactor, together with the use of a dual-pressure ammonia synthesis configuration [53].

As a variant of the traditional system (Figure 2.11), the Linde Ammonia Concept (LAC™) [54] is a well-consolidated technology for ammonia synthesis from NG: it introduces a pressure swing absorption (PSA) and an air separation unit (ASU) to

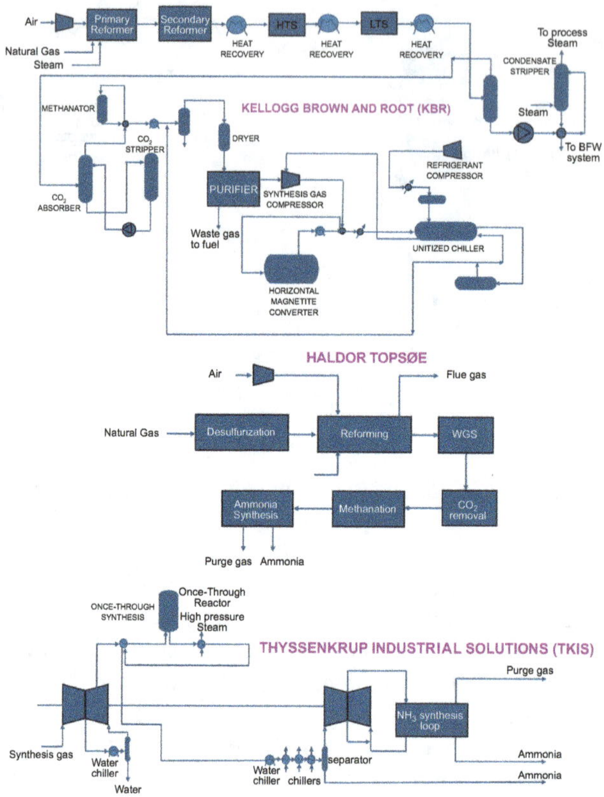

Figure 2.13 Ammonia production technologies

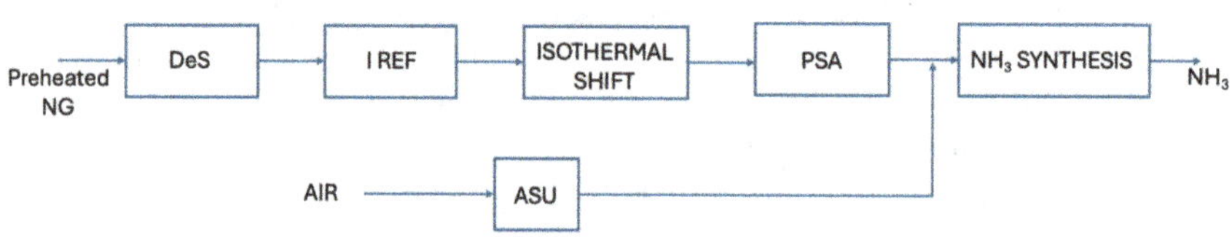

Figure 2.14 Linde ammonia concept scheme (derived from [53])

reach a high purity of H_2 and N_2, respectively, inside the ammonia reactor (Figure 2.14). The first plant using that technology was in Barola (India), constructed in 1998 and with an ammonia production of about 1,350 metric tons per day (MTD), followed by installations in China and Australia [54].

Despite most ammonia plants worldwide using natural gas steam-reformed hydrogen to synthesize NH_3, some countries, particularly China, the largest ammonia producer, rely on coal gasification [53]. This method is, in fact, the second most commonly used for NH_3 production [55]. In such plants, an ASU is crucial: N_2 is used in the synthesis reactor, while O_2 is needed for coal gasification.

The resulting synthesis gas is a mixture of H2, CO, CO_2, and methane. After mixing, the gas is processed in cleaning units (sour gas shift unit and acid gas removal unit) before being sent to the ammonia reactor, where nitrogen separated in the ASU is added [53].

Traditional ammonia production from fossil fuels without carbon capture and storage (CCS) technologies is associated with what is known as "gray ammonia," which generates approximately 150 g of CO_2 equivalent emissions per MJ of energy produced during the process. Introducing CCS technology can significantly reduce GHG emissions to around 16 g of CO_2 equivalent per MJ [56]. There are many existing plants that produce blue ammonia (from fossil fuels with CCS), which will be discussed in the next sections.

2.5 Ammonia innovative production process

As mentioned in the earlier section, the H–B process coupled to steam methane reforming is nowadays the main industrial method for ammonia production. However, even if it has been optimized over time, due to elevated GHG emissions, the industries need to pursue a more sustainable production technology. It must be underlined that, with 150 million tons of ammonia yearly production, this sector is responsible for around 500 Mt of CO_2-eq emissions annually, representing over 1% of global GHG emissions [28]. Moreover, if methane leakage and other indirect emissions related to the supply of natural gas are considered, the total GHG emissions would be more than half Gt. Implying a CCS system allows for a strong reduction in CO_2 emission [4,36]: it enables to reach almost carbon neutrality. In this sense, to reach government requirements in terms of GHG emissions, a de-carbonization of the ammonia production process is necessary, especially due to the expected increase in NH_3 demand because of population growth. Figure 2.15 shows three global ammonia production scenarios, as explained in [57]: (i) the stated policies scenario, which follows the current industries trend; (ii) the sustainable development scenario, which, instead, considers introducing technologies comparable with the Paris Agreement; (iii) the net zero emission by 2050 scenario, which is the one compatible to reach world net zero emissions by 2050.

To reach a zero-emission target for the ammonia production process, other technologies must be used (i.e. water electrolysis feeding the system by renewables, biomass gasification, or performing a methane pyrolysis) [4]. Among the possible solutions, the most feasible in the short term is the H_2 generation via electrolysis [4,58]. Methane pyrolysis, nowadays, is not at commercial-scale installations. In this case, hydrogen is produced by using very high-temperature heat provided by electrical plasma to split methane into H_2 and C atoms, without burning it [59]. Biomass, instead, as well as having the great limitation of being feasible only in areas adjacent to large forest reserves, has the negativity of having a high environmental impact in terms of deforestation [60].

Focusing on hydrogen generation via electrolysis, this solution minimizes GHG emissions but has some drawbacks. It requires high-purity water to avoid

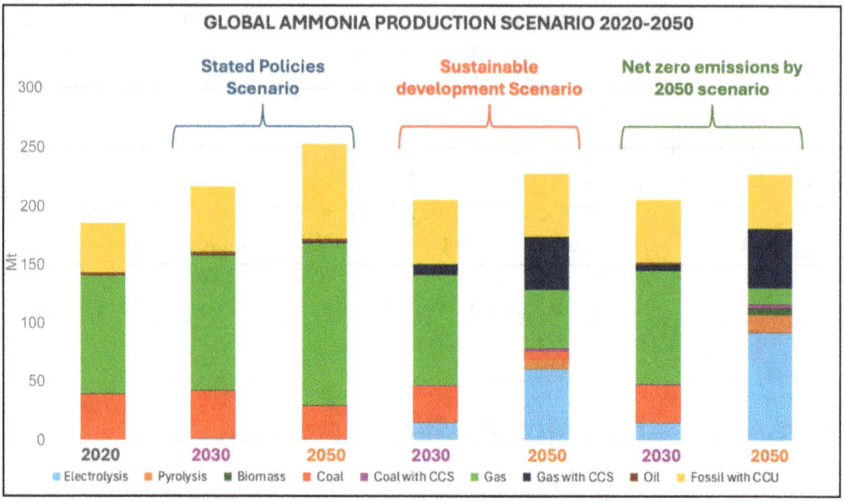

Figure 2.15 Global NH_3 production scenario 2020–2050 (data derived from [57])

damaging the electrolyzer and is highly affected by electricity costs. Additionally, electrolyzers have a high capital expenditure (CAPEX) [61], and the system is eco-friendly only if it is fully powered by renewable energy sources, which necessitates large-scale storage to manage the intermittency of these sources.

In the layout, the electrolyzer is coupled with an ASU for N_2 production before entering the H–B loop. During operation, O_2 is a by-product of both water electrolysis and air cryogenic separation.

Electrolysis can be carried out at low temperatures using alkaline electrolysis cells (AECs) or proton exchange membrane electrolysis cells (PEMECs), or at high temperatures using solid oxide electrolysis cells (SOECs).

Electrolyzer-based systems show higher energy consumption due to the substantial electricity required for water dissociation (>90% of the total inlet energy) compared to natural gas-based systems. However, in terms of global efficiency, electrolyzer-based systems are the best solution. SOEC technology achieves the best performance with lower energy consumption (3.7 kWh/Nm^3 H_2 vs. 4–5 kWh/Nm^3 H_2) because a portion of the required energy is supplied as heat. Furthermore, SOECs are well-suited for ammonia synthesis loop, as the exothermic H–B reactor can provide the steam needed by the SOECs. The main drawbacks of SOEC technology include its lower readiness level compared to AEC and PEMEC, as well as its anticipated lifetime and high CAPEX coupled with stack replacement costs [61,62].

AEC and PEMEC both have high readiness levels, but PEMEC has the advantage of not using an electrolyte (usually Na-based or K-based), allowing for higher purity H_2 production [63]. Additionally, PEMEC has promising prospects for decreasing capital costs [61] and operates at higher efficiency.

Table 2.5 provides a comparison among high-temperature electrolyzers, low-temperature electrolyzers, and the H–B process fed by natural gas.

Table 2.5 Ammonia production: technologies comparison [30,63,66–68]

	Methane to ammonia	**El-based AEC/ PEM**	**El-based SOEC**
TRL	9	8	5
System energy consumption	7.6–9.5 kWh/kg NH_3 (the lowest value without CCS, the highest one with CCS)	9–10 kWh/kg NH_3	7.5–8.5 kWh/kg NH_3
System energy efficiency	Around 60%	AEC: >60% PE-MEC: 68–71%	SOEC: 70–75%
CO_2 emission	>1.7–2.2 kg CO_2/kg NH_3	None if RESs are considered	None if RESs are considered

2.6 Overview of the main ammonia production site

In 2021, there were 490 active ammonia production plants, divided into 64 countries, with a total production of about 150 Mt/y, mainly for fertilizers production [67]. As explained before, China is the biggest producer in the world, with about 30% of the global production (Figure 2.16) [3], followed by Russia, the USA, and India with 9.5% each.

Despite the larger global production of NH_3, China has only small production companies on its territory, as reported in Figure 2.17. As explained in [68], the production of NH_3 is mainly coal-based, as in China there are no NG reservoirs, resulting in a consistent CO_2 emission (about 4 tons of CO_2 emitted every ton of NH_3), also due to the limited use of CCS technology. However, it must be underlined that, nowadays, China has the possibility to produce green ammonia through an electrolysis-based system fed by renewables, taking advantage of existing factories for electrolyzers' production at a relatively low cost compared to US/European cost manufacturers. In fact, China is involved in three renewable ammonia projects [68]: (i) the Da'an project, with 180,000 tons per year of green ammonia; (ii) the Baotou project, with 390,000 t/y by 2025; (iii) the Chifeng project with 20,000 t/y in 2023 (phase 0) till 900,000 in 2029 (phase 3).

The major ammonia production companies are mainly concentrated in North America, Europe, the Middle East, Russia, India, and Australia: Figures 2.18 and 2.19 show their distribution in the world and their installed capacities in Mt/y of NH_3, respectively [67].

The greatest ammonia production company is the CF Industries (Illinois, USA), with a total capacity of 10.5 Mt/y (5% of the global production). The company's registered office is in Deerfield, Illinois (USA), but it has nine installed plants all over the world. The biggest one is the Donaldsonville (Louisiana) plant, which is the largest ammonia production complex in the world (4.3 Mt/y of installed capacity) [5], followed by Medicine Hat (Albania), Port Neal (Iowa), Verdigris (Oklahoma), with an ammonia production installed capacity of about 1.2

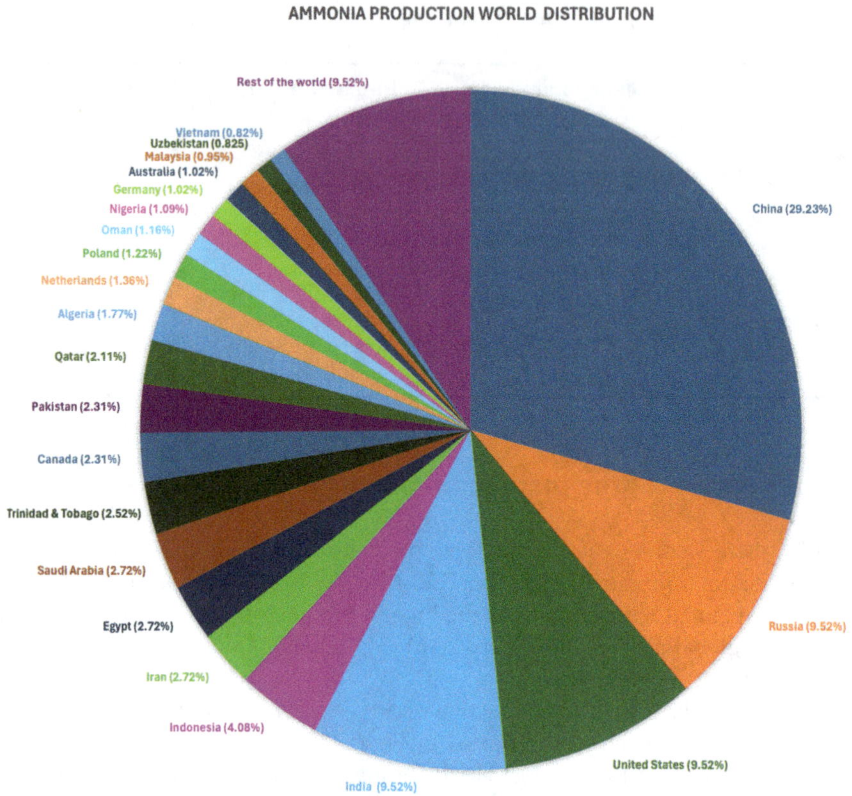

Figure 2.16 Ammonia production world distribution (data derived from [3])

Mt/y. Smaller CF Industries' plants are the ones of Billingham (UK), Yazoo City (Mississippi), Courtright (Ontario), Woodward (Oklahoma), and Point Lisas (Trinidad), with an ammonia yearly production below 500,000 tons. CF Industries' NH_3 production comes from the traditional H–B process but they plan to reach a 25% reduction in CO_2-equivalent emission by 2030 and net-zero carbon emission by 2050 [69], by installing a new electrolyzer for green ammonia production and by following the CCS project [70].

The second largest ammonia production company is YARA (Oslo, Norway), with 7.7 Mt/y of ammonia production from the traditional H–B process using fossil gases. Moreover, YARA company promotes green ammonia production by electrifying its factory at Herøya [71].

The third ammonia producer in the world is Nutrien (Canada), with about 7 million tons of installed ammonia capacity. It synthesizes ammonia from natural gas, to be used mainly for nitrogen production (Nutrien is the third nitrogen producer in the world with a production of 11 Mt/y) [7,72]. Nutrien, besides, is active in carbon capture and CO_2 abatement, with a low-carbon ammonia production of 1.2 million tons in 2023 [67].

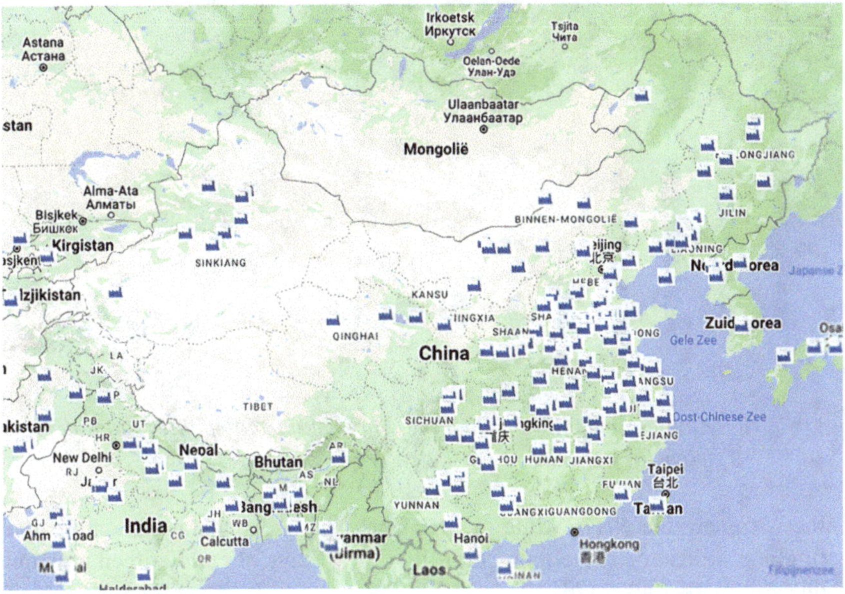

Figure 2.17 China ammonia production factory. Reprinted, with permission, from [68].

Figure 2.18 Ammonia major world production companies (data derived from [67])

The last largest production company is OCI-Nitrogen (Chemelot), which is the second largest in Europe with an ammonia yearly production of 6.4 Mt. This company is also one of the leaders in biogas production and it is active in renewable and low-carbon ammonia synthesis [73]. OCI-Nitrogen uses biogas in Dutch and US plants and electrolyzers in Egyptian facilities to produce hydrogen for ammonia

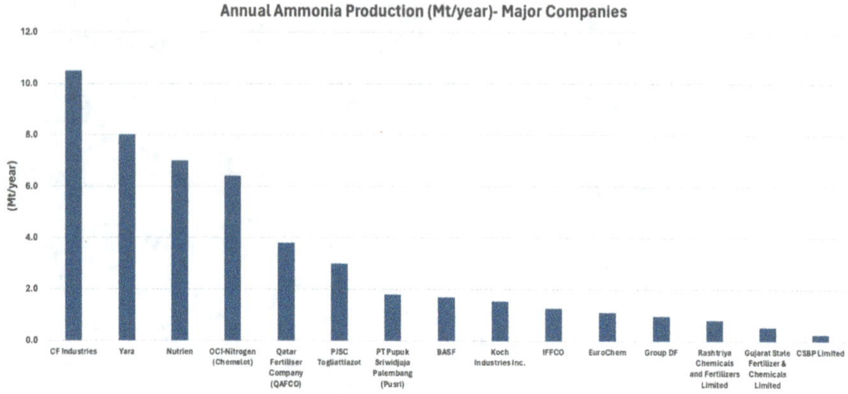

Figure 2.19 Annual ammonia production (Mt/y)-major companies (data derived from [67])

synthesis (renewable NH_3), while in US plants it utilizes natural gas with carbon capture and utilization (blue ammonia). It is, nowadays, the only green ammonia producer at market level [73].

Smaller companies are Qatar Fertilizer Company (QAFCO) (Qatar) [8] and PJSC Togliattiazot (Russia), with an ammonia yearly production of about 3.8 and 3 Mt/y, respectively.

As briefly introduced in the previous session, China has in mind three big projects on green ammonia production [68]. The first one is the Da'an project, by the Jilin Electric Power company [74], which wants to ensure a green ammonia production of about 180,000 t/y by producing hydrogen via alkaline and PEM electrolysis. It will install 39 alkaline units (total installed power of about 5 MW), provided by LONGi Hydrogen and SANY Hydrogen companies, and 50 PEM units. The second project aims to install a new renewable ammonia plant in Baotou (Mongolia) with an expected production of 390,000 t/y. The plant will be based on wind and solar RESs and will be in operation by 2025 [75,76]. The last main project is a multi-phase project which aims to install a green ammonia production plant at Chifeng of 1.5 Mt/y by 2029. The first phase should already be concluded, with the installation of a pilot plant of 20,000 t/y production by September 2023, followed by the installation of a 300,000 t/y plant by December 2024 (actually under CF Industries is planning projects for blue ammonia production). Regarding the first, the company has recently signed an agreement with ExxonMobil for the sequestration of 2 Mt of CO_2 per year from the Donaldsonville complex: with the starting of CCS, expected by 2025, CF Industries will be able to produce up to 1.7 Mt/y of blue ammonia. The company has in mind another blue ammonia installation in Louisiana, but it is under development [70].

Another significant green ammonia project is the Allied Green Ammonia initiative, in collaboration with Topsoe [77]. This project involves hydrogen generation by solid oxide electrolyzers (SOEs) and aims to produce 900,000 tons of green ammonia per year by 2028. YARA, the second largest ammonia production company, has electrified its factory at Herøya [71]. This facility features a 24 MW electrolyzer demonstration plant capable of producing 20,500 tons of green ammonia per year in 2023 [78]. Additionally, YARA, in partnership with Ørsted, plans to use renewable hydrogen from wind-powered electrolyzers (100 MW) to produce ammonia at its Sluiskil facility by 2025 [79]. Furthermore, YARA, together with BASF, is developing a blue ammonia project on the US Gulf Coast, aiming for a 1.2–1.4 million tons per year ammonia plant [80].

In 2023, Nutrien company produced about 1.2 Mt of blue ammonia, thanks to their facilities in Geismar and Joffre (Germany and Canada, respectively) [81], through carbon capture utilization and storage facilities. Furthermore, OCI Nitrogen, as the fourth world ammonia productor, has in mind to buy green hydrogen from New Fortress Energy (water electrolysis production) and to produce about 160,000 t/y of green ammonia, in addition to their existing Egyptian facilities, in its plant in Beaumont (Texas) by 2026 [82]. As for blue ammonia production, they are building a new facility in Texas, which will start the production in 2025 [73]. Another project for green ammonia production is the one from 600 MW solar park in Chile by Mejillones Ammonia Energy company, with an expected production of 300,000 t/y by 2027 at the first stage of the project [83]. As one of the biggest green ammonia projects, ACME and IHI (Japan) have in mind to build a new plant in Odisha (India) capable to reach 1.3 Mt/y of NH_3 production by renewables [84], and exporting it through the Gopalpur port to Japan. The production is expected to start in 2027.

Finally, two blue ammonia projects are described in [85]: the Pedirka blue project, which aims to install a 1 Mt/y capacity plant in the Northern Territory (Australia), and the Project Phoenix, which aims to install a production plant in Evanston, Wyoming (USA).

The described projects are reported in Figure 2.20, together with some potential new sites for green ammonia production.

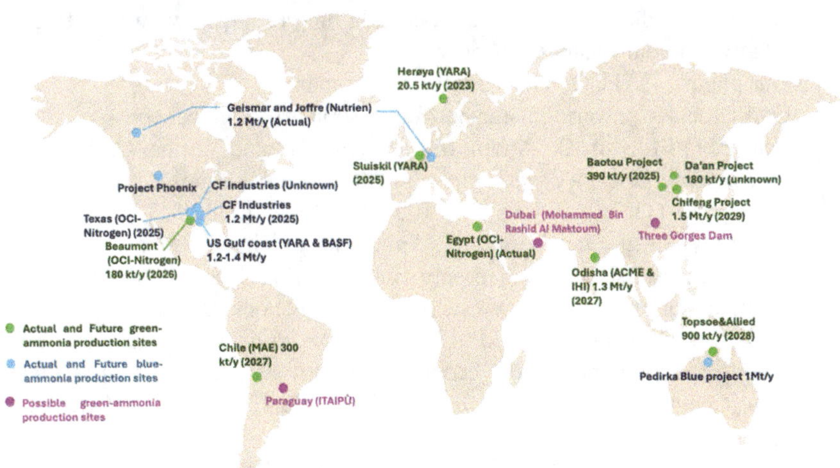

Figure 2.20 Actual and future green/blue ammonia production sites and possible new sites

Green ammonia production is related to RES capability of the site: the future 5,000 MW photovoltaic plant in Dubai [86] (planned by 2050), the Three Gorges Dam [87] and the Itaipù hydroelectric plant [88], which are the first and second world's largest hydroelectric existing plants respectively, can be some of the possible sources for green ammonia future production.

References

[1] Ojelade, O.A., Zaman, S.F., and Ni, B.-J., Green ammonia production technologies: a review of practical progress. *Journal of Environmental Management*. 2023;342:118348. doi:https://doi.org/10.1016/j.jenvman.2023.118348.

[2] Chen, S., Perathoner, S., Ampelli, C., and Centi, G., Chapter 2 – electrochemical dinitrogen activation: to find a sustainable way to produce ammonia, In Albonetti, S., Perathoner, S., and Quadrelli, E.A. (eds.), *Studies in Surface Science and Catalysis*, vol. 178. Amsterdam: Elsevier; 2019. pp. 31–46, https://doi.org/10.1016/B978-0-444-64127-4.00002-1.

[3] US Geological Survey, *Ammonia Production Worldwide in 2023, by Country (in 1,000 metric tons). Statista.* Statista Inc.; 2024 (accessed 4 March 2024). Ammonia production by country 2023. Available online: https://www.statista.com/statistics/1266244/global-ammonia-production-by-country/.

[4] IEA, *Ammonia Technology Roadmap*. Paris: IEA; 2021. Available online: https://www.iea.org/reports/ammonia-technology-roadmap, License: CC BY 4.0.

[5] Ammonia production | Available online: https://www.cfindustries.com/what-we-do/ammonia-production.

[6] Yara International ASA's 2022 Integrated Report. Available online: https://www.yara.com/siteassets/investors/057-reports-and-presentations/annual-reports/2022/yara-integrated-report-2022.pdf.

[7] Nitrogen products & locations. Available online: https://www.nutrien.com/what-we-do/our-business/nitrogen.

[8] Our plants | QAFCO | Qatar Fertiliser Company.

[9] OCI Global – powering a cleaner future sooner (oci-global.com).

[10] Dincer, I., Erdemir, D., Aydin, M.I., Karasu, H., and Vezina, G. Ammonia. In: *Ammonia Energy Technologies. Lecture Notes in Energy*, vol 91. Cham: Springer; 2022. https://doi.org/10.1007/978-3-031-13532-3_1.

[11] Padappayil, R.P., and Borger, J., Ammonia toxicity. In: *StatPearls [Internet]*. Treasure Island, FL: Treasure Island, FL: StatPearls Publishing; 2024. Available online: https://www.ncbi.nlm.nih.gov/books/NBK546677/.

[12] Morlanés, N., Katikaneni, S.P., Paglieri, S.N., *et al.*, A technological roadmap to the ammonia energy economy: current state and missing technologies, *Chemical Engineering Journal*. 2021;408:127310. https://doi.org/10.1016/j.cej.2020.127310.

[13] Tawalbeh, M., Murtaza, S.Z., Al-Othman, A., Alami, A.H., Singh, K., and Olabi, A.G., Ammonia: a versatile candidate for the use in energy storage systems. *Renewable Energy*. 2022;194:955–977.

[14] Christensen, C.H., Sørensen, R.Z., Johannessen, T., *et al.*, Metal ammine complexes for hydrogen storage. *Journal of Materials Chemistry*. 2005;15 (38):4106–4108.

[15] Elishav, O., Mosevitzky Lis, B., Valera-Medina, A., and Grader, G.S., "Storage and distribution of ammonia." In Valera-Medina, A., and Banares-Alcantara, R. (eds.), *Techno-Economic Challenges of Green Ammonia as an Energy Vector*. Amsterdam: Elsevier; 2020, pp. 85–103. https://doi.org/10.1016/B978-0-12-820560-0.00005-9.

[16] Appl, M. "Ammonia, 1. Introduction." In *Ullmann's Encyclopedia of Industrial Chemistry, Dannstadt-Schauernheim.* Weinheim: Wiley-VCH Verlag GmbH & Co. KGaA; 2012. pp. 107–133.

[17] Dybkjaer, I., "Ammonia production processes." In Nielsen, A. (ed.), *Ammonia*. Berlin: Springer; 1995. pp. 199–327.

[18] Hignett, T.P., "Transportation and storage of ammonia." In *Fertilizer Manual*. Dordrecht: Springer; 1985. pp. 73–82.

[19] Klerke, A., Christensen, C.H., Nørskov, J.K., and Vegge, T., "Ammonia for hydrogen storage: challenges and opportunities." *Journal of Materials Chemistry*. 2008;18(20):2304–2310.

[20] Rouwenhorst, K.H., Ham, A.G.V.d., Mul, G., and Kersten, S.R., "Islanded ammonia power systems: technology review & conceptual process design." *Renewable and Sustainable Energy Reviews*. 2019;114:109339.

[21] Zhang, T., Miyaoka, H., Miyaoka, H., Ichikawa, T., and Kojima, Y., Review on ammonia absorption materials: metal hydrides, halides, and borohydrides. *ACS Applied Energy Materials*. 2018;1:232–242. https://doi.org/10.1021/acsaem.7b00111.

[22] Malmali, M., Le, G., Hendrickson, J., Prince, J., McCormick, A.V., and Cussler, E.L., "Better absorbents for ammonia separation." *ACS Sustainable Chemistry and Engineering*. 2018:6(5):6536–6546.

[23] Osman, O., Sgouridis, S., and Sleptchenko, A., "Scaling the production of renewable ammonia: a techno-economic optimization applied in regions with high insolation." *Journal of Cleaner Production*. 2020;271:121627.

[24] Aoki, T., Ichikawa, T., Miyaoka, H., and Kojima, Y., "Thermodynamics on ammonia absorption of metal halides and borohydrides." *Journal of Physical Chemistry*. 2014;118:18412–18416.

[25] Kojima, Y., and Yamaguchi, M., Ammonia storage materials for nitrogen recycling hydrogen and energy carriers. *International Journal of Hydrogen Energy*. 2020;45:10233–10246. https://doi.org/10.1016/j.ijhydene.2020.01.145.

[26] Elmøe, T.D., Sørensen, R.Z., Quaade, U., Christensen, C.H., Nørskov J.K., and Johannessen, T., A high-density ammonia storage/delivery system based on Mg(NH3)6Cl2 for SCR-DeNOx in vehicles. *Chemical Engineering Science*. 2006;61:2618–2625. https://doi.org/10.1016/j.ces.2005.11.038.

[27] Egerer, J., Grimm, V., Niazmand, K., and Runge, P., The economics of global green ammonia trade–"shipping Australian wind and sunshine to Germany." *Applied Energy*. 2023;334:120662.

[28] Bonnet-Cantalloube, B., Espitalier-Noël, M., de Carvalho, P.F., Fonseca, J., and Pawelec, G., *Clean Ammonia in the Future Energy System*. Hydrogen Europe; 2023. Available online: https://hydrogeneurope.eu/wp-content/uploads/2023/03/2023.03_H2Europe_Clean_Ammonia_Report_DIGITAL_FINAL.pdf.

[29] Laval, A., Hafnia, H.T., and Vestas, S.G., Ammonfuel-an industrial view of ammonia as a marine fuel. *Hafnia*. 2020;7:32–59.

[30] Bird, F., Clarke, A., Davies, P., and Surkovic, E., *Ammonia: Zero-carbon Fertiliser, Fuel and Energy Store.* 2020. Available online: https://royalsociety.org/-/media/policy/projects/green-ammonia/green-ammonia-policy-briefing.pdf.

[31] Jinwoo Kim, C.H., and Seo, Y., End-to-end value chain analysis of isolated renewable energy using hydrogen and ammonia energy carrier. *Energy Conversion and Management*. 2022;254:115247. https://doi.org/10.1016/j.enconman.2022.115247.

[32] Yadav, A. and Jeong, B., Safety evaluation of using ammonia as marine fuel by analysing gas dispersion. *Journal of International Maritime Safety Environmental Affairs and Shipping*. 2022;6(2):99–116. doi:10.1080/25725084.2022.2083295.

[33] Kawasaki delivered LPG-powered AXIS RIVER LPG/NH3 carrier (July 2023). Available online: https://www.vesselfinder.com/news/26467-Kawasaki-Delivered-LPG-powered-AXIS-RIVER-LPGNH3-Carrier.

[34] Magnusson, S.T., and Murphy-Cannella, D., Ammonia – what is it and how can it be applied to the energy and shipping sectors? (August 2021). Available online: https://www.econnectenergy.com/articles/ammonia-what-is-it-and-how-can-it-be-applied-to-the-energy-sector#:~:text=Ammonia%20can%20be%20transported%20by%20three%20different%20ship,semi-refrigerated%20%284-8%20barg%2F-10C%29%20and%20under%20pressure%20%2817%20barg%2F45C%29.

[35] Ammonia: a technology brief on ammonia as a fuel for maritime transport. Available online: https://theliquidgrid.com/maritime-decarbonization/maritime-zero-emission-fuels/marine-ammonia/#:~:text=Ammonia%20has%20been%20carried%20as%20a%20cargo%20aboard,liquid%20aboard%20ships%20to%20keep%20the%20pressure%20low.

[36] Rouwenhorst, K.H.R., Krzywda, P.M., Benes, N.E., Mul, G., and Lefferts, L., *Ammonia Production Technologies*. Amsterdam: Elsevier; 2021. https://doi.org/10.1016/b978-0-12-820560-0.00004-7.

[37] Fully refrigerated tankers that carry LPG, ammonia & vinyl chloride. Available online: http://liquefiedgascarrier.com/Fully-Refrigerated-Ships.html.

[38] Shakirov, A., *LNG shipping cost.* 2021. Available online: https://www.insidelogistics.ca/opinions/lng-shipping-costs/#:~:text=As%20a%20result%2C%20LNG%20shipping%20costs%20dropped%20significantly,%240.5%20mmBtu%20%28from%20%240.9%20mmBtu%20in%20January%202020%29.

[39] Salmon, N., and Bañares-Alcántara, R., Green ammonia as a spatial energy vector: a review. *Sustainable Energy Fuels*. 2021;5:2814. doi:10.1039/D1SE00345C.

[40] Gezerman, A.O., Industrial scale ammonia pipeline transfer system and exergy analysis, *Kemija U Industriji*. 2021;70:711–716. https://doi.org/10.15255/kui.2020.080.

[41] NH3 Fuel Association. The Logical Path Forward, 2019. Available online: https://nh3fuelassociation.org/comparisons/ (accessed 6 December 2024).

[42] Togliatti-Odessa ammonia pipeline will become operational. Canada: N. p., 1979. Available online: https://www.osti.gov/etdeweb/biblio/6195696.

[43] Ammonia pipeline "Tolyatti-Odessa" (SE "Ukrhimtransammiak"). Available online: https://se.ua/en/projects/ammonia-pipeline-tolyatti-odessa-se-ukrhim-transammiak/.

[44] Bridge, A., R8638: Ammonia wagon, Limerick Jct. Available online: https://www.geograph.ie/photo/2678387 (accessed 13 November 2024).

[45] Anhydrous Ammonia Transport Tank. Available online: https://www.bnhgastanksindia.com/anhydrous-ammonia-transport-tanks.

[46] Amhamed, A.I., Shuibul Qarnain, S., Hewlett, S., *et al.*, Ammonia production plants—a review. *Fuels*. 2022;3(3):408–435. https://doi.org/10.3390/fuels3030026.

[47] Haber, F., and Le Rossignol, R., *Production of Ammonia.* U.S. Patent 1202995A, 13 August 1909. Available online: https://patents.google.com/patent/US1202995A/en?oq=(U.S+patent+1202995) (accessed 16 January 2021).

[48] Wright, L., and Pickford, A., *Ammonia Synthesis System*. U.S. Patent 3721532A, 8 February 1971. Available online: https://patents.google.com/patent/US3721532A/en (accessed 27 January 2021).

[49] Da Rosa, A.V., *Process for Production of Ammonia.* U.S. Patent 4107277A, 15 August 1978. Available online: https://patents.google.com/patent/US4107277A/en (accessed 16 January 2022).

[50] Becker, C.L., *Low Energy Ammonia Synthesis Process.* U.S. Patent 4148866, 4 October 1976. Available online: https://patents.google.com/patent/US4153673A/en (accessed 16 January 2022).

[51] Shires, P.J., Cassata, J.R., Mandelik, B.G., and Dijk, C.P., *Preparation of Ammonia Synthesis Gas.* U.S. Patent 4479925, 30 October 1984.

[52] Modak, J.M., Haber process for ammonia synthesis. *Resonance*. 2011; 16:1159–1167. 10.1007/s12045–011-0130-0.

[53] Pattabathula, V., and Richardson, J., Introduction to ammonia production. *Chemical Engineering Progress*. 2016;112(9):69–75.

[54] Linde, Linde Engineering, Ammonia, LAC™ Process. Available online: https://www.linde-engineering.com/en/process-plants/hydrogen_and_synthesis_gas_plants/gas_products/ammonia/index.html.

[55] Bicer, Y., Dincer, I., Vezina, G., *et al.* Impact assessment and environmental evaluation of various ammonia production processes. *Environmental Management.* 2017;59:842–855. https://doi.org/10.1007/s00267-017-0831-6.

[56] Satista, Average GHG intensity of ammonia production for the shipping industry in 2022, by ammonia type, December 8, 2023. Available online:

https://www.statista.com/statistics/1367299/ghg-emissions-production-grey-ammonia-blue-ammonia/#:~:text=A%20typical%20process%20of%20ammonia,the%20grey%20ammonia%20production%20process.

[57] IEA, Global ammonia production by technology and scenario 2020–2050. Available online: https://www.iea.org/data-and-statistics/charts/global-ammonia-production-by-technology-and-scenario-2020-2050.

[58] Ghavam, S., Vahdati, M., Wilson, I.A., and Styring, P., Sustainable ammonia production processes. *Frontiers in Energy Research*. 2021;9:34.

[59] Patlolla, S.R., Katsu, K., Sharafian, A., Wei, K., Herrera, O.E., and Mérida, W., A review of methane pyrolysis technologies for hydrogen production. *Renewable and Sustainable Energy Reviews*. 2023;181:113323.

[60] Cardoso, J.S., Silva, V., Chavando, J.A.M., Eusébio, D., Hall, M.J., and Costa, M., Small-scale biomass gasification for green ammonia production in Portugal: a techno-economic study. *Energy & Fuels*. 2021;35(17):13847–13862.

[61] Bermudez, J.M., Evangelopoulo S., and Pavan F., *Electrolysers Analysis – IEA Tracking Report*. Available online: https://www.iea.org/energy-system/low-emission-fuels/electrolysers.

[62] Schmidt, O., Gambhir, A., Staffell, I., Hawkes, A., Nelson, J., and Few, S., Future cost and performance of water electrolysis: an expert elicitation study. *International Journal of Hydrogen Energy*. 2017;42:30470–30492.

[63] Guo, Y., Li, G., Zhou, J., and Liu, Y., Comparison between hydrogen production by alkaline water electrolysis and hydrogen production by PEM electrolysis. *IOP Conference Series: Earth and Environmental Science*. 2019;371(4):042022.

[64] Moses, P.G., *Hydrogen at Scale*. 2022. Available online: https://energiforsk.se/media/30824/hydrogen-production-in-practice-poul-georg-moses-haldor-topsoe.pdf.

[65] Smith, C., Hill, A.K., and Torrente-Murciano, L., Current and future role of Haber–Bosch ammonia in a carbon-free energy landscape. *Energy and Environmental Science*. 2020;13:331–344. https://doi.org/10.1039/c9ee02873k.

[66] Grigoriev, S.A., Fateev, V.N., Bessarabov, D.G., and Millet, P., Current status, research trends, and challenges in water electrolysis science and technology. *International Journal of Hydrogen Energy*. 2020;45:26036–26058. https://doi.org/10.1016/j.ijhydene.2020.03.109.

[67] AMMPOWER – Major Ammonia Producing companies and their capacities, November 21, 2022. Available online: https://www.iamm.green/ammonia-producing-companies/.

[68] Ammonia Energy Association. China: scaling-up "flexible" ammonia production powered by renewable energy, June 28, 2023. Available online: https://www.ammoniaenergy.org/articles/china-scaling-up-flexible-ammonia-production-powered-by-renewable-energy/.

[69] CF Industries, *Environment*. Available online: https://www.cfindustries.com/sustainability/environment.

[70] CF Industries, *Carbon Capture & Sequestration: The Safe, Proven Pathway to Rapid Reduction of CO_2 Emissions*. Available online: https://www.cfindustries.com/globalassets/cf-industries/media/documents/ccs-factsheet.pdf.

[71] YARA, Yara begins electrifying the factory at Herøya, December 17, 2021. Available online: https://www.yara.com/corporate-releases/yara-begins-electrifying-the-factory-at-heroya/#hegra-english.

[72] Nutrien, Global sustainability report, December 31, 2023. Available online: https://nutrien-prod-asset.s3.us-east-2.amazonaws.com/s3fs-public/uploads/2024-03/Nutrien_Global_Sustainability_Report_on2023v2.pdf.

[73] OC-Nitrogen, annual report 2022. Available online: https://oci-global.com/storage/2023/04/OCI-Annual-Report-2022-vf2.pdf.

[74] Ammonia Energy, New electrolysis-based ammonia projects in China, May 7, 2023. Available online: https://www.ammoniaenergy.org/articles/new-electrolysis-based-ammonia-projects-in-china/.

[75] Ammonia Energy, Topsoe: dynamic renewable ammonia production in China, January 31, 2023. Available online: https://www.ammoniaenergy.org/articles/topsoe-dynamic-renewable-ammonia-production-in-china/.

[76] Ammonia Energy, Flexible ammonia synthesis: shifting the narrative around hydrogen storage, April 27, 2023. Available online: https://www.ammoniaenergy.org/articles/flexible-ammonia-synthesis-shifting-the-narrative-around-hydrogen-storage/.

[77] Topsoe, November 15, 2023, Topsoe signs first Australian green ammonia contract with Allied Green Ammonia. Available online: https://www.topsoe.com/press-releases/topsoe-signs-first-australian-green-ammonia-contract-with-allied-green-ammonia.

[78] Herøya Industipark, The world's largest and first green hydrogen plant starts production at Herøya in 2023. Available online: https://www.heroya-industripark.no/en/news/the-worlds-largest-and-first-green-hydrogen-plant-starts-production-at-heroeya-in-2023.

[79] YARA, Ørsted and Yara seek to develop groundbreaking green ammonia project in the Netherlands, October 5, 2020. Available online: https://www.yara.com/corporate-releases/orsted-and-yara-seek-to-develop-groundbreaking-green-ammonia-project-in-the-netherlands/.

[80] YARA, BASF and Yara to evaluate low-carbon blue ammonia project at U.S. Gulf Coast, June 29, 2023. Available online: https://www.yara.com/corporate-releases/basf-and-yara-to-evaluate-low-carbon-blue-ammonia-project-at-u.s.-gulf-coast/.

[81] Nutrien, Low-Carbon Ammonia. Available online: https://www.nutrien.com/sustainability/strategy/low-carbon-fertilizer.

[82] Chemical & Engineering news, OCI to produce green ammonia and methanol in Texas, September 22, 2023. Available online: https://cen.acs.org/energy/hydrogen-power/OCI-produce-green-ammonia-methanol/101/web/2023/09.

[83] Ammonia Energy Association, Mejillones Ammonia Energy: renewable ammonia in Chile, March 13, 2024. Available online: https://www.ammoniaenergy.org/articles/mejillones-ammonia-energy-renewable-ammonia-in-chile/.
[84] Ammonia Energy Association, ACME to supply renewable ammonia to IHI in Japan, February 6, 2024. Available online: https://www.ammoniaenergy.org/articles/acme-to-supply-renewable-ammonia-to-ihi-in-japan/.
[85] Ammonia Energy Association, Blue ammonia in the Northern Territory & Wyoming, January 27, 2022. Available online: https://www.ammoniaenergy.org/articles/blue-ammonia-in-the-northern-territory-wyoming/.
[86] C40 cities, Dubai's 'Mohammed Bin Rashid Al Maktoum' 5,000MW Solar Park Aims to Save 6.5 Million tCO2e Annually, April 2019. Available online: https://www.c40.org/case-studies/dubai-s-mohammed-bin-rashid-al-maktoum-5-000mw-solar-park-aims-to-save-6-5-million-tco2e-annually/.
[87] USGS science for a changing world, Three Gorges Dam: The World's Largest Hydroelectric Plant, June 6, 2018. Available online: https://www.usgs.gov/special-topics/water-science-school/science/three-gorges-dam-worlds-largest-hydroelectric-plant#:~:text=The%20Three%20Gorges%20Dam%20on,the%20world%27s%20biggest%20hydroelectric%20facility.
[88] Power Technology, The Itaipu Hydroelectric Dam Project, Brazil, July 31, 2020. Available online: https://www.power-technology.com/projects/itaipu-hydroelectric/.

Chapter 3

Measurement techniques for ammonia combustion systems

Y. Li[1], *X. Shi*[1], *Y. Zhang*[1] *and W. Li*[1]

3.1 Introduction

Although most measurement techniques for hydrocarbon fuels can be applied to ammonia-fueled systems, ammonia combustion diagnostics have become increasingly a hot topic due to the presence of nitrogen elements. This emphasizes the need for quantitative measurement of key nitrogen-containing components such as NH_3, NH_2, NH, NO, NO_2, and N_2O. This chapter summarizes major measurement techniques for ammonia combustion systems, categorized into three types: spectroscopy diagnostics, optical diagnostics, and sampling analysis. The principles and applications of various measurement techniques are briefly introduced, focusing on the unique aspects of measurement techniques oriented for ammonia-fueled systems.

Due to their relatively simple principles and long-standing application across various disciplines, conventional methods like hot-wire anemometry, thermocouples, pressure sensors, and photomultiplier tubes are not further discussed in this chapter.

3.2 Spectroscopy diagnostics

3.2.1 Chemiluminescence

Chemiluminescence, the phenomenon observed when the vibronically excited product of an exoergic reaction relaxes to its ground state with photon emissions, can be defined in simplistic terms as chemical reactions that emit light [1]. The chemical reaction produces energy in sufficient amounts (approximately 300 $kJ \cdot mol^{-1}$ and 150 $kJ \cdot mol^{-1}$ for blue- and red-light emissions, respectively), inducing the transition of an electron from its ground state to an excited state. This transition is often accompanied by vibrational and rotational changes in the molecule. The return of the electron to its ground state with photon emission is called

[1]School of Mechanical Engineering, Shanghai Jiao Tong University, China

chemiluminescence. The excited molecule can also lose energy through chemical reactions, collisional deactivation, internal conversion, or inter-system crossing [2].

The fraction of molecules emitting a photon upon returning to the ground state is the quantum yield (ϕ_{cl}). It is the product of three ratios:

$$\phi_{cl} = \phi_c \cdot \phi_e \cdot \phi_f$$

where ϕ_c is the fraction of reacting molecules giving an excitable molecule and accounts for the yield of the chemical reaction; ϕ_e is the fraction of such molecules in an electronically excited state and relates to the efficiency of the energy transfer; and ϕ_f is the fraction of these excited molecules that return to the ground state by emitting a photon (Figure 3.1).

The emission characteristics of the chemiluminescent process are determined by the sensitizer and the efficiency of the energy transfer, which complies with the Förster law [3]:

$$E = \frac{R_0^6}{d^6 + R_0^6}$$

where E is the energy transfer efficiency ranging from 0 (no transfer) to 1 (complete transfer), d represents the distance in Å between the donor and acceptor molecules, and the Förster distance R_0 (in Å, also) is given by the equation:

$$R_0 = 9700\left(J \cdot K^2 \cdot \phi_{cl} \cdot n^{-4}\right)^{1/6}$$

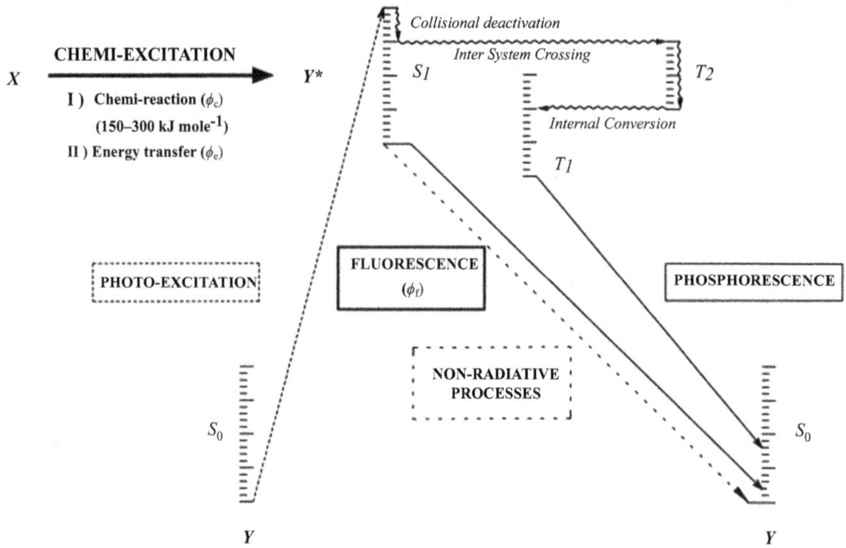

Figure 3.1 Diagram placing chemiluminescence among the most typical photophysical processes. Reprinted, with permission, from [2]. Copyright 2000 Elsevier.

where J is the spectral overlap integral between donor emission and acceptor absorption spectra, K is the orientation factor for dipole–dipole interaction, and n is the refractive index of the medium between the donor and the acceptor [4–7].

With non-intrusiveness and relative simplicity, chemiluminescence detection is well suitable for sensing flame characteristics in harsh environments when conventional *in situ* monitoring becomes impractical. In hydrocarbon flames, the self-chemiluminescence and the intense chemiluminescence emissions from radicals like OH*, CH_2O*, and CH* are widely used to sense relevant flame characteristics. In ammonia combustion, chemiluminescence methods focus on both flame self-chemiluminescence and key radical and molecule chemiluminescence, particularly nitrogen-containing radicals, including OH*, NH*, NH_2*, NH_3*, NO*, and NO_2*.

Flame self-chemiluminescence is the sum of chemiluminescence from various substances in the flame, distributed over a broad wavelength range. It allows for qualitative analysis of ammonia flame morphology and stability but cannot discern specific chemiluminescence species. In contrast, chemiluminescence from individual molecules or radicals, which emit at distinct wavelengths, can characterize different flame properties. Table 3.1 and Figure 3.2 illustrate the spectral transitions and individual lines of these species, as well as UV and UV + visible chemiluminescence spectra in ammonia combustion.

OH* is widely used in hydrogen-containing fuel systems, with spectral contributions in the 300–325 nm range [10]. In ammonia flames, OH* helps determine flame structure [11–14], flame front [15], equivalence ratio [16], NO emission [17], and local/global heat release rate [18]. It also serves as a benchmark for comparing chemiluminescence models with experimental results [10].

NH_2* intensity characterizes combustion intensity [19] and correlates positively with the equivalence ratio [20]. The visible emission from the ammonia flame front is attributed to NH_2* radicals, while in the post-flame zone, NO_2* chemiluminescence dominates in lean flames and NH_2* chemiluminescence dominates in rich flames [21]. NH_2* is the main contributor to visible ammonia flame emissions [22] and heat release rate marker [23]. Vigueras-Zúñiga *et al.* [24] found that using OH, CH, and NH_2 better correlates with high heat release rates in methane/ammonia blends. Similarly, NO_2*, NO*, NH*, and NH_2* intensities also originated as the validation parameters between kinetic modeling and ammonia laminar experiments [8].

Table 3.1 Spectral transitions and individual lines of the species in ammonia flames [8]

Species	Transition	Wavelength, nm
NO(A)	$A^2\Sigma^+ – X^2\Pi$	226
OH*	$A^2\Sigma^+ – X^2\Pi$	309
NH*	$A^3\Pi – X^3\Sigma^-$	336
NO_2*	$\tilde{A}^2B_2 – \tilde{X}^2A_1$	
NH_2*	$A^2A_1 – X^2B_1$	

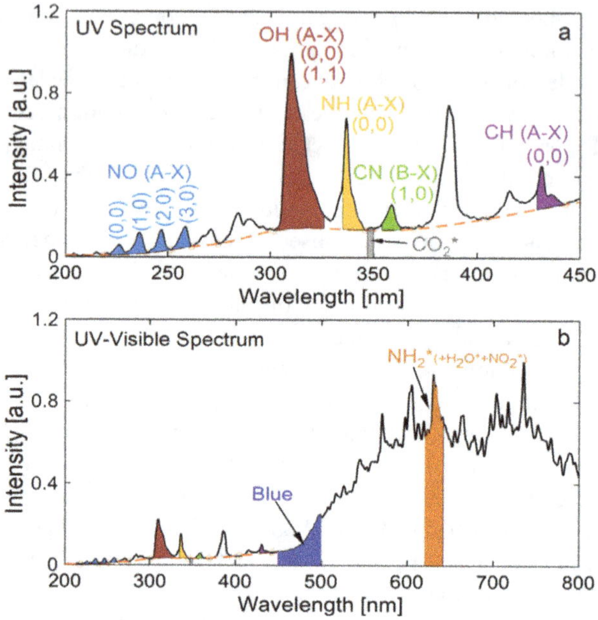

Figure 3.2 *(a) UV and (b) UV + visible chemiluminescence spectra for stoichiometric 60%NH_3/40%H_2/air swirl flame. Reprinted, with permission, from [9]. Copyright 2023 Elsevier.*

Ratios of chemiluminescence intensities between multiple species, such as OH*/NO* and CN*/NH*, are used to assess equivalence ratios and combustion conditions [16]. The CN*/NH* ratio correlates with combustion temperature and NOx concentrations in co-firing systems, providing a suitable control parameter [25]. Ratios like CN*/NO* and NH*/CH* can also be utilized to calculate the equivalence ratios [26].

To summarize, chemiluminescence imaging of ammonia flames is relatively straightforward, typically requiring only a camera and specific filters, providing abundant information on ammonia combustion characteristics. However, the acquired chemiluminescence images result from a three-dimensional path integral and lack spatial resolution. While Abel inversion can offer quantitative properties for laminar and stable flames, turbulent, asymmetric flames are primarily characterized qualitatively or semi-quantitatively, limiting broader applications in ammonia combustion analysis.

3.2.2 Absorption spectroscopy techniques

3.2.2.1 Frequency modulation spectroscopy (FMS)

Frequency modulation spectroscopy (FMS) is an optical heterodyne spectroscopy technique capable of sensitive and rapid detection of absorption or dispersion features [27]. In FMS, the optical frequency of a laser is sinusoidally modulated using an electro-optic modulator (EOM) at frequencies equal to or greater than the halfwidth at half maximum of the absorption transition lineshape. This typically requires modulation frequencies ranging from 100 MHz to GHz rates for infrared (IR) transitions at 1 atm. The modulation effectively shifts absorption and dispersion information to the harmonics of the modulation frequency, allowing reduced lower-frequency noise.

As shown in Figure 3.3, the EOM modulates the refractive index by adjusting the applied voltage, which alters the phase or frequency of the laser beam. Without wavelength-specific absorption, the exiting electric field forms a balanced triplet, canceling beat signals at the photodetector. However, if the triplet resonates with an absorption transition, the resulting phase shift and attenuation create an imbalance, producing a detectable beat signal at the modulation frequency f_m. This beat signal can be extracted via homodyne detection (e.g. with a lock-in amplifier or double-balanced mixer), and compared with measured reference or simulated signals to infer gas properties. By scanning the carrier frequency of the laser across an absorption feature, FMS spectra can be measured [28]. FMS allows for highly sensitive measurements, down to absorbances of 10^{-8} [29], making it effective for detecting narrow absorption transitions [28].

In ammonia combustion, research utilizing FMS technology is relatively limited. Song *et al.* [31,32] used FMS in shock tube experiments for sensitive detection of NH_2 to determine the overall reaction rates (e.g., $NH_2 + NO \rightarrow$ products) over a wide range of temperatures. Since HCN is a primary cyanide species in combustion processes and contributes to NOx formation, Stuhr *et al.* [33] established a mid-IR frequency modulation spectroscopy (MIR-FMS) technique for quantitative HCN detection behind shockwaves.

The FMS technique boasts strong resilience against interference, making it suitable for stable measurements in complex environments. However, it requires

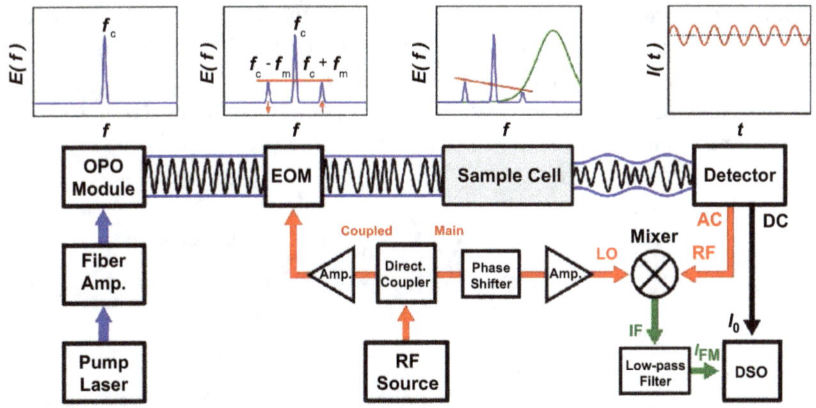

Figure 3.3 A typical experimental setup for an FMS experiment. Reprinted, with permission, from [30]. Copyright 2019 Optica Publishing Group.

high stability in the reference signal for frequency modulation, and its signal processing is intricate, somewhat limiting its application in ammonia combustion research.

3.2.2.2 Wavelength-modulation spectroscopy (WMS)

a) WMS technique

Molecules can absorb photons at specific wavelengths, leading to transition from a lower to higher energy levels. The amount of light absorbed at a given wavelength is proportional to the fraction of molecules in the absorbing quantum state. Laser absorption spectroscopy (LAS) leverages this relationship for quantitative measurements of gas temperature and composition. WMS, one of the LAS techniques, enhances the measurement accuracy of gas conditions in various environments.

As shown in Figure 3.4, WMS involves modulating the laser wavelength at frequency f_m while scanning across a molecule's absorption transition. This modulation embeds discrete frequencies into the detector signal, shifting absorption information to harmonics of f_m, which can be extracted via digital lock-in filters. WMS signals derive gas conditions through calibration-free WMS models. WMS operates with a modulation frequency much lower than the absorption linewidth, typically from a few kHz to MHz, distinguishing it from FMS.

Studies have utilized WMS techniques for ammonia concentration detection under conventional conditions. For instance, Cong and Sun [35–37] improved the conventional WMS with logarithmic-transformed data processing for trace NH_3 detection, making it applicable for ammonia combustion contexts. Tommasi *et al.* [38] reported a compact analyzer for NH_3 monitoring, addressing the problem of the line profile distortions. Du *et al.* [39] developed a low-cost and *in situ* optical sensor for multiparameter measurements in SCR processes, while Peng *et al.* [40] present a highly sensitive NH_3 sensor for high-temperature applications.

In summary, WMS offers high sensitivity, selectivity, background suppression, and robustness for gas concentration measurements but requires careful system implementation and calibration, and associated high costs. Despite these challenges, WMS remains a powerful technique for precise and accurate gas sensing in various industrial, environmental, and scientific applications. Especially, it effectively measures ammonia combustion-related species, including NH_3, NOx, and H_2O, from parts per billion (ppb) to parts per million (ppm) levels.

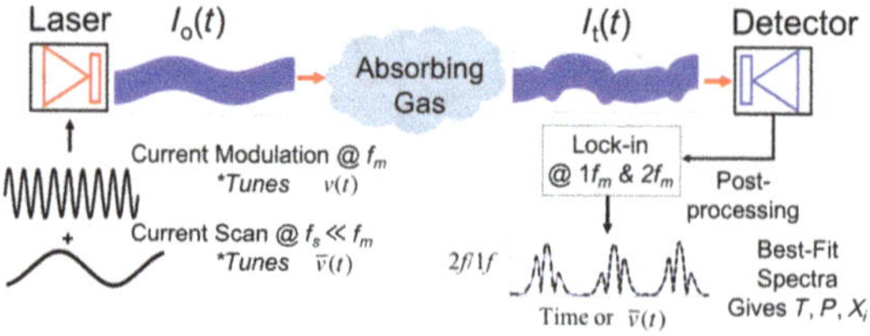

Figure 3.4 Schematic of laser-absorption measurements using WMS. Reprinted, with permission, from [34].

b) Tunable diode laser absorption spectroscopy (TDLAS)

TDLAS is a widely used WMS technique for measuring the concentration of gases such as NH_3, and H_2O in a gaseous mixture. This technique is based on the characteristics of distributed feedback (DFB) diode lasers which have narrow line widths and tunability. As shown in Figure 3.5, a TDLAS gas detector system is capable of obtaining spectra of single molecule absorption lines within the characteristic absorption spectrum region [41]. TDLAS sensors commonly exploit the laser's fast tuning capability to modulate the wavelength, causing it to sweep back and forth across an absorption feature at a precise modulation frequency [42]. When narrow-band light passes through a gas cell, Beer's law describes the exact relationship between IR light intensity and gas concentration, as follows:

$$I = I_0 \cdot e^{kP}$$

where I is the intensity of light striking the detector, I_0 is the initial intensity of radiation, k is a system-dependent constant, and P is the concentration of the gas to be measured [42].

Compared with WMS techniques, TDLAS typically employs laser wavelength modulation and direct absorption spectroscopy (DAS) techniques, offering higher sensitivity and selectivity. TDLAS is mainly used for gas concentration measurements, making it ideal for detecting various gas concentrations and temperatures in ammonia flames. As shown in Figure 3.6, many critical gas species, including NH_3, exhibit absorption in the IR spectral region. TDLAS detects whose molecules are rovibrationally excited in the IR line spectrum by lasers [43]. Similarly, combustion products such as NO, NO_2, and H_2O show absorption in the visible and near-ultraviolet regions. TDLAS measures gas concentration and temperature, with the latter often inferred from the detection of special components like H_2O, following

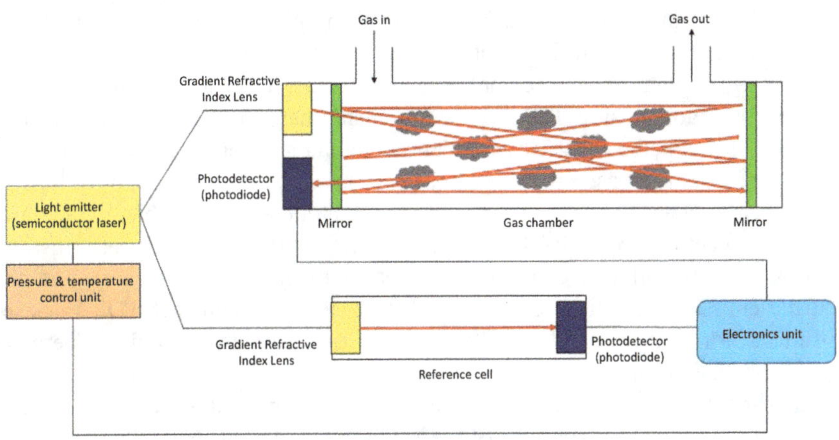

Figure 3.5 Schematic of a TDLAS gas detector system. Reprinted, with permission, from [43]. Copyright 2019 Elsevier.

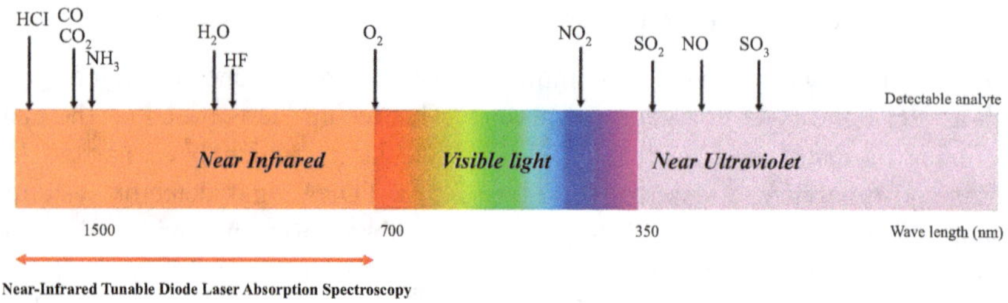

Figure 3.6 Detectable analytes in the electromagnetic spectrum. Reprinted, with permission, from [43]. Copyright 2019 Elsevier.

certain principles [44]. For NH_3 detection, the wavelength spectrum range from 1450 to 1560 nm (approximately 6400–6900 cm^{-1}) is typically chosen [81].

The TDLAS technique mainly includes DAS and WMS techniques. DAS features a straightforward experimental setup but limited sensitivity. In contrast, WMS avoids low-frequency noise interference by superimposing high-frequency modulated signals. WMS is further divided into Calibration (C-WMS) and Calibration-Free (CF-WMS) approaches. CFWMS uses the fitting method to derive concentration, eliminating the need for periodic calibration and providing much convenience. Meanwhile, normalizing the second harmonic (2f) signal using the primary harmonic (1f) signal can effectively eliminate effects from light intensity fluctuations, detector gain, and other factors [44].

TDLAS is widely used for industrial ammonia measurements. Biondo *et al.* [45] presented an *in situ* multispecies portable emission measurement system (PEMS) capable of measuring tailpipe emissions, including H_2O, CO_2, CO, NO, NH_3, and CH_4 without the necessity of online calibration. Lu *et al.* [46] developed an efficient, portable, real-time, and high-precision NH_3 remote sensor system adopting the 2f/1f signal processing technique. The detection limit matches Chinese safety standards for indoor NH_3 concentration of 40 ppm, suitable for detecting ammonia leaks in industrial production processes.

In high temperatures and complex environments closer to ammonia combustion, several studies using TDLAS measurements have been conducted. Sharma *et al.* [47] developed a harsh environment ammonia slip sensor based on TDLAS for reliable *in situ* measurements for gas turbine exhaust in misalignment-prone harsh environments. Thoma *et al.* [48] proved open-path tunable diode laser absorption spectroscopy (OP-TDLAS) as a viable alternative to Fourier transform spectroscopy for pollutant concentration measurements, applied to ammonia and methane emission detection. Cui *et al.* [49] developed a compact TDLAS-based instrument for real-time NH_3 (1–20 ppm) measurements in high-temperature applications using the 1512.2 nm absorption line. Guo *et al.* [44] reported a laser sensor that simultaneously measures NH_3, NO, and temperature at high temperatures and humidity, detecting H_2O absorption peaks at 6609.85 cm^{-1} and 6612.02 cm^{-1}, and performing temperature calculations using two-line thermometry through H_2O absorption, as shown in Figure 3.7.

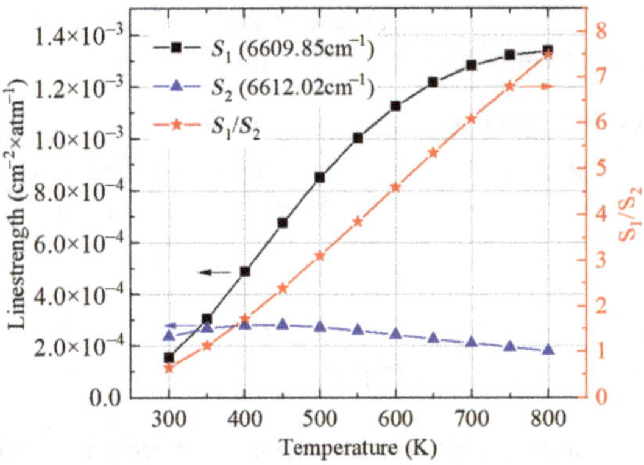

Figure 3.7 Intensity of the two ammonia spectral lines and their ratio with temperature. Reprinted, with permission, from [44]. Copyright 2024 Elsevier.

Lately, Zhu *et al.* performed time-resolved NO measurements of $NH_3/O_2/Ar$ [50], and $NH_3/H_2/O_2/Ar$ [51] mixtures using TDLAS in a shock tube covering a wide temperature range with various pressures and equivalence ratios. This study investigated the mechanisms of NH_3 oxidation and NO formation, contributing to NOx pollutant evolution research. They also developed a mid-IR quantum cascade laser (Mid-IR QCL)-based TDLAS spectrometer for ammonia detection, capable of operating in low- and high-frequency (140 Hz and 10 kHz) scan modes and fixed-wavelength mode. This technique enables the measurement of NH_3 line parameters and absolute NH_3 quantification, providing spectrally resolved NH_3 absorption cross-sections across pressures of 1.46–3.25 bar and temperatures of 1000–1800 K, enhancing the understanding of NH_3 evolution behavior in high-temperature environments [50].

Apart from ammonia combustion, TDLAS is applied in various applications, such as monitoring gas emissions, including NH_3, NO, and H_2O selective catalytic reduction (SCR) systems. Research has focused on developing gas sensors using TDLAS techniques [44,52–54] for online monitoring of SCR outlet flue gas and optimizing SCR performance in thermal power plants. Besides, improvements have been made to provide new or enhanced line intensities and increase signal immunity for ammonia absorption [43,55–58], resulting in lower uncertainty and more accurate ammonia combustion measurements.

In summary, TDLAS offers high sensitivity, quick response, field online application, multipoint remote monitoring, and selectivity for target trace gases, making it widely employed for detecting atmospheric trace gases from ppb to ppm levels. However, it has some drawbacks. In different temperature conditions, the amplitude of the second harmonic detected by a lock-in amplifier fluctuates with ambient temperature variations for the same gas concentration, and noise from the light source or optical system reduces detectability. The requirement for a long optical path length and consequently large stationary configuration (e.g. long path multipass cells) challenges compact and portable gas sensor applications [43].

Additionally, TDLAS lacks temporal and 2D/3D spatial resolution, limiting its adaptability for 2D/3D ammonia combustion measurements.

3.2.2.3 Cavity-ringdown spectroscopy (CRDS)

CRDS is a linear, DAS. The absorption spectroscopy is generally governed by the Beer-Lambert law for the measurement of sample concentration, which is given by

$$I_{\text{out}} = I_{\text{in}} \cdot \exp(-\alpha \cdot L)$$

where I_{in} represents the incident light intensity and I_{out} is the transmitted light intensity after passing through a sample with absorption coefficient, α, and path length, L.

For samples with low absorption coefficients, the sensitivity of DAS is limited by the accurate measurement of changes in transmitted intensity, i.e. $\delta I = (I_{\text{in}} - I_{\text{out}}) / I_{\text{in}}$, which is indeed greatly influenced by intensity fluctuations of the light source. The issue is particularly critical for gaseous samples, as they possess the lowest absorption coefficient, leading to the transmitted signal being mostly obscured by system noise. Enhancing detection sensitivity can be achieved by increasing the sample path length, thereby improving the signal-to-noise ratio (SNR). CRDS enhances light-matter interactions by at least a thousandfold within a small sample length [59].

CRDS utilizes a high-finesse and stable optical cavity formed by two (up to four) high-reflectivity (R $\geq$ 99.99%) mirrors with specific physical parameters (as shown in Figure 3.8). When laser radiation is directed along the central axis of the optical cavity, it is trapped inside the cavity and undergoes multiple reflections between the mirrors before decaying. Therefore, the sample path length as traversed by the light is effectively enhanced by almost a thousandfold within the optical cavity, which eventually enhances the sensitivity of the system. Moreover, the key measurement parameter in the CRDS technique is ring-down time (τ), which signifies the photon lifetime inside the optical cavity, accounting for all intrinsic and sample-related losses. Hence, τ provides time-domain measurements of sample concentration, independent of laser intensity fluctuations [59].

TCRDS contain various forms of light enhancement or hybrid techniques, including optical frequency comb (OFC) cavity-enhanced absorption spectroscopy,

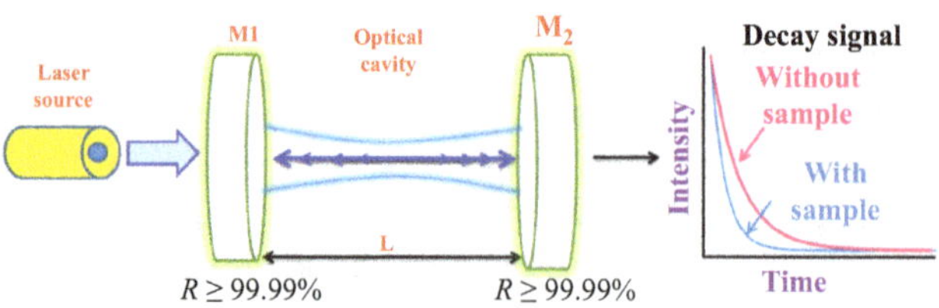

Figure 3.8 Schematic of the basic principle of the cavity ring-down technique. Reprinted, with permission, from [59]. Copyright 2021 American Chemical Society.

two-photon CRDS, Faraday rotation CRDS (FR-CRDS), swept-frequency CRDS, cavity-enhanced photoacoustic spectroscopy (CE-PAS), and evanescent wave CRDS (EW-CRDS), etc.

CRDS is mainly used for fine measurement of gas species. Paldus *et al.* [60] achieved a detection limit of 0.25 ppbv for ammonia using a 16 mW CW-QCL DFB laser. Thorpe *et al.* [61,62] improved systems using a broad-bandwidth OFC or a mode-locked erbium-doped fiber laser source, enabling investigation of a very large spectral bandwidth (1.45–1.65 μm) for various molecular absorptions, including Ar, C_2H_2, O_2, H_2O, and NH_3. Lately, He *et al.* [63] achieved detection limits of around 11 ppbv for NH_3 in N_2 and 300 ppbv for NH_3 in the air using a fiber-coupled rapidly CW-CRDS technique.

CRDS is also suitable for industrial research, though less explored in ammonia combustion. He *et al.* [63,64] developed a rapidly swept CW-CRDS approach for quantitative trace-level spectroscopic sensing of CO_2, H_2O, CH_4, and NH_3 at ambient conditions, achieving a detectable mixing ratio of ~11 ppbv. Cavity ringdown spectroscopy has also been used to measure NH and NH_2 radicals in the downstream Ar-NH_3 plasma jets, investigating the dissociation of NH_3 [65].

Overall, CRDS offers superior SNR and ultra-high detection sensitivity, ranging from parts-per-billion (10^{-9}) to parts-per-trillion (10^{-12}). It provides time-domain signal measurement without the requirement of secondary calibration. Nevertheless, it requires high-reflectivity mirrors to form an optical cavity, with reflectivity coatings effective within specific discrete and narrow wavelength ranges.

3.2.3 Fluorescence spectroscopy techniques

Fluorescence, in general, refers to spontaneous emission of light. Laser-induced fluorescence (LIF) involves a two-step process: absorption of a laser photon, followed by delayed emission of light, as shown in Figure 3.9(a). LIF uses continuous wave (CW) or pulsed lasers, tunable for specific absorption spectrum features [28].

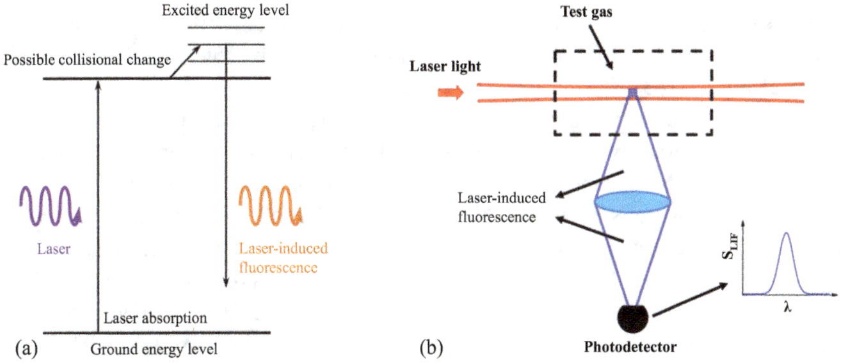

Figure 3.9 (a) Simple model for LIF and (b) schematic of LIF experimental setup. Reprinted, with permission, from [67].

In the realm of ammonia combustion research, LIF serves as a powerful tool for studying various chemical species such as NH_3, OH, NO, and NH. This technique involves exciting these target species to high-energy states using lasers and then monitoring the emitted fluorescence as the excited-state species return to their ground states. The fluorescence signal is detected at an angle to the excitation beam, allowing for detailed spatial measurements along the laser path (Figure 3.9(b)), termed planar laser-induced fluorescence (PLIF). The ability to provide omnidirectional measurements makes LIF invaluable for studying temperature, pressure, species concentration, and velocity with spatial resolutions as fine as 1 mm^3 [66].

OH-LIF is the most commonly employed technique in studies involving pure ammonia [68–70] and co-firing systems such as NH_3/H_2 [71–73], and NH_3/CH_4 [74–77]. OH plays a crucial role in ammonia flames, where its distribution can characterize the flame front, combustion intensity, heat release, and other critical features. Typically, OH fluorescence is excited using wavelengths around 283 nm by a dye laser pumped by an Nd:YAG laser, with an emission wavelength of about 310 nm.

Apart from OH-LIF, NO-LIF and NH-LIF are used to measure radical concentrations and flame structures [78]. The excitation wavelengths for NO-LIF and NH-LIF typically range around 225 nm and 303/336 nm, respectively, using Nd:YAG and dye lasers combined with various dye mixtures. For instance, NO laser-saturated fluorescence (LSF) has been used in non-stretched premixed $NH_3/CH_4/NO$/air flame to measure NO concentrations in the post-flame zone [79]. Brackmann *et al.* [80] conducted quantitative picosecond time-resolved NO-LIF measurements in $NH_3/CH_4/N_2/O_2$ premixed flames, while Xu *et al.* [81] adopted NH-PLIF to study flame/turbulence interactions in NH_3/air premixed flames at high Karlovitz numbers.

Advancements in excitation laser technologies for NO-LIF have significantly improved performance capabilities. For example, frequency-mixing crystals can generate 225.5 nm from the fundamental 676.5 nm and corresponding frequency-doubled 338 nm, facilitating quantitative NO-LIF in NH_3/H_2/air non-premixed counterflow flames [82]. Besides, a 1-D NO-LIF method that employs excitation near 236 nm to avoid ammonia absorption was combined with 1-D Raman spectroscopy, revealing a large amount of NO (approximately 1500 ppm) in the $NH_3/H_2/N_2$ jet flames [83].

Beyond these, other LIF techniques like NH_3-LIF, NH_2-LIF, CN-LIF, and polycyclic aromatic hydrocarbons (PAH)-LIF are employed for comprehensive measurements in NH_3-doped flames. Femtosecond two-photon LIF (*fs*-TPLIF) has been used to measure NH_3 in NH_3/N_2 mixtures [84] and laminar premixed NH_3/air flames [85], observing two-head bands of NH_3 around 565 nm, demonstrating its feasibility for *in situ* NH_3 combustion measurements. NH_2-LIF, excited at 532 nm using three Nd:YAG lasers with temporal stretching to 500 ns, provides the NH_2 concentration in laminar counterflow non-premixed and premixed $NH_3/H_2/N_2$/air flames, with signal sampled between 620 and 630 nm. Although NO_2-LIF interferes with NH_2-LIF, its effects are negligible on the fuel side [86]. CN B–X (1–0) lines near 359 nm, excited using a Nd:YAG-pumped dye laser and a solid-state alexandrite laser, combined with detection of the (1–1) band near 387 nm, capture CN concentrations in NH_3 doped premixed CH_4/air flames, offering around four times stronger signals than conventional Nd:YAG pumped dye laser systems. This method will be useful for studying blended ammonia/hydrocarbon fuel flames [87]. Lastly, PAH-LIF, excited at 266 nm, was used to detect PAH content and study PAH formation in NH_3/C_2H_4 [88–94] and NH_3/n-heptane [95] flames, revealing the inhibition effects of NH_3 addition on soot formation.

Due to the high cost of lasers, researchers try to use a single Nd:YAG pumped laser system with different dye mixtures to obtain various LIF signals separately. Given the strong interactions among the radicals in flames, simultaneous two- or three-species LIF techniques are frequently employed. In ammonia/hydrocarbon fuel cofiring systems, CH-/OH-LIF, CH_2O-/OH-LIF, or NH_3-/CH-LIF are used. Simultaneous CH-LIF (excited at 314 nm) and OH-LIF are used to measure concentrations in $NH_3/C_2H_4/N_2$/air turbulent flame, showing that CH and OH decrease with ammonia substitution, mainly upstream of soot formation regions [96]. Two-photon NH_3-LIF and CH-LIF, with excitation at 304.8 nm and 383.3 nm, are effective for NH_3 detection in both laminar and turbulent NH_3/CH_4/air flames up to 5 bar (Figure 3.10) [97].

In NH_3 and NH_3/H_2 flames, two-species LIFs include NH-/NO-LIF [98–102], NO-/OH-LIF [103,104], NH-/OH-LIF [105], and NH_3-/NH-LIF [106], and three-species LIFs include OH-/NH-/NO-LIF [107], NH_3-/NH-/NO-LIF (Figure 3.11) [108], and NO-/OH-*ns*PLIF/H-*fs* two-photon absorption LIF [109] are widely implemented, although these apparatus systems are more complex. Findings reveal that local NO concentration heavily depends on local OH levels in non-premixed NH_3/air flame [103], with overlapped NH and OH layers in turbulent premixed

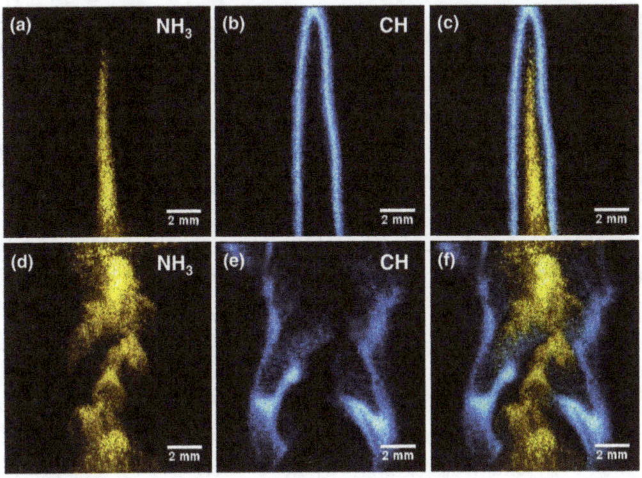

Figure 3.10 Single-shot NH_3 and CH LIF images simultaneously acquired in (a–c) laminar and (d–f) turbulent NH_3/CH_4/air/flames. Reprinted, with permission, from [97]. Copyright 2014 Springer Nature.

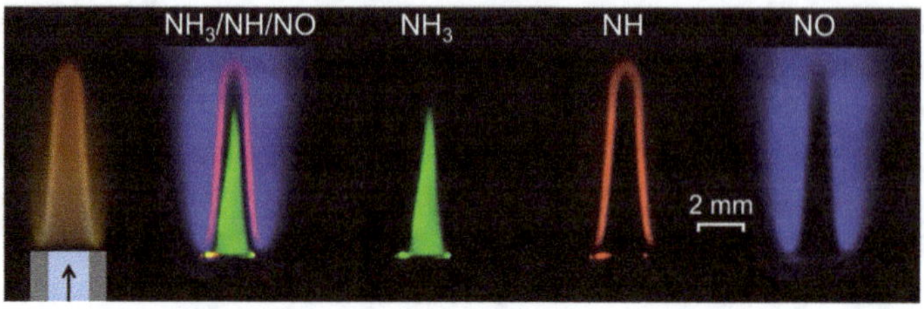

Figure 3.11 Three-species PLIF images in premixed NH_3/H_2/air flame. Reprinted, with permission, from [108]. Copyright 2023 Elsevier.

NH_3/air flame [114]. NH-PLIF images effectively characterize the fuel consumption layer since NH formation is associated with NH_3 consumption. NO spans in a wide region across preheat, reaction, and post-flame zones, forming mainly in the reaction zone but transported by turbulence and convection to the post-flame zone [99]. NO formation also occurs in the inner premixed flames and the diffusion flame, marking burnt product locations [100,111], while NH profiles can serve as flame front markers [100,101].

Two-species or three-species LIF signals are commonly used for qualitative analysis, while some studies [83,98,107] have determined quantitative profiles through integrated LIF theory, giving detailed insights into radical distribution in flames. To manage the high cost of laser systems, researchers employ frequency-mixing units to extend multiple-laser functionalities. For instance, Wang *et al.* [102] employed 303.545 nm to excite NH-PLIF, and 236.214 nm to excite NO-LIF, which was generated by combining residual beam energy at 303.545 nm mixed with the fundamental of the pump laser at 1064 nm. Wang *et al.* [108] also developed a PLIF technique using a single-dye laser for simultaneous, single-shot PLIF imaging of the reactant NH_3, radical NH, and pollutant NO in NH_3/H_2 flames. In this system, wavelengths near 304 nm and 237 nm are available at the same time with a single dye laser by combining outputs of the frequency-doubling and frequency mixing units. Similarly, Wang *et al.* [106] simplified experimental complexities by co-exciting two-photon NH_3-LIF and single-photon NH-LIF at 304.79 nm simultaneously.

In flames of hydrocarbon fuels, $[CH_2O][OH]$ is usually served as a heat release rate surrogate (HRRS), leading to extensive research using simultaneous CH_2O-/OH-LIF imaging. Conversely, in ammonia flames, $[NH_2][NO]$ is found to be a promising HRRS [110,111]. Thus, NH-/NO-LIF results help understand heat release features in ammonia flames.

Additionally, OH-LIF and NO-LIF techniques not only quantify species concentrations but also derive temperature information using the two-line LIF method. Rahinov *et al.* [112] and Bell *et al.* [78] have used OH-PLIF and NO-PLIF for temperature measurements in NH_3/CH_4/air flames, achieving approximately 50 K error margins. Recent studies, such as the single-shot 2D temperature measurements using thermally assisted OH PLIF excited in the $A^2\Sigma^+ - X^2\Pi$ (1–0) band from

Dulin *et al.* [113], offer to detect a temperature range of 1500–1900 K with an root mean square (RMS) error below 80 K, which is crucial for identifying high-temperature zones in flames.

Overall, LIF provides versatility and effectiveness in advancing the understanding of ammonia combustion through non-destructive and real-time/spatial measurements of species concentration, heat release, and temperature. However, its complex instrumentation and susceptibility to background fluorescence limit its applicability. Ongoing research focuses on simplifying systems and enhancing signal strength to expand the role of LIF techniques in ammonia combustion studies.

3.2.4 Scattering spectroscopy techniques

3.2.4.1 Spontaneous Raman scattering (SRS)

When an incident photon interacts with a crystal lattice or molecule, it can be scattered either elastically or inelastically. In elastic scattering, known as Rayleigh scattering, the energy of the scattered photon equals that of the incident photon. In inelastic scattering, or the Raman effect [114], the energy of the scattered photon differs. Stokes Raman scattering occurs when the scattered photon has lower energy (red-shifted), while anti-Stokes Raman scattering occurs when it has higher energy (blue-shifted). The Raman signal, as shown in Figure 3.12, is independent of collisional quenching and varies linearly with laser power and molecule number density [115].

The SRS technique is used to measure temperature and species concentrations, providing valuable insights into turbulence/chemistry interactions and the thermochemical structure of hydrogen/air and hydrocarbon/air flames. However, its application to ammonia flames faces challenges such as a lack of detailed high-temperature spectra, strong flame luminosity, and fluorescence interference. Tang *et al.* [116] extend this technique for ammonia flames, developing a novel approach to estimate the temperature dependence of the Raman signal and fluorescence

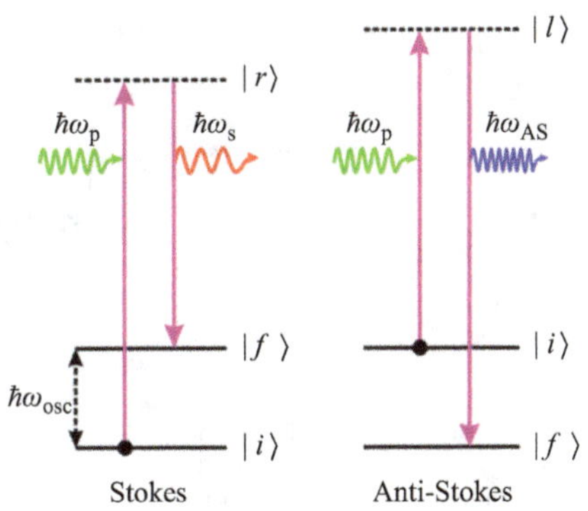

Figure 3.12 Energy transfer process in Stokes (left) and anti-Stokes (right) Raman scattering. Reprinted, with permission, from [114]. Copyright 2019 Springer Nature.

interference in high-pressure ammonia counterflow diffusion flames. Yang *et al.* [117] measured ammonia Raman spectra between 299 K and 760 K using a continuous wave 532 nm laser, normalizing the spectra to study the temperature behavior. Guo *et al.* [118] investigated the flame morphology and structure in ammonia/ethylene and ammonia/acetylene laminar diffusion flames using SRS technology. Najafi *et al.* [119] studied the spatial distribution of temperature and major species, including CH_4, NH_3, CO_2, H_2O, N_2, CO, O_2, and H_2, in ammonia flames through SRS. These studies enhance our understanding of ammonia combustion and demonstrate the potential of Raman spectroscopy for detailed analysis of flame characteristics.

SRS technique uniquely provides instantaneous, spatially resolved measurements of temperature and most major species simultaneously using a single laser beam without tuning. These simultaneous measurements are particularly important in non-premixed combustion, where understanding local stoichiometry is essential [115]. However, this technique is not widely used in ammonia combustion, which requires further development and improvement to overcome existing challenges.

3.2.4.2 Coherent anti-Stokes Raman spectroscopy (CARS)

CARS spectroscopy is an important tool for measuring temperature and species concentrations in combustion. As shown in Figure 3.13, three laser beams converge at a common intersection point. This includes an oscillating polarization at the Raman frequency $W_{\text{pump}} = W_{\text{probe}}$, where W_{pump} and W_{probe} are the frequencies of

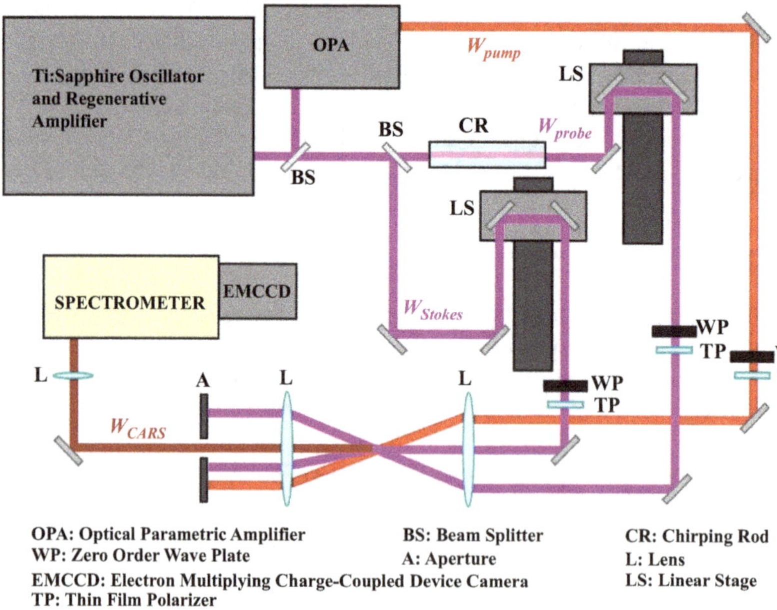

Figure 3.13 An experimental setup for CARS (four-wave mixing). Reprinted, with permission, from [120]. Copyright 2015 Elsevier.

the beams. The CARS signal frequency is $W_{CARS} = 2W_{pump} - W_{Stokes}$, with W_{Stokes} being the frequency of the third laser beam. The CARS signal propagates as a laser-like beam, following the phase-matching condition, and is then directed to a spectrograph. The resulting spectrum provides temperature information.

Though CARS offers high Raman conversion efficiency and moderate resolution, it is rarely used in ammonia combustion measurement. In premixed NH_3/air flame from a porous-plug burner, rotational CARS has been used to measure temperatures and relative oxygen concentrations [107]. The limitations include its single-point measurement capability, lacking spatial and time-resolved resolution, and the need for a precise experimental setup.

3.3 Optical diagnostics

3.3.1 Schlieren method

The physical principle of the Schlieren imaging method is rooted in Snell's Law, which describes how light changes direction when passing through different mediums. In a homogeneous medium like a vacuum, light travels uniformly at a constant velocity. When encountering inhomogeneous media such as moving fluids, light rays refract and deviate from their original path, resulting in Schlieren patterns [121]. A basic Schlieren setup typically implements lenses arranged in a straight line, as shown in Figure 3.14. A parallel light beam passes through the Schlieren test subject, and then passes through some lens. The resulting image is focused onto the knife-edge cut-off and projected onto a screen or photographic sensor.

In ammonia combustion studies, the Schlieren method is widely used in the laminar burning velocity measurement in spherical flames [123–129], and for visualizing the structure of ammonia spray flames [72,130–134]. Schlieren setups are straightforward to implement. However, limited information can be obtained, and precise data on specific species distributions are not available.

3.3.2 Scattering techniques

Rayleigh scattering occurs when the dimension of the scatter is much smaller than the wavelength of the incident electromagnetic radiation, showing a pronounced

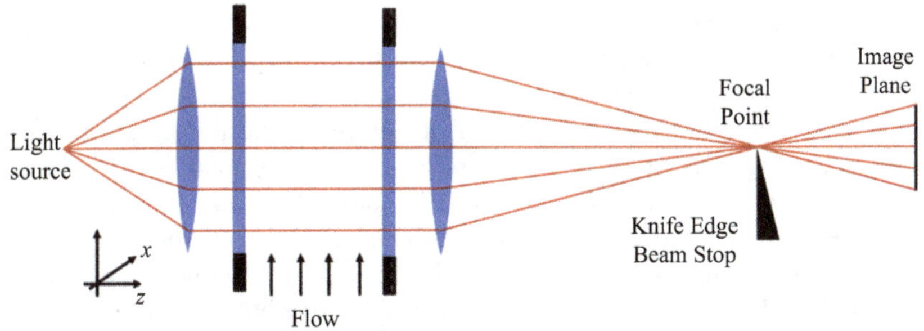

Figure 3.14 A typical Schlieren optics system. Reprinted, with permission, from [122].

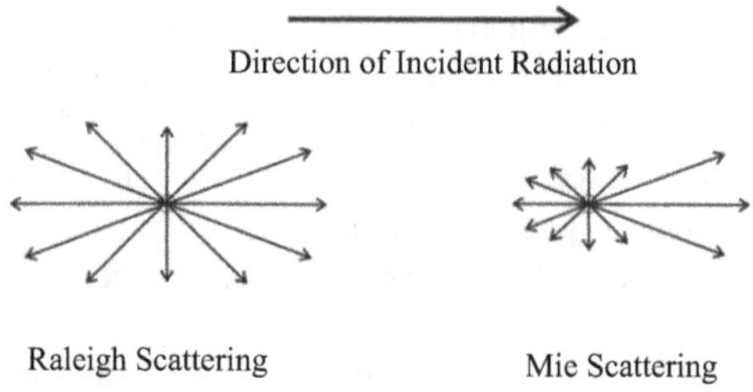

Figure 3.15 Schematic of Raleigh scattering and Mie scattering. Reprinted, with permission, from [135].

wavelength dependency. Mie scattering, on the other hand, occurs when the dimension of the scatter is much larger than the wavelength of the incident electromagnetic radiation [135]. Figure 3.15 illustrates these two scattering phenomena.

In ammonia combustion studies, particle imaging velocimetry based on Mie scattering signals is the most common optical method for velocity measurement. Particles like titanium dioxide (TiO_2) or aluminum trioxide (Al_2O_3), featuring an average diameter of several micrometers, are usually utilized for seeding in the flow, and a 532 nm laser serves as the light source for particle scattering [73,75,136,137]. This method provides two-dimensional or three-dimensional velocity distribution in ammonia flames, facilitating detailed analysis of the combustion jet core area, vortex, etc.

3.4 Sampling and analysis methods

3.4.1 Gas chromatography (GC)

GC is a physical separation method primarily used to separate various chemical compounds introduced into the facility as a mixture and to quantitatively determine their relative proportions. Combined with other physicochemical methods, GC can also provide qualitative (structural) information about the separated substances. The method is limited to relatively volatile (low-molecular-weight) compounds. Separation occurs based on the relative affinity of various mixture components to the stationary phase (solid or liquid), while the mobile phase (gas) migrates them through the system. GC is a dynamic separation method, where the separation of components occurs in a heterogeneous phase system [138]. Figure 3.16 illustrates the operating process of the GC system.

GC is commonly used for analyzing major species in the ammonia combustion process, including NH_3, H_2, O_2, N_2. Ripepi *et al.* [140] developed a real-time GC-MS (mass spectrometry) method for sensitive isotope NH_3 quantification without requiring external sample manipulations. Zaffaroni *et al.* [141] proposed an *in situ* GC method for ammonia detection applicable to various electrochemical devices under a wide range of conditions. This *in situ* GC method quantifies ammonia present in the gas phase in 5 minutes, with a lower detection limit of about 150 ppb.

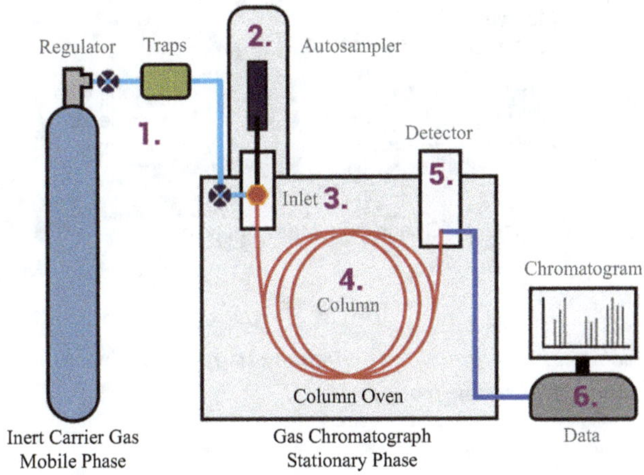

Figure 3.16 Schematic of a GC system. Reprinted, with permission, from [139].

Liu *et al.* [142] analyzed the major ammonia-rich combustion reactants (NH_3) and products (H_2 and N_2) using a fast sampling system and GC, observing that H_2 production increases with the initial temperature, and initially increases and then decreases with the initial pressure. GC offers high resolution and effective separation of components in complex mixtures, but its standalone use is limited by detection sensitivity and the range of detectable compounds. Therefore, many studies now combine GC with mass spectrometry (MS) to enhance analytical capabilities, which will be introduced in the following section.

3.4.2 Mass spectrometry

MS is a technique used for quantitative analysis of small gas molecules, providing high sensitivity and specificity. MS measurements are performed in the gas phase on ionized analytes. Molecules are ionized and then introduced into an electrical field, where they are sorted by the mass-to-charge ratio (m/z). An MS consists of an ion source, a mass analyzer measuring the m/z of the ionized analytes, and a detector that registers the number of ions at each m/z value [143]. The functional components of a quadrupole MS are shown in Figure 3.17.

MS can detect a broad range of ammonia combustion products, including NO, NO_2, N_2O, H_2O, O_2, and N_2 with high resolution. Synchrotron vacuum ultraviolet photoionization mass spectrometry (SVUV-PIMS) was implemented to investigate NH_3/CH_3OH oxidation in a jet-stirred reactor (JSR) [145], with important species (e.g. NO, NO_2, N_2O, N_2, H_2, and H_2O) been effectively detected and quantified. Besides, online SVUV-PIMS and GC techniques were applied to study the oxidation process of ammonia or blends of ammonia and other fuels [146–149], providing sufficient information about species mole fraction for further analysis. The coupling of GC and MS provides a powerful analytical tool with high sensitivity and specificity for quantitative and qualitative analysis. However, it also has limitations, such as high costs, and the need for specialized training.

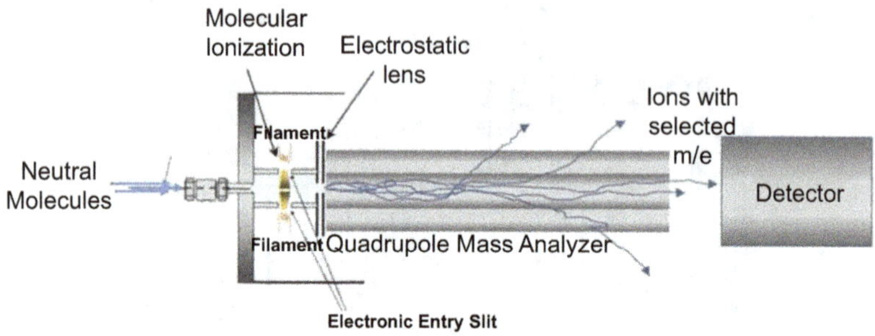

Figure 3.17 Schematic of functional components of a quadrupole MS. Reprinted, with permission, from [144].

3.4.3 Gas analyzer

Gas analyzers, typically based on electrochemical sensors, can detect various species. In ammonia combustion studies, gas analyzers are used more widely than GC and MS due to their low cost and simplicity. Gas analyzers can measure the mole fractions of CO, NO, NO_2, NH_3, H_2, and other species in combustion exhaust [14,70,150–152]. The measurement accuracy is about $\pm$ 1% F.S., which is relatively lower than that of spectroscopy-based techniques like Fourier-transform infrared spectroscopy (FTIR).

3.4.4 Fourier-transform infrared spectroscopy

FTIR is a technique for measuring mole fractions of various gas compounds, based on principles of absorption spectroscopy. Many compounds absorb IR energy at specific wave numbers (or wavelength) proportionally to their concentration. In an FTIR spectrometer, some IR radiation is absorbed by the sample, while some is transmitted. The resulting molecular absorption and transmission response can be used to identify the components and their concentrations. Compared to other IR techniques, FTIR can measure multiple components in real-time using an interferometer that collects a broad range of wavelengths [153]. The basic operation principle of FTIR spectroscopy is shown in Figure 3.18.

FTIR is capable of simultaneously measuring NH_3, N_2O, CO_2, CO, NO, NO_2, CH_4, etc., making it an attractive tool for testing ammonia combustion exhaust. Several studies have utilized FTIR to analyze NOx and residual NH_3 emissions [137,154–157]. FTIR has advantages such as high sensitivity and precision, rapid data acquisition, and a high SNR, which are beneficial for exhaust emissions detection. However, it also has limitations including potential interferences from H_2O and CO_2, high cost, and the complexity of data interpretation.

3.4.5 Quantum cascade lasers (QCL)

Electrons may occupy states in one of two continuous energy bands (valence (low energy electrons) and conduction (high energy electrons) bands). Conventional semiconductor laser diodes generate light by jumping between bands.

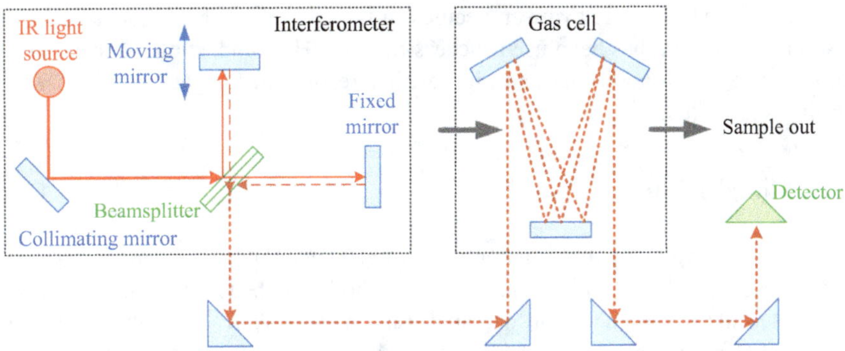

Figure 3.18 Schematic operation principle of FTIR spectroscopy. Reprinted, with permission, from [153]. Copyright 2021 MDPI.

However, a QCL does not use semiconductors, but a series of thin layers that form a superlattice. This superlattice creates a varying electric potential, which in turn leads to a varying probability of electrons occupying different energy positions along the system. This is referred to as one-dimensional multiple quantum well confinement and leads to the splitting of the band of permitted energies into a number of discrete electronic sub-bands. The hundreds of nanometer-scale semiconductor layers allows to create a specific electronic density of states which leads to optical gain and lasing between cascaded electron sub-bands.

Additionally, in semiconductor laser diodes, electrons and holes are annihilated after recombining across the band gap. However, in a unipolar QCL, once an electron has undergone a transition and emitted a photon in one period of the superlattice, it can tunnel into the next period of the structure where another photon can be emitted. The process gives rise to the name "cascade," which can make quantum efficiency greater than unity with higher power outputs.

The particularity of this concept is that the emitted wavelength does only depend on the layer thicknesses, not on the constituent materials. As a consequence, the same technology is capable of producing lasers in a very broad spectral range (λ from 3 to 20 μm). This is however a complex technology, that requires a very demanding control of all the fabrication steps, from the quantum design, to the multi-layer material growth, the device fabrication, the thermal management, and the optical control of the emitted light [158].

Companies such as Emerson [159] have employed the technology to create fast, high-resolution spectroscopy to detect and identify a range of molecules in the mid-IR wavelength range (2,500–12,000 nm). Coupled with tunable diode laser (TDL) spectroscopy, a single instrument is able to provide greater insight and monitoring in both the near and mid-IR range of spectroscopic light, delivering data on emissions. These systems include up to six high-resolution lasers to measure both the near and mid-IR spectral regions for real-time, optimal gas measurement and analysis down to sub ppm concentrations of species such as NH_3, NO, N_2O,

NO_2, O_2, and H_2O, all relevant molecules in the oxidation of ammonia via combustion. The system has been used successfully in NH_3 combustion test campaigns [160], hence providing a fast and high-quality resolution for gas emissions when burning such a fuel.

3.5 Conclusion

This chapter introduces three major measurement techniques for ammonia combustion systems. It is evident that some methods, such as chemiluminescence, laser-induced fluorescence, and gas analyzers, have already been widely used in ammonia combustion studies, while others are still in development. No single measurement technique can capture all the necessary data on temperature, composition, flow field, and other parameters for ammonia combustion. Therefore, combining various measurement techniques is crucial. Additionally, the measurement techniques for ammonia combustion systems will continue to evolve, becoming more high-speed, multi-species, and precisely quantified, with improved time-resolved and spatial resolutions.

References

[1] I. Weeks. *Chemiluminescence Immunoassay*, Elsevier, Amsterdam, 1992.

[2] C. Dodeigne, L. Thunus, and R. Lejeune. Chemiluminescence as diagnostic tool. A review. *Talanta* 51 (2000) 415–439.

[3] T. Förster. Zwischenmolekulare energiewanderung und fluoreszenz. *Annalen Der Physik* 2 (1948) 55–75.

[4] A.K. Campbell, and A. Patel. A homogeneous immunoassay for cyclic nucleotides based on chemiluminescence energy transfer. *Biochemical Journal* 216 (1983) 185–194.

[5] D.J. Campbell, and R. Hunter. Drop-cone penetration in situ and on minimally disturbed soil cores. *Journal of Soil Science* 37 (1986) 153–163.

[6] T. Förster. 10th Spiers memorial lecture. Transfer mechanisms of electronic excitation. *Discussions of the Faraday Society* 27 (1959) 7–17.

[7] L. Stryer. Fluorescence energy transfer as a spectroscopic ruler. *Annual Review of Biochemistry* 47 (1978) 819–846.

[8] A.A. Konnov. An exploratory modelling study of chemiluminescence in ammonia-fuelled flames. Part 1. *Combustion and Flame* 253 (2023) 112788.

[9] S. Mashruk, X. Zhu, W.L. Roberts, T.F. Guiberti, and A. Valera-Medina. Chemiluminescent footprint of premixed ammonia-methane-air swirling flames. *Proceedings of the Combustion Institute* 39 (2023) 1415–1423.

[10] G. Capriolo, G. Issayev, X. Zhu, J. Vargas, and T.F. Guiberti. An experimental and kinetic study of $OH(A^2\Sigma^+)$ formation and quenching in ammonia-hydrogen-air flames. *Combustion and Flame* 260 (2024) 113258.

[11] D. Chen, J. Li, X. Li, Y. Guo, H. Huang, and N. Kobayashi. Experimental studies on the OH^* chemiluminescence and structure characteristics in NH_3/H_2 and NH_3/cracked gas swirl flames. *International Journal of Hydrogen Energy* 52 (2024) 1370–1379.

[12] D.E. Thomas, C. Wadkar, C.F.W. Goertemiller, and W.F. Northrop. Structure and nitric oxide formation in laminar diffusion flames of ammonia-hydrogen and air. *Fuel* 362 (2024) 130764.

[13] V.N. Viswamithra, and S.K. Menon. A distributed fuel injection approach to suppress lean blow-out and NO_x emissions in a methane-ammonia-fueled premixed swirl combustor. *Journal of Engineering for Gas Turbines and Power-Transactions of the Asme* 144 (2022) 061015.

[14] X. Zhu, A.A. Khateeb, T.F. Guiberti, and W.L. Roberts. NO and OH^* emission characteristics of very-lean to stoichiometric ammonia-hydrogen-air swirl flames. *Proceedings of the Combustion Institute* 38 (2021) 5155–5162.

[15] D. Pugh, P. Bowen, A. Valera-Medina, A. Giles, J. Runyon, and R. Marsh. Influence of steam addition and elevated ambient conditions on NO_x reduction in a staged premixed swirling NH_3/H_2 flame. *Proceedings of the Combustion Institute* 37 (2019) 5401–5409.

[16] X. Zhu, W.L. Roberts, and T.F. Guiberti. UV-visible chemiluminescence signature of laminar ammonia-hydrogen-air flames. *Proceedings of the Combustion Institute* 39 (2023) 4227–4235.

[17] X. Zhu, A.A. Khateeb, W.L. Roberts, and T.F. Guiberti. Chemiluminescence signature of premixed ammonia-methane-air flames. *Combustion and Flame* 231 (2021) 111508.

[18] V. Scharl, T. Lackovic, and T. Sattelmayer. Characterization of ammonia spray combustion and mixture formation under high-pressure, direct injection conditions. *Fuel* 333 (2023) 126454.

[19] G.T. Kim, J. Park, S.H. Chung, and C.S. Yoo. Effects of non-thermal plasma on turbulent premixed flames of ammonia/air in a swirl combustor. *Fuel* 323 (2022) 124227.

[20] M. Suarez, M.K. Hay, K. Naude, and W.D. Kulatilaka. NH_2^* chemiluminescence in premixed $NH_3/H_2/N_2$-air flames. *AIAA SciTech 2024 Forum*, 2024, https://arc.aiaa.org/doi/abs/10.2514/6.2024-0391.

[21] W. Weng, M. Alden, and Z. Li. Visible chemiluminescence of ammonia premixed flames and its application for flame diagnostics. *Proceedings of the Combustion Institute* 39 (2023) 4327–4334.

[22] V. Scharl, and T. Sattelmayer. Spectroscopic investigation of diesel-piloted ammonia spray combustion. *Fuel* 358 (2024) 130201.

[23] A. Karan, G. Dayma, C. Chauveau, and F. Halter. Insight into the inner structure of stretched premixed ammonia-air flames. *Proceedings of the Combustion Institute* 39 (2023) 1743–1752.

[24] M.O. Vigueras-Zuniga, M.E. Tejeda-del-Cueto, S. Mashruk, M. Kovaleva, C.L. Ordonez-Romero, and A. Valera-Medina. Methane/ammonia radical formation during high temperature reactions in swirl burners. *Energies* 14 (2021) 6624.

[25] A. Leipertz, R. Obertacke, and F. Wintrich. Industrial combustion control using UV emission tomography. *Symposium (International) on Combustion* 26 (1996) 2869–2875.

[26] T.F. Guiberti, N.N. Shohdy, S. Cardona, X. Zhu, L. Selle, and C.J. Lapeyre. Chemiluminescence- and machine learning-based monitoring of premixed ammonia-methane-air flames. *Applications in Energy and Combustion Science* 16 (2023) 100212.

[27] G.C. Bjorklund, M.D. Levenson, S. Jose, and C. Ortiz. Frequency modulation (FM) spectroscopy. *Applied Physics B* 32 (1983) 145–152.

[28] R.K. Hanson, R.M. Spearrin, and C.S. Goldenstein. *Spectroscopy and Optical Diagnostics for Gases*, Springer, New York, 2015.

[29] J.A. Silver. Frequency-modulation spectroscopy for trace species detection: theory and comparison among experimental methods. *Applied Optics* 31 (1992) 707–717.

[30] M. Stuhr, N. Faßheber, and G. Friedrichs. Single-tone mid-infrared frequency modulation spectroscopy for sensitive detection of transient species. *Optics Express* 27 (2019) 26499–26512.

[31] S. Song, R.K. Hanson, C.T. Bowman, and D.M. Golden. Shock tube determination of the overall rate of NH_2+NO→products at high temperatures. *Proceedings of the Combustion Institute* 28 (2000) 2403–2409.

[32] S.H. Song, R.K. Hanson, C.T. Bowman, and D.M. Golden. A shock tube study of the product branching ratio of the NH_2+NO reaction at high temperatures. *Journal of Physical Chemistry A* 106 (2002) 9233–9235.

[33] M. Stuhr, and G. Friedrichs. Mid-infrared frequency modulation detection of HCN and its reaction with O atoms behind shock waves. *Journal of Physical Chemistry A* 126 (2022) 9485–9496.

[34] The Goldenstein Group. Wavelength-modulation spectroscopy. https://goldensteingroup.com/our-research/wavelength-modulation-spectroscopy/ (accessed 29 June 2024).

[35] M. Cong, and D. Sun. Detection of ammonia using logarithmic-transformed wavelength modulation spectrum. *4th Annual International Workshop on Materials Science and Engineering (IWMSE)*, Xian, China, 2018, pp. 012132.

[36] M. Cong, and D. Sun. Recovery of ammonia absorption spectroscopy using asynchronous differential detection. *Conference on Optical Sensing and Imaging Technologies and Applications*, Beijing, China, 2018, pp. 108460.

[37] M. Cong, and D. Sun. Design of a gas sensor based on asynchronous double beam structure and balanced photodetector. *Infrared Physics & Technology* 93 (2018) 20–24.

[38] E. De Tommasi, G. Casa, and L. Gianfrani. An intensity-stabilized diode-laser spectrometer for sensitive detection of NH_3. *IEEE Transactions on Instrumentation and Measurement* 56 (2007) 309–312.

[39] Z. Du, Y. Yan, J. Li, S. Zhang, X. Yang, and Y. Xiao. In situ, multi-parameter optical sensor for monitoring the selective catalytic reduction process of diesel engines. *Sensors and Actuators B-Chemical* 267 (2018) 255–264.

[40] W.Y. Peng, R. Sur, C.L. Strand, R.M. Spearrin, J.B. Jeffries, and R.K. Hanson. High-sensitivity in situ QCLAS-based ammonia concentration sensor for high-temperature applications. *Applied Physics B-Lasers and Optics* 122 (2016) 188.

[41] Y. He, Y. Zhang, R. Kan, L. Wang, K. You, and J. Liu. Study on mash gas monitoring with distributed multipoint fiber optic sensors system in coal mine. 2012 *Symposium on Photonics and Optoelectronics*, Shanghai, China, 2012, pp. 1–4.

[42] S.K. Chaulya, and G.M. Prasad. Gas sensors for underground mines and hazardous areas. In: *Sensing and Monitoring Technologies for Mines and Hazardous Areas*. Elsevier, Amsterdam, 2016, pp. 161–212.

[43] D. Kwak, Y. Lei, and R. Maric. Ammonia gas sensors: a comprehensive review. *Talanta* 204 (2019) 713–730.

[44] S. Guo, J. Li, Y. Wei, *et al.* A multi-laser hybrid absorption sensor for simultaneous measurement of NH_3, NO, and temperature in denitrification flue gas. *Infrared Physics & Technology* 136 (2024) 105034.

[45] L. Biondo, H. Gerken, L. Illmann, *et al.* Advantages of simultaneous in situ multispecies detection for portable emission measurement applications. *SAE International Journal of Sustainable Transportation, Energy, Environment, & Policy* 2 (2021) 161–171.

[46] H. Lu, C. Zheng, L. Zhang, *et al.* A remote sensor system based on TDLAS technique for ammonia leakage monitoring. *Sensors* 21 (2021) 2448.

[47] R. Sharma, S. Maity, A. Bekal, S. Vartak, A. Sridharan, and C. Mitra. Optical sensors for harsh environment applications. *Sensors for Extreme Harsh Environments II 9491*, 2015, pp. 29–35.

[48] E.D. Thoma, R.C. Shores, E.L. Thompson, *et al.* Open-path tunable diode laser absorption spectroscopy for acquisition of fugitive emission flux data. *Journal of the Air & Waste Management Association* 55 (2005) 658–668.

[49] H. Cui, J. Li, F. Wang, G. Lv, W. Wang, and J. Fan. High-temperature determination of ammonia by tunable diode laser absorption spectroscopy (TDLAS). *Instrumentation Science & Technology* 52 (2024) 221–235.

[50] D. Zhu, S. Agarwal, L. Seifert, B. Shu, R. Fernandes, and Z. Qu. NH_3 absorption line study and application near 1084.6 cm^{-1}. *Infrared Physics & Technology* 136 (2024) 105058.

[51] D. Zhu, L. Ruwe, S. Schmitt, B. Shu, K. Kohse-Hoeinghaus, and A. Lucassen. Interactions in ammonia and hydrogen oxidation examined in a flow reactor and a shock tube. *Journal of Physical Chemistry A* 127 (2023) 2351–2366.

[52] X. Guo, F. Zheng, C. Li, *et al.* A portable sensor for in-situ measurement of ammonia based on near-infrared laser absorption spectroscopy. *Optics and Lasers in Engineering* 115 (2019) 243–248.

[53] W. Wang, and Z. Zhang, Study on trace detection method of ammonia escape based on TDLAS. *4th IEEE International Symposium on Computer, Consumer and Control (IS3C)*, Taiwan, China, 2018, pp. 294–297.

[54] L. Zhang, F. Wang, L. Yu, J. Yan, and K. Cen. The research for trace ammonia escape monitoring system based on tunable diode laser absorption spectroscopy. *Spectroscopy and Spectral Analysis* 35 (2015) 1639–1642.

[55] W. Nie, R. Kan, Z. Xu, *et al.* Measurements of line strengths for some lines of ammonia in 6611–6618 cm^{-1}. *Acta Physica Sinica* 66 (2017) 054207.

[56] A. Pogany, O. Werhahn, and V. Ebert. Measurement of ammonia line intensities in the 1.5 μm region by direct tunable diode laser absorption spectroscopy. *Journal of Quantitative Spectroscopy & Radiative Transfer* 276 (2021) 107884.

[57] K. You, Y. Zhang, Y. He, and L. Wang. Automatic gain adjustment modular in open path detection based on laser technology. *Journal of Atmospheric and Environmental Optics* 7 (2012) 269–275.

[58] Y. Zhao, X. Zhao, Z. Wang, R. Zhang, and Y. Wang. A new method for eliminating background signal drift to improve the detection precision in continuous harmonic detection. *Spectroscopy and Spectral Analysis* 35 (2015) 3224–3229.

[59] A. Maity, S. Maithani, and M. Pradhan. Cavity ring-down spectroscopy: recent technological advancements, techniques, and applications. *Analytical Chemistry* 93 (2020) 388–416.

[60] B.A. Paldus, C.C. Harb, T.G. Spence, *et al.* Cavity ringdown spectroscopy using mid-infrared quantum-cascade lasers. *Optics Letters* 25 (2000) 666–668.

[61] M.J. Thorpe, D.D. Hudson, K.D. Moll, J. Lasri, and J. Ye. Cavity-ringdown molecular spectroscopy based on an optical frequency comb at 1.45-1. 65 μm. *Optics Letters* 32 (2007) 307–309.

[62] M.J. Thorpe, K.D. Moll, R.J. Jones, B. Safdi, and J. Ye. Broadband cavity ringdown spectroscopy for sensitive and rapid molecular detection. *Science* 311 (2006) 1595–1599.

[63] Y. He, C. Jin, R. Kan, *et al.* Remote open-path cavity-ringdown spectroscopic sensing of trace gases in air, based on distributed passive sensors linked by km-long optical fibers. *Optics Express* 22 (2014) 13170–13189.

[64] Y. He, R. Kan, F.V. Englich, W. Liu, and B.J. Orr. Simultaneous multi-laser, multi-species trace- level sensing of gas mixtures by rapidly swept continuous-wave cavity-ringdown spectroscopy. *Optics Express* 18 (2010) 20059–20071.

[65] P.J. van den Oever, J.H. van Helden, C.C.H. Lamers, *et al.* Density and production of NH and NH_2 in an Ar-NH_3 expanding plasma jet. *Journal of Applied Physics* 98 (2005) 093301.

[66] S. Wang, C.L. Strand, and R.K. Hanson. Spectrally-resolved absorption and laser-induced fluorescence of high-temperature gases. *AIAA SciTech Forum*, San Diego, California, 2019.

[67] Hanson Research Group. Laser-induced fluorescence. https://hanson.stanford.edu/our-approaches/imaging-and-photophysics/laser-induced-fluorescence (accessed 29 June 2024).

[68] E.C. Okafor, M. Tsukamoto, A. Hayakawa, *et al.* Influence of wall heat loss on the emission characteristics of premixed ammonia-air swirling flames

interacting with the combustor wall. *Proceedings of the Combustion Institute* 38 (2021) 5139–5146.
[69] J. Choe, and W. Sun. Experimental investigation of non-equilibrium plasma-assisted ammonia flames using NH_2*chemiluminescence and OH planar laser-induced fluorescence. *Proceedings of the Combustion Institute* 39 (2023) 5439–5446.
[70] Y. Tang, D.J. Xie, B.L. Shi, N.F. Wang, and S.Q. Li. Flammability enhancement of swirling ammonia/air combustion using AC powered gliding arc discharges. *Fuel* 313 (2022) 122674.
[71] A. Alfazazi, A.M. Elbaz, J. Li, S. Abdelwahid, H.G. Im, and B. Dally. Characteristics of ammonia-hydrogen nonpremixed bluff-body-stabilized flames. *Combustion and Flame* 258 (2023) 113066.
[72] T. Li, N. Wang, Z. Zhang, *et al.* A comparison between turbulent non-premixed jet flames of CH_4 and the 50%NH_3+50%H_2 blend. *Combustion and Flame* 246 (2022) 112477.
[73] X. Shi, T. Lian, Y. Zhang, *et al.* Enhanced ammonia combustion by partial pre-cracking strategy in a gas turbine model combustor: flame macro-structures, lean blowout characteristics and exhaust emissions. *Applications in Energy and Combustion Science* 17 (2024) 100247.
[74] L. Ji, J. Wang, G. Hu, R. Mao, W. Zhang, and Z. Huang. Experimental study on structure and blow-off characteristics of NH_3/CH_4 co-firing flames in a swirl combustor. *Fuel* 314 (2022) 123027.
[75] J. Sun, Q. Huang, Y. Tang, and S. Li. Stabilization and emission characteristics of gliding arc-assisted NH_3/CH_4/air premixed flames in a swirl combustor. *Energy & Fuels* 36 (2022) 8520–8527.
[76] S. Wang, G. Wang, A.M. Elbaz, T.F. Guiberti, and W.L. Roberts. Experimental investigation of the structure and NO emissions from swirling lean premixed NH_3/CH_4/air flames and their correlation with OH. *Energy & Fuels* 37 (2023) 13341–13353.
[77] M. Zhang, Z. An, L. Wang, *et al.* The regulation effect of methane and hydrogen on the emission characteristics of ammonia/air combustion in a model combustor. *International Journal of Hydrogen Energy* 46 (2021) 21013–21025.
[78] J.B. Bell, M.S. Day, J.F. Grcar, *et al.* Detailed modeling and laser-induced fluorescence imaging of nitric oxide in a NH_3-seeded non-premixed methane/air flame. *Proceedings of the Combustion Institute* 29 (2002) 2195–2202.
[79] B. Li, Y. He, Z. Li, and A.A. Konnov. Measurements of NO concentration in NH_3-doped CH_4 + air flames using saturated laser-induced fluorescence and probe sampling. *Combustion and Flame* 160 (2013) 40–46.
[80] C. Brackmann, J. Bood, J.D. Naucler, A.A. Konnov, and M. Alden. Quantitative picosecond laser-induced fluorescence measurements of nitric oxide in flames. *Proceedings of the Combustion Institute* 36 (2017) 4533–4540.
[81] L. Xu, Q. Fan, X. Liu, *et al.* Flame/turbulence interaction in ammonia/air premixed flames at high karlovitz numbers. *Proceedings of the Combustion Institute* 39 (2023) 2289–2298.

[82] D.E. Thomas, K.P. Shrestha, F. Mauss, and W.F. Northrop. Extinction and NO formation of ammonia-hydrogen and air non-premixed counterflow flames. *Proceedings of the Combustion Institute* 39 (2023) 1803–1812.

[83] G. Wang, H. Tang, C. Yang, G. Magnotti, W.L. Roberts, and T.F. Guiberti. Quantitative laser-induced fluorescence of NO in ammonia-hydrogen-nitrogen turbulent jet flames at elevated pressure. *Proceedings of the Combustion Institute* 39 (2023) 1465–1474.

[84] D. Zhang, Q. Gao, B. Li, J. Liu, and Z. Li. Instantaneous one-dimensional ammonia measurements with femtosecond two-photon laser-induced fluorescence (fs-TPLIF). *International Journal of Hydrogen Energy* 44 (2019) 25740–25745.

[85] J.X. Liu, J. Gao, B. Li, D.Y. Zhang, Y.F. Tian, and Z.S. Li. Ammonia measurements with femtosecond two-photon laser-induced fluorescence in premixed NH_3/air flames. *Energy & Fuels* 34 (2020) 1177–1183.

[86] H. Tang, D. Ezendeeva, and G. Magnotti. Simultaneous measurements of NH_2 and major species and temperature with a novel excitation scheme in ammonia combustion at atmospheric pressure. *Combustion and Flame* 250 (2023) 112639.

[87] B. Zhou, C. Brackmann, Z. Li, and M. Alden. Development and application of CN PLIF for single-shot imaging in turbulent flames. *Combustion and Flame* 162 (2015) 368–374.

[88] A.M. Bennett, P. Liu, Z. Li, *et al.* Soot formation in laminar flames of ethylene/ammonia. *Combustion and Flame* 220 (2020) 210–218.

[89] W.R. Boyette, A.R.W. Macfarlane, S.A. Steinmetz, M.J. Dunn, W.L. Roberts, and A.R. Masri. On the combined effects of compositional inhomogeneity and ammonia addition to turbulent flames of ethylene. *Proceedings of the Combustion Institute* 39 (2023) 4355–4364.

[90] X. Cheng, Y. Li, Y. Xu, Y. Liu, and B. Wang. Study of effects of ammonia addition on soot formation characteristics in n-heptane co-flow laminar diffusion flames. *Combustion and Flame* 235 (2022) 111683.

[91] Y. Liu, X. Cheng, Y. Li, L. Qiu, X. Wang, and Y. Xu. Effects of ammonia addition on soot formation in ethylene laminar diffusion flames. *Fuel* 292 (2021) 120416.

[92] Y. Li, Y. Zhang, G. Yang, *et al.* A comparative study on PAH characteristics of ethanol and ammonia as fuel additives in a premixed flame. *Journal of the Energy Institute* 101 (2022) 56–66.

[93] Y. Li, Y. Zhang, R. Zhan, Z. Huang, and H. Lin. Effects of ammonia addition on PAH formation in laminar premixed ethylene flames based on laser-induced fluorescence measurement. *Energy* 213 (2020) 118868.

[94] F. Ren, X. Cheng, Z. Gao, Z. Huang, and L. Zhu. Effects of NH_3 addition on polycyclic aromatic hydrocarbon and soot formation in C_2H_4 co-flow diffusion flames. *Combustion and Flame* 241 (2022) 111958.

[95] Y. Li, Y. Zhang, R. Zhan, Z. Huang, and H. Lin. Experimental and kinetic modeling study of ammonia addition on PAH characteristics in premixed n-heptane flames. *Fuel Processing Technology* 214 (2021) 106682.

[96] W.R. Boyette, S.A. Steinmetz, T.F. Guiberti, M.J. Dunn, W.L. Roberts, and A.R. Masri. Soot formation in turbulent flames of ethylene/hydrogen/ ammonia. *Combustion and Flame* 226 (2021) 315–324.

[97] C. Brackmann, O. Hole, B. Zhou, Z.S. Li, and M. Alden. Characterization of ammonia two-photon laser-induced fluorescence for gas-phase diagnostics. *Applied Physics B-Lasers and Optics* 115 (2014) 25–33.

[98] C. Brackmann, E.J.K. Nilsson, J.D. Naucler, M. Alden, and A.A. Konnov. Formation of NO and NH in NH_3-doped CH_4+N_2+O_2 flame: experiments and modelling. *Combustion and Flame* 194 (2018) 278–284.

[99] Q. Fan, X. Liu, X. Cai, *et al.* Structure and scalar correlation of ammonia/air turbulent premixed flames in the distributed reaction zone regime. *Combustion and Flame* 241 (2022) 112090.

[100] R.C. Rocha, S. Zhong, L. Xu, *et al.* Structure and laminar flame speed of an ammonia/methane/air premixed flame under varying pressure and equivalence ratio. *Energy & Fuels* 35 (2021) 7179–7192.

[101] X. Cai, Q. Fan, J. Wang, Z. Huang, and Z. Li. Effect of differential diffusion on the structure of high karlovitz number premixed ammonia/hydrogen/air flames. *Journal of Engineering Thermophysics* 43 (2022) 2025–2032.

[102] G. Wang, H. Shi, W.L. Roberts, and T.F. Guiberti. Simultaneous imaging of NO and NH in an ammonia-hydrogen-nitrogen flame using a single dye laser. *Combustion and Flame* 245 (2022) 112355.

[103] K.D.K.A. Somarathne, E.C. Okafor, A. Hayakawa, *et al.* Emission characteristics of turbulent non-premixed ammonia/air and methane/air swirl flames through a rich-lean combustor under various wall thermal boundary conditions at high pressure. *Combustion and Flame* 210 (2019) 247–261.

[104] E.C. Okafor, K.D.K.A. Somarathne, R. Ratthanan, *et al.* Control of NO_x and other emissions in micro gas turbine combustors fuelled with mixtures of methane and ammonia. *Combustion and Flame* 211 (2020) 406–416.

[105] Q.S. Fan, X. Liu, L.L. Xu, *et al.* Flame structure and burning velocity of ammonia/air turbulent premixed flames at high Karlovitz number conditions. *Combustion and Flame* 238 (2022) 111943.

[106] Z. Wang, X. Li, L. Li, Z. Zhao, B. Zhou, and X. Gan. Strategy for simultaneous multi-scalar imaging in turbulent NH_3/H_2 premixed flames using a single laser system. *Combustion and Flame* 242 (2022) 112185.

[107] C. Brackmann, V.A. Alekseev, B. Zhou, *et al.* Structure of premixed ammonia plus air flames at atmospheric pressure: laser diagnostics and kinetic modeling. *Combustion and Flame* 163 (2016) 370–381.

[108] G. Wang, S. Wang, and T.F. Guiberti. Simultaneous planar laser-induced fluorescence measurement of reactant NH_3, radical NH, and pollutant NO in ammonia-hydrogen flames using a single dye laser. *Combustion and Flame* 256 (2023) 112981.

[109] Y. Fan, Z. Wang, Y. Wang, M. Lee, W.D. Kulatilaka, and Y. Suzuki. Species structures in preheated ammonia micro flames. *Proceedings of the Combustion Institute* 39 (2023) 4427–4436.

[110] C. Song, H. Wang, M. Cheng, T. Tian, K. Luo, and J. Fan. A DNS study of heat release rate surrogates with unity and non-unity exponents for ammonia/air premixed flames. *International Journal of Hydrogen Energy* 48 (2023) 16470–16480.

[111] X. Zhu, T.F. Guiberti, R. Li, and W.L. Roberts. Numerical study of heat release rate markers in laminar premixed ammonia-methane-air flames. *Fuel* 318 (2022) 123599.

[112] I. Rahinov, A. Goldman, and S. Cheskis. Absorption spectroscopy diagnostics of amidogen in ammonia-doped methane/air flames. *Combustion and Flame* 145 (2006) 105–116.

[113] V. Dulin, D. Sharaborin, R. Tolstoguzov, *et al.* Assessment of single-shot temperature measurements by thermally-assisted OH PLIF using excitation in the $A^2\Sigma^+–X^2\Pi$ (1-0) band. *Proceedings of the Combustion Institute* 38 (2021) 1877–1883.

[114] R.R. Jones, D.C. Hooper, L. Zhang, D. Wolverson, and V.K. Valev. Raman techniques: fundamentals and frontiers. *Nanoscale Research Letters* 14 (2019) 2231.

[115] R.W. Dibble. The spontaneous Raman scattering technique applied to nonpremixed flames of methane. *Combustion and Flame* 67 (1987) 189–206.

[116] H. Tang, C. Yang, G. Wang, T.F. Guiberti, and G. Magnotti. Raman spectroscopy for quantitative measurements of temperature and major species in high-pressure non-premixed $NH_3/H_2/N_2$ counterflow flames. *Combustion and Flame* 237 (2022) 111840.

[117] C. Yang, D. Ezendeeva, T. Yu, and G. Magnotti. Temperature dependent Raman spectra of ammonia ranging from 3150 cm^{-1} to 3810 cm^{-1} for combustion applications. *Optics Express* 29 (2021) 33234–33244.

[118] L. Guo, C. Yu, W. Sun, *et al.* Study on effects of ethylene or acetylene addition on the stability of ammonia laminar diffusion flame by optical diagnostics and chemical kinetics. *Applied Energy* 362 (2024) 123032.

[119] S.B.N. Najafi, A.V. Mokhov, and H.B. Levinsky. Investigation of the stability, radiation, and structure of laminar coflow diffusion flames of CH_4/NH_3 mixtures. *Combustion and Flame* 244 (2022) 112282.

[120] C.N. Dennis, C.D. Slabaugh, I.G. Boxx, W. Meier, and R.P. Lucht. Chirped probe pulse femtosecond coherent anti-Stokes Raman scattering thermometry at 5kHz in a gas turbine model combustor. *Proceedings of the Combustion Institute* 35 (2015) 3731–3738.

[121] A. Mazumdar. Principles and techniques of schlieren imaging systems. Technical Report CUCS-016-13. Department of Computer imaging Science, Columbia University, 2013. Available from: https://academiccommons.columbia.edu/doi/10.7916/D8TX3PWV.

[122] W. Frei. Creating Schlieren-Type Visualizations in COMSOL Multiphysics®. (2022). https://www.comsol.com/blogs/creating-schlieren-type-visualizations-in-comsol-multiphysics (accessed 29 June 2024).

[123] B. Mei, S. Ma, Y. Zhang, X. Zhang, W. Li, and Y. Li. Exploration on laminar flame propagation of ammonia and syngas mixtures up to 10 atm. *Combustion and Flame* 220 (2020) 368–377.

[124] E.C. Okafor, Y. Naito, S. Colson, *et al.* Experimental and numerical study of the laminar burning velocity of CH_4-NH_3-air premixed flames. *Combustion and Flame* 187 (2018) 185–198.

[125] E.C. Okafor, Y. Naito, S. Colson, *et al.* Measurement and modelling of the laminar burning velocity of methane-ammonia-air flames at high pressures using a reduced reaction mechanism. *Combustion and Flame* 204 (2019) 162–175.

[126] B. Mei, S. Ma, X. Zhang, and Y. Li. Characterizing ammonia and nitric oxide interaction with outwardly propagating spherical flame method. *Proceedings of the Combustion Institute* 38 (2021) 2477–2485.

[127] B.W. Mei, X.Y. Zhang, S.Y. Ma, *et al.* Experimental and kinetic modeling investigation on the laminar flame propagation of ammonia under oxygen enrichment and elevated pressure conditions. *Combustion and Flame* 210 (2019) 236–246.

[128] B.W. Mei, J.G. Zhang, X.X. Shi, Z.Y. Xi, and Y.Y. Li. Enhancement of ammonia combustion with partial fuel cracking strategy: laminar flame propagation and kinetic modeling investigation of $NH_3/H_2/N_2$/air mixtures up to 10 atm. *Combustion and Flame* 231 (2021) 111472.

[129] J.G. Zhang, B.W. Mei, W. Li, *et al.* Unraveling pressure effects in laminar flame propagation of ammonia: a comparative study with hydrogen, methane, and ammonia/hydrogen. *Energy & Fuels* 36 (2022) 8528–8537.

[130] Q. Cheng, K. Ojanen, Y. Diao, O. Kaario, and M. Larmi. Dynamics of the ammonia spray using high-speed schlieren imaging. *SAE International Journal of Advances and Current Practices in Mobility* 1 (2022) 1138–1153.

[131] R. Payri, J. Garcia-Oliver, G. Bracho, and J. Cao. Experimental characterization of direct injection liquid ammonia sprays under non-reacting diesel-like conditions. *Fuel* 362 (2024) 130851.

[132] Q. Zhang, S. Shan, Z. Zhou, Z. Wang, and K. Cen. Jet penetration characteristics and NO combustion emission process of ammonia/methane fuel in a flue gas environment. *Physics of Fluids* 34 (2022) 093301.

[133] M. Kohansal, M. Kiani, S. Masoumi, S. Nourinejad, M. Ashjaee, and E. Houshfar. Experimental and numerical investigation of NH_3/CH_4 mixture combustion properties under elevated initial pressure and temperature. *Energy & Fuels* 37 (2023) 10681–10696.

[134] D. Wei, H. Fang, L. Hu, Y. Rong, and H. Zhou. Effect of acoustic excitation on the combustion and emission characteristics of methane-ammonia-air swirling flame. *Fuel* 352 (2023) 129117.

[135] R. Nave. Mie Scattering. http://hyperphysics.phy-astr.gsu.edu/hbase/atmos/blusky.html (accessed 29 June 2024).

[136] S. Zitouni, P. Brequigny, and C. Mounaim-Rousselle. Turbulent flame speed and morphology of pure ammonia flames and blends with methane or hydrogen. *Proceedings of the Combustion Institute* 39 (2023) 2269–2278.

[137] R. Zhang, L. Chen, H. Wei, *et al.* Experimental investigation on reactivity-controlled compression ignition (RCCI) combustion characteristic of n-heptane/ammonia based on an optical engine. *International Journal of Engine Research* 24 (2023) 2478–2488.

[138] M. Lee, M. Novotny, and K. Bartle. Gas chromatography/mass spectrometric and nuclear magnetic resonance determination of polynuclear aromatic hydrocarbons in airborne particulates. *Analytical Chemistry*, 48 (1976) 1566–1572.

[139] PerkinElmer Inc. Gas chromatography explained: what it is and how it works. https://blog.perkinelmer.com/posts/gas-chromatography-explained-what-it-is-and-how-it-works/ (accessed 29 June 2024).

[140] D. Ripepi, R. Zaffaroni, M. Kolen, J. Middelkoop, and F.M. Mulder. Operando isotope selective ammonia quantification in nitrogen reduction studies via gas chromatography-mass spectrometry. *Sustainable Energy & Fuels* 6 (2022) 1945–1949.

[141] R. Zaffaroni, D. Ripepi, J. Middelkoop, and F.M. Mulder. Gas chromatographic method for in situ ammonia quantification at parts per billion levels. *ACS Energy Letters* 5 (2020) 3773–3777.

[142] W. Liu, Y. Qi, R. Zhang, Q. Zhang, and Z. Wang. Hydrogen production from ammonia-rich combustion for fuel reforming under high temperature and high pressure conditions. *Fuel* 327 (2022) 124830.

[143] B. Calabrese. Experimental platforms for extracting biological data: mass spectrometry, microarray, next generation sequencing. In: S. Ranganathan, M. Gribskov, K. Nakai, and C. Schönbach (eds.), *Encyclopedia of Bioinformatics and Computational Biology*. Academic Press, Oxford, 2019, pp. 126–129.

[144] MKS Instruments. Mass Spectrometry. https://www.mks.com/n/mass-spectrometry (accessed 29 June 2024).

[145] Y. Zhuang, R. Wu, X. Wang, R. Zhai, and C. Gao. An experimental and modeling study on the oxidation of ammonia-methanol mixtures in a jet stirred reactor. *Fuel* 356 (2024) 129628.

[146] R. Tang, Q. Xu, J. Pan, *et al.* An experimental and modeling study of ammonia oxidation in a jet stirred reactor. *Combustion and Flame* 240 (2022) 112007.

[147] J. Pan, R. Tang, Z. Wang, *et al.* An experimental and modeling study on the oxidation of ammonia and n-heptane with JSR. *Proceedings of the Combustion Institute* 39 (2023) 477–485.

[148] Z. Sun, Y. Deng, S. Song, J. Yang, W. Yuan, and F. Qi. Experimental and kinetic modeling study of the homogeneous chemistry of NH_3 and NO_x with CH_4 at the diluted conditions. *Combustion and Flame* 243 (2022) 112015.

[149] Y. Deng, Z. Sun, W. Yuan, J. Yang, Z. Zhou, and F. Qi. Exploring NH_3 and NO_x interaction chemistry with CH_4 and C_2H_4 at moderate temperatures and various pressures. *Frontiers in Energy Research* 10 (2022) 828836.

[150] Y. Chen, B. Zhang, Y. Su, C. Sui, and J. Zhang. Effect and mechanism of combustion enhancement and emission reduction for non-premixed pure ammonia combustion based on fuel preheating. *Fuel* 308 (2022) 122017.

[151] H. Shi, S. Liu, C. Zou, *et al.* Experimental study and mechanism analysis of the NO_x emissions in the NH_3 MILD combustion by a novel burner. *Fuel* 310 (2022) 122417.
[152] Y. Tang, D. Xie, C. Wei, B. Huang, B. Shi, and N. Wang. Effect of microsecond repetitively pulsed discharges on lean blow-off limit and emission of rapidly-mixed ammonia/air swirling flames. *Applications in Energy and Combustion Science* 14 (2023) 100140.
[153] B. Giechaskiel, and M. Clairotte. Fourier Transform Infrared (FTIR) spectroscopy for measurements of vehicle exhaust emissions: a review. *Applied Sciences-Basel* 11 (2021) 7416.
[154] R. Ju, J. Wang, M. Zhang, *et al.* Stability and emission characteristics of ammonia/air premixed swirling flames with rotating gliding arc discharge plasma. *Energy* 277 (2023) 127649.
[155] M. Kosowska-Golachowska, A. Luckos, and A. Kijo-Kleczkowska. Pollutant emissions during oxy-fuel combustion of biomass in a bench scale CFB combustor. *Energies* 15 (2022) 706.
[156] M. Zhang, Z. An, X. Wei, J. Wang, Z. Huang, and H. Tan. Emission analysis of the CH_4/NH_3/air co-firing fuels in a model combustor. *Fuel* 291 (2021) 120135.
[157] M. Zhang, W. Xu, R. Wang, *et al.* Wall heat loss effect on the emission characteristics of ammonia swirling flames in a model gas turbine combustor. *Combustion and Flame* 256 (2023) 112955.
[158] Quantum laser cascade technology. [Online]. Available from: https://mirsense.com/qcl-technologies/ [Accessed 18.July.2024]
[159] Rosemount continuous laser gas analyzer, Emerson. [Online]. Available from: https://www.emerson.com/documents/automation/brochure-rosemount-quantum-cascade-laser-continuous-gas-analyzers-en-73002.pdf [Accessed 18. July.2024]
[160] S. Mashruk, E.C. Okafor, M. Kovaleva, *et al.* Evolution of N_2O production at lean combustion condition in NH_3/H_2/air premixed swirling flames. *Combustion and Flame* 224 (2022) 112299.

Chapter 4
Combustion systems for ammonia-fueled gas furnaces and boilers

M. de Joannon[1], *G.B. Ariemma*[1], *M.V. Manna*[1], *P. Sabia*[1], *G. Sorrentino*[1] *and R. Ragucci*[1]

4.1 Introduction

According to the International Energy Agency [1], industrial heat constitutes a significant portion of both industrial energy demand and global energy consumption, making up two-thirds and almost one-fifth, respectively. The share of energy consumption for different sectors is reported in Figure 4.1. It is clearly shown that about 32% of this energy is consumed by industry and almost 74% of this energy is spent for heat production. Here these percentages have to be considered as an indication of gross numbers due to continuously and swiftly change of scenario that energy sector is experiencing, leaded by two main factors: the ongoing energy transition and the unstable geopolitical equilibria. As expected, it is mainly related to the use of coal, oil and natural gas and consequently to the CO_2 emissions, as shown in Figure 4.2.

There are numbers of evaluations and scenarios for CO_2 emission from this sector and all of them converge in the identification of a level higher than 30% of the total CO_2 emissions from the industrial sector, as shown in Figure 4.1.

In the central scenario of the World Energy Outlook [1], industrial heat demand is projected to increase across all temperature levels and high-temperature heat (as from boilers) is expected to contribute a quarter of the total growth in heat demand in industry by 2040. The current need for industrial heat predominantly depends on fossil fuels, biomass, and electricity, with minimal utilization of renewable resources in specific sectors.

Further considerations are shown in Figure 4.2, where the CO_2 emission impact is specified for different industrial sectors. The main comment rising from the analysis of Figure 4.2 is that five heavy industries account for 80% of industrial CO_2 emissions. They come from the heat production for reaching the high temperatures characteristics of these processes, as well as from the process itself,

[1]Istituto di Scienze e Tecnologie per l'Energia e la Mobilità Sostenibili, Consiglio Nazionale delle Ricerche, Napoli, Italy

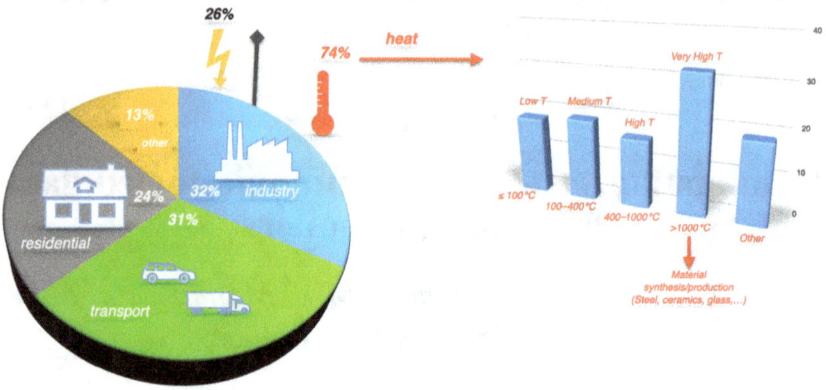

Figure 4.1 Heat production sharing among different sectors and the characteristic temperature for heat in industrial sector. Reprinted, with permission, from [140].

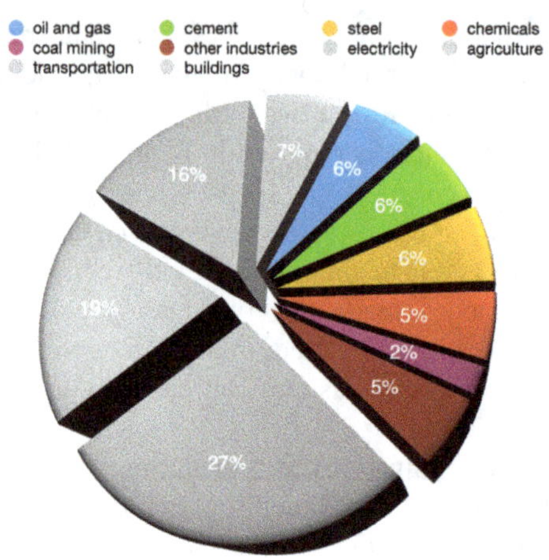

Figure 4.2 CO_2 emission impact of different sectors. Reprinted, with permission, from [141].

intended as the CO_2 deriving from the transformation of the matter along the process chain. In both cases, such a CO_2 source is hard to reduce.

There are different altenatives to decarbonize the hard-to-abate industrial system. In general, the first practical step is aimed to increase the global process efficiency, avoiding/minimizing heat losses all over the segments of production chain, with the aim to reduce the fuel consumption, and subsequently the CO_2 release. In this context, the implementation of robust and flexible monitoring/control strategies through digital twin and soft sensing approaches may cover a fundamental role.

Other possible routes can come through co-firing low-carbon fuels with conventional fuels in standard boiler/furnaces as a potential short-term viable approach for mitigating CO_2 emissions. The intrinsic short-term nature derives from minor adjustments/retrofitting of existing infrastructures. In China, boiler plants that have been already exercised with co-firing biomass and coal demonstrate that a high biomass/coal ratio is feasible [2,3]. NO_x emissions were lowered by ammonia released from biomass pyrolysis/combustion with the only drawback related to sever fouling propensity from biomass combustion.

Renewable heat sources can be cost-effective and practical solutions, though the integration into low-energy demanding system is much suitable with respect to hard-to-abate systems. In addition, despite bio-energy could fulfill the low/high-temperature heat demands, its availability is local and daily/seasonally intermittent, and its economic and environmental viability relies on regional factors.

Given this short-term solution background, to seriously address the decarbonization of industrial sectors, innovative technologies have to be conceived and a substantial change of the production chain should be taken into account. Of course, this approach requires a relevant economic effort in term of R&D of advanced processes/technologies with a medium/long-term intrinsic nature.

This is the case of carbon capture and sequestration (CCS) up to carbon capture, sequestration and utilization with the conversion of CO_2 into valuable and marketable products (chemicals, fuels). The spread-out of these technologies are then hindered by the prohibitive costs/extra energy required for CO_2 liquefaction/transportation, along with the necessity to find proper geological storage sites. For instance, the CCS concept has been demonstrated in small plants; thus, still scientific efforts should be paid for its technological maturation.

Another valuable alternative comes through the electrification. Herein, it has to be taken into account that the hard-to-abate industry is very difficult to electrify, because of the high temperature involved in many processes. Unquestionably, this is the case of steel, glass, as well as ceramic industries and cement that can have sections working at temperatures well beyond 1000 °C. To give at a glance the typical temperatures required for several hard-to-abate industries, temperature ranges are reported in Figure 4.3, in relation to the most energy-intensive steps along specific material production chain.

Figure 4.3 Typical working temperature range for several industrial production materials

Indeed, the typical working temperatures are reported for the production of aluminum/alloy, cement, glass, steel and iron, and ceramics. In the case of aluminum/alloy, the highest temperatures are required for the alumina smelting (≅950 °C), where alumina is poured into a reduction cell to separate aluminum from oxygen. Lower temperatures (≅350 °C–500 °C) are then required for the aluminum extrusion process. In the case of cement, the highest temperatures are reached in the kiln, where the clinker is melted at temperatures approximately equal to 1300 °C–1450 °C. The pre-calciner section demands for lower temperatures (≅950 °C). In the glass industry, the raw materials (silica sand, calcium oxide, soda, magnesium), prior a purification steps, are fed to a furnace where they are molten at temperatures around 1600 °C. This material is destinated to a "floated" section at about 1000 °C, to form a 5–6 mm layer at 1000 °C. The steel and iron industries then requires temperature between 900 °C and 1700 °C, while ceramic ones require temperature from 1000 °C to 2200 °C, going from conventional potteries to advanced ceramic materials.

Though the low temperature (below 150 °C) and medium temperature heat (between 150 °C and 400 °C) segments can be electrified, they represent only the 33% of the total heat, whereas the remaining part is heat with characteristic temperatures higher than 400 °C, with a 32% of the total heat at T>1000 °C. Other electrically higher temperatures heating solutions, such as induction and microwave heating, are being developed, but currently still at a relatively low technological readiness level (TRL).

Three main issues related to the fully electrification of the industrial sectors are:

- High temperature involved especially in material synthesis and treatments that cannot be reached without combustion process
- The need to strongly increase the availability of electricity, that should be at least duplicated, with the impact also on the related infrastructures
- The need to rebuild a huge number of industrial sites that are generally conceived and commissioned to last decades, with new typology of heating systems

In addition, even though a rapid development of these technological sectors can be foreseen, the main hindering factor has to be correlated to the energy cost [4,5]. To abate this barrier, the whole renewable energy chain should be developed.

Another possible electrification route for the industrial sectors comes through an indirect approach where energy (from renewable sources) can be used to produce H_2 from water electrolysis to destine to direct oxidation systems [4].

Given lower energy demand, electrification can constitute a practical solution for other sectors, for instance for domestic application, even though the accessibility to electricity and its infrastructures has to be carefully evaluated. Indeed, it has to be taken into account that there are almost 750 million people in the world without accessibility to the electric energy. This number increases if those with a limited access to electricity are considered. The energy mix proposed in the heating district concept can be considered among the most efficient solution for household heat. Nevertheless, they are not yet a concrete reality to be considered the obvious solution. On the other hand, the critical aspect of CO_2 emission for heat generation is its utilization in industries.

Starting from renewable energy, the other chance to decarbonize the industrial sectors come through the use of new energy carriers, intended as long-time renewable chemical storage. Using hydrogen as an alternative fuel initially appears to be the most promising and environmentally friendly solution, because it is not-toxic, has a high calorific value, and its combustion product is water. The main limitations to the use of hydrogen are transportation and storage, since it needs to be compressed and stored in special-polymeric material (to avoid steel embrittlement), high pressures vessel or liquified at cryogenic conditions, resulting in non-economic solutions. In addition, its strong reactivity, along with the high adiabatic temperature and hence a strong tendency to thermal-NOx, pose some limitation in combustion systems for both safety and emissions.

In view of these limitations, ammonia is gaining a considerable acceptance as hydrogen carrier and as fuel itself in heat and power sectors. The main advantages with respect to hydrogen consist in firmly established and economical infrastructures for storage and transportation, as widely used as fertilizer or as desulfurizing/De-NOx agent of combustion flue gas in boilers/furnaces. In addition, ammonia has a high volumetric energy density, high octane number, and good anti-explosive properties [6]. On the down side, ammonia combustion could potentially result in elevated NOx emissions (as consequence of the augmented nitrogen content in the fuel), unburned ammonia, and CO emissions when co-fired.

This chapter is focused on the analysis of the potentialities and of the state of the art of ammonia utilization in the main hard-to-abate industrial sectors. The fuel flexibility needed to accommodate the locally and timely available fuel is much easily attainable with respect to aviation and engine sectors, by applying advanced combustion technologies, such as MILD/flameless or Oxy-fuel combustion, which are already relatively widely deployed in hard-to-abate sectors, due to their high efficiency. Therefore, in the following a general description of solutions evaluated in literature and applied in practical applications will be discussed, looking at standard burners, MILD/flameless, and Oxy-fuel technology. Then, the chapter will follow with a description of the different processes in the hard-to-abate sectors illustrating the potentialities to involve the utilization of ammonia as relevant energy carrier in the decarbonizing of the sector.

4.2 Overview of strategies for ammonia utilization in furnaces and boilers

The utilization of new energy carriers in combustion-based energy systems require the adaptation of combustion unit to the characteristics of the molecules to be used as fuel. The energy transition is bringing into play a certain number of carriers with very different characteristics, especially referring to combustion intensity. As already widely discussed in Chapter 1, among them, ammonia has a low combustion intensity, characterized by low flame velocity and adiabatic flame temperature (AFT) as well as limited flammability range with respect to fossil fuels, and even more with respect to hydrogen.

Therefore, ammonia utilization in burners for furnaces and boilers requires a careful evaluation of the potentiality and appropriate strategies to make effective ammonia in this field taking into account also an easily integration of such carrier in existing plants.

Different strategies can be used and have been evaluated for ammonia utilization in systems of interest in this chapter. They rely on fluid-dynamic and/or chemical interventions that help to overcome the low combustion intensities of ammonia.

When these strategies are adopted and successful with respect to ammonia oxidation process, still attention has to be paid to the NO_x emissions, in terms of fuel and thermal NO_x, while N_2O formation has to be thoroughly contained as a greenhouse gas with a global warming potential in between 265 and 298 times larger than CO_2 itself and an environmental life time of 131 ± 10 years. Likewise, the oxidation processes have to be optimized also to avoid ammonia slip in the atmosphere.

In relation to the hard-to-abate industry, the ammonia oxidation causes a change of the atmosphere with respect to the combustion of conventional fuels (lower adiabatic temperature, lower radiation, different residence times) that have to be discussed in relation to material manufacturing processes. In addition, the basic nature of ammonia itself along with high corrosive properties has to be addressed to preserve material properties. This is a critical aspect that will be discussed case by case.

4.2.1 Ammonia/fuels mixtures

As discussed many times in this book, the use of pure ammonia in conventional combustion systems is limited because the stabilization of ammonia flames is not an easy task, resulting in poor combustion properties in ammonia. The conventional approach to overcome this hindering issue while enabling decarbonation is to replace part of fossil fuels with ammonia, with a partial reduction of CO_2 emissions.

Utilization of ammonia with other fuels used as enhancers, such as alcohols, hydrogen, liquid hydrocarbon, has been widely studied mainly for ICEs and GTs, where higher quality fuels are used. Basically, high reactive fuels release radical species with respect to ammonia, thus are able to modulate the scarce ammonia combustion properties. Within this contingency strategy, many practical interventions envisage the use of partially cracked (catalytic or thermal cracking) ammonia to hydrogen [7,8], with the aim to burn $NH_3/H_2/N_2$ mixtures, where H_2 can aid the oxidation process, given its high reactivity.

Traditional combustion systems rely on the stabilization of conventional premixed or diffusive flames. While considering a pre-mixed flame, the parameter that can be associated to the global reaction rate is the laminar burning velocity (LBV). Indeed, to give at a glance the effect of the enhancement of ammonia reactivity when mixed with conventional fuels, the LBV can be analyzed. In Figure 4.4 (semi-log scale), the laminar flame speed is reported for ammonia/methane, ammonia/ethanol, ammonia/methanol, and ammonia/H_2 blends (secondary axis) by Tornatore *et al.* [9].

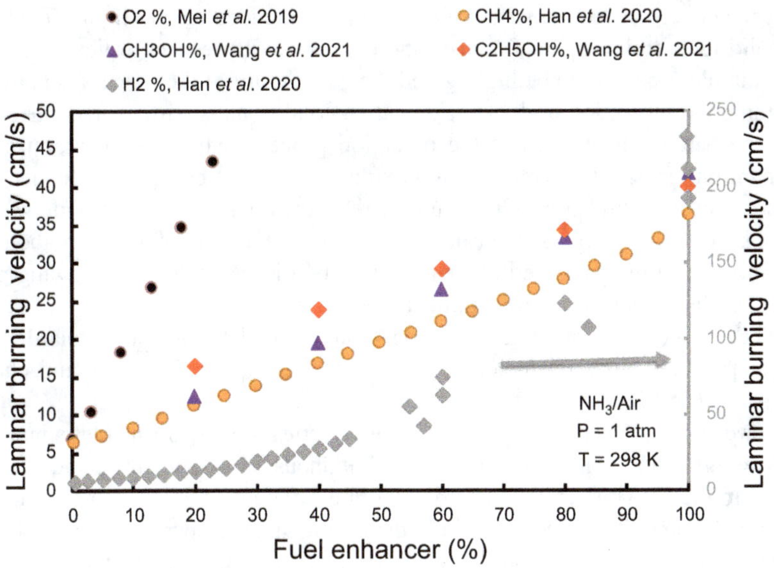

Figure 4.4 Laminar flame speed for ammonia/fuel-enhancer blends. Reprinted, with permission, from [9].

It is possible to note as the addition of H_2 or CH_4 [10], CH_3OH or C_2H_5OH [11] notably increases the NH_3 LBV, from a typical value of 7 cm/s for pure ammonia. Up to a relative concentration equal to 20%, the enhancing effect is almost equal for all the analyzed molecules. Afterward, a further concentration increment leads to a higher LBV but the slope of the curve changes toward lower value, reaching almost a plateau. Only in the case of H_2, the LBV increases almost linearly with the H_2 concentration.

Other strategies to increase the LBV relies on the increase of mixture inlet temperature or the modification of the oxidation environment, with the addition of O_2 [12] (as surplus with respect to air conditions), ozone [13], or hydrogen peroxides [14]. In Figure 4.4, the LBV enhancement for ammonia is reported also with O_2 addition to the system (up to a concentration of 20%). It is worth noting that this strategy is particular effective to enhance ammonia LBV, thus supporting the implementation of Oxy-fuel processes.

Han *et al.* [10] and Lhuillier *et al.* [15] reported data pre-heating the mixture up to 448 K and 473 K, respectively. Restricting the analysis to the stoichiometric NH_3/air mixture, S_L increases almost linearly with T, up to 14 cm/s for $T = 448$ K.

The main drawback of the ammonia/fuel blend strategy to reach a stable oxidation process is that it relies on higher release of both fuels and thermal-NOx production, due to the increase in the radical pool and temperature. In the case of ammonia/carbon-based fuel blends, the minimization of NOx has to be discussed also with respect to CO emissions, as the conditions to minimize these pollutant categories occurs under different operating conditions in terms of mixture equivalent ratio, while fuel-rich conditions to minimize NOx would pose main issues also related to ammonia slip and HCN formation.

The concept of ammonia/fuel blends can be extended also to liquid (for ICEs) and solid fuels (relevant for furnaces and boilers). With respect to liquid or solid/ammonia blends, it has to be highlighted the complexity of the process, and indeed the ammonia interaction under coal oxidative conditions is particularly intriguing, since the heterogeneousness of the oxidation process, where liquid atomization/vaporization or solid devolatilization/oxidation has to be taken into account. Ammonia/coal combustion is particularly relevant for furnaces and boilers; thus, herein a brief paragraph is dedicated to this topic. Further information about the utilization of ammonia mixed other fuels will be discussed in the following paragraphs, with respect to specific applications.

The coal combustion multi-step process involving heating and devolatilization of coal particles, char formation, oxidation of devolatilized species and char is schematized in Figure 4.5.

For example, Wang *et al.* [17] performed a series of experimental tests in a classical one-dimensional flame furnace with a bituminous coal. Ammonia could be added directly in the coal burner or fed separately into different positions. Authors demonstrated that ammonia changes drastically the coal ignition temperature (Figure 4.6(a)), in relation to the ammonia content. They identified an ammonia content lower than

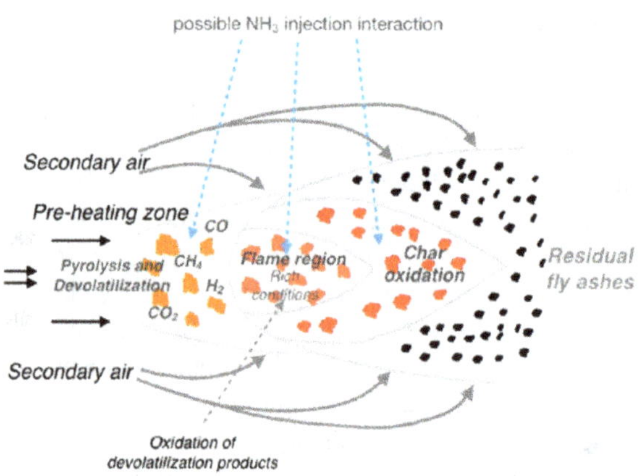

Figure 4.5 Ammonia-coal interaction during combustion. Reprinted, with permission, from [142].

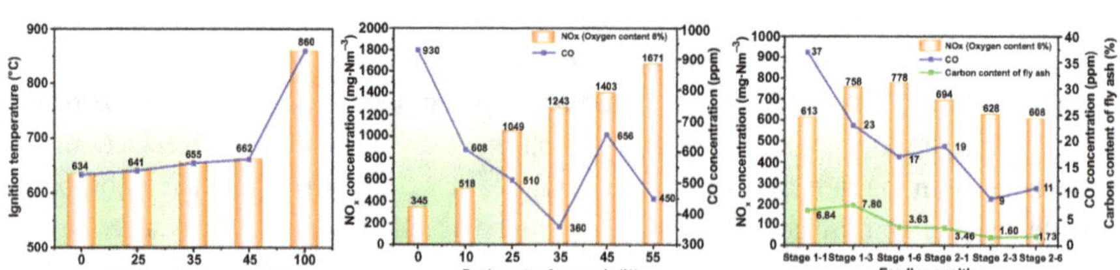

Figure 4.6 (a) Ignition delay, (b) NOx/CO of ammonia-coal mixture as a function of ammonia content; (c) NO_x, CO, and carbon content of fly ashes in staged ammonia fed. Reprinted, with permission, from [17].

45%, to not drastically modify the ignition temperature with respect to the pure coal combustion. The main issues again were related to NO formation (Figure 4.6(b)). For instance, for an ammonia content equal to 25%, the addition of ammonia led to a drastic increment of NO, with strong fluctuation of CO, indicating a non-ideal combustion process with plausible instabilities. As a contingency strategy, the authors proposed ammonia fuel staging for the primary and secondary air duct. They demonstrated that fuel staging is beneficial both in terms of NO_x and CO reduction.

In addition, to clarify the complexness of the process, authors claimed that the char characteristics are affected by the presence of ammonia, as the nitrogen content in the char changes with ammonia mixing ratio, because nitrogen, released from the ammonia oxidation, is adsorbed or embedded into the char itself [17].

4.2.2 MILD/flameless and oxyfuel combustion

The use of ammonia as energy carrier is mainly limited by its poor combustion characteristics. As widely discussed, its low flame velocity and narrow flammability limits make the standard combustion process difficult to be sustained. The use of a fuel enhancer to improve its propensity to burn has been already discussed. The other options to overcome these issues are the implementation of MILD combustion or oxyfuel combustion, which are suitable advanced technologies mainly for furnaces and burners.

The strength of MILD combustion relies on the paradigm it is grounded, radically different from those of standard combustion processes. Indeed, the local conditions of fuel and oxidant mixture are characterized by high dilution levels and pre-heating. The level of local dilution has to be high enough to make the fuel/oxidant mixture composition fall outside the flammability limits. At the same time, the pre-heating of the mixture should guarantee a local temperature higher than autoignition temperature. In such a way it has been demonstrated that the process evolves through an igni-diffusion process of mixture kernels that ignites, reacts, and disappears evolving independently from each other and fill the entire volume of the combustion chamber. As a consequence, there is no stabilization of a standard flame making the process independent on the most critical combustion parameter of ammonia as the flame velocity. On the contrary, the process is flameless as shown in Figure 4.7 of a 10 kW combustion chamber [18] and does not rely on the diffusive or deflagrative structure typical of diffusion or premixed flame. Overall, the two relations that identify MILD combustion are reported in Figure 4.7(a) alongside the composition and temperature plan where the regions corresponding to the different combustion regimes are identified.

The local dilution and temperature are generally obtained by high internal recirculation, with recirculation rate (recirculated mass rate/fresh mass rate) higher than 5. The reference combustion chamber configurations used are those based on a reverse or cyclonic flows. Their explicative sketch is reported in Figure 4.7(c).

Based on local autoignition, MILD combustion is able to stabilize a wide palette of fuels with very different combustion characteristics, as summarized in Figure 4.7, resulting in a real fuel flexible technology.

With respect to ammonia, it has been demonstrated that MILD combustion is able to stabilize pure ammonia combustion also producing a limited amount of NOx with respect to standard combustion processes, as shown in Figure 4.8, where

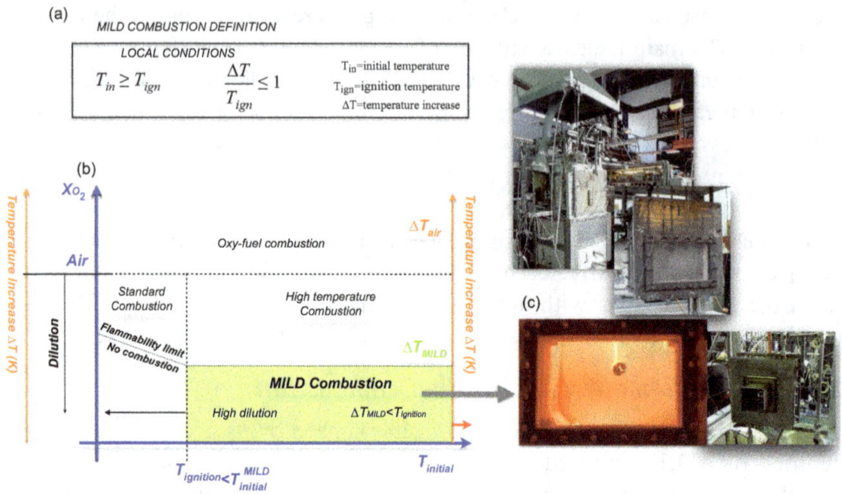

Figure 4.7 MILD combustion: (a) definition and (b) map of combustion technologies, (c) 10 kW burner

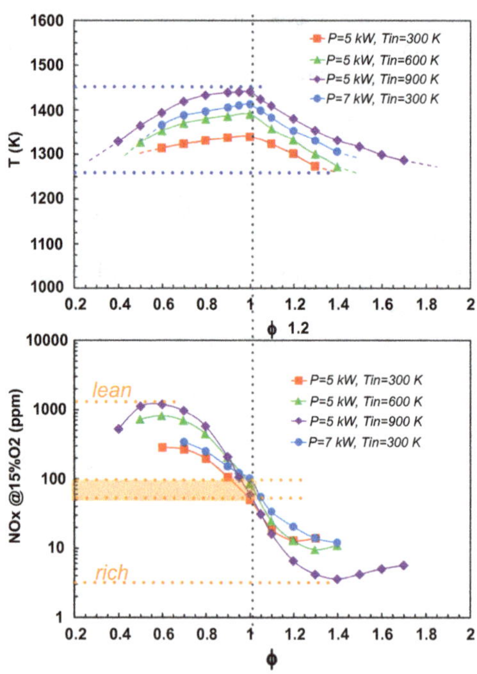

Figure 4.8 Temperature and NOx profile in ammonia MILD combustion in cyclonic burner (adapted from [83])

temperature and NOx emission obtained by Sorrentino et al. in a power flexible cyclonic burner have been reported.

It has been also demonstrated that the use of a fuel enhancer such as hydrogen, methane [19], and alcohol [20] extends the range of process stability in MILD combustion conditions, though an increase of NOx emission is measured, as reported in Figure 4.9 for ammonia methane blends.

A possible further mitigation for NOx emission can be obtained by adding water to the fuel (Figure 4.10). Due to MILD combustion characteristics, the process keeps its stability also diluting ammonia with water, leading to a reduction up to almost 50% of NOx with respect to already low level of MILD combustion. Moreover, the presence of water allows to face also with one of the issues encountered in burning ammonia, as the decrease of radiative heat transfer from the reactive volume due to the absence of CO_2. Indeed, looking at the emissivity of

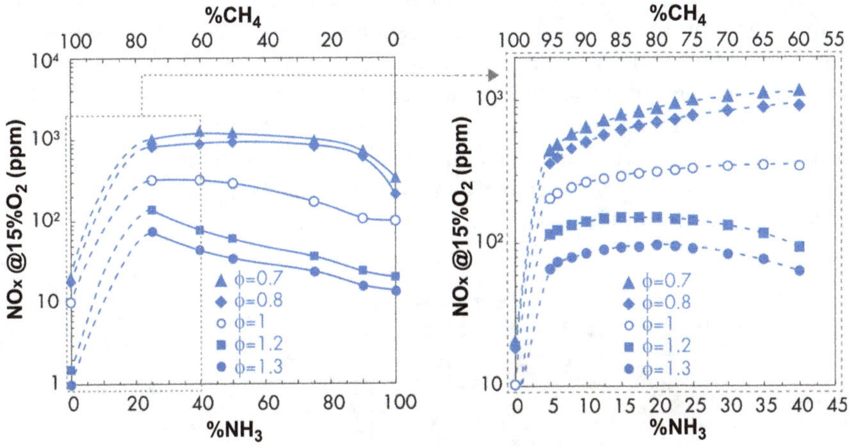

Figure 4.9 NOx emissions measured for ammonia/methane blends in MILD combustion conditions. Reprinted, with permission, from [20].

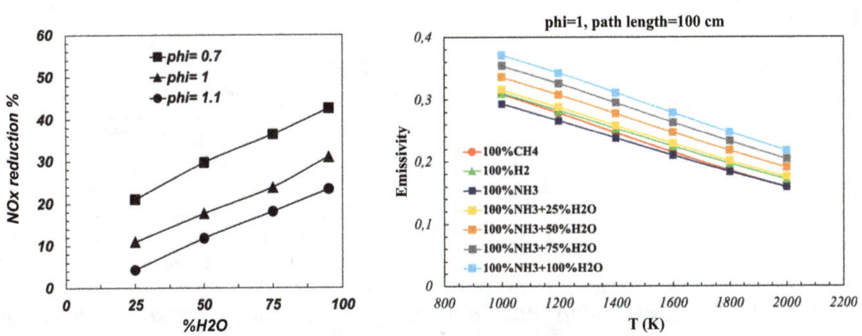

Figure 4.10 Effect of water addition on ammonia MILD combustion: NOx reduction and emissivity

exhaust gases derived from combustion of ammonia in presence of different water percentage in Figure 4.10, it is possible to see that by feeding the same weight percentage of ammonia and water in MILD conditions, the emissivity of the flue gases reaches the emissivity of methane burnt gases.

The increase in emissivity is also obtained by burning ammonia under oxyfuel combustion conditions. This consolidated combustion technology involves the use of oxygen-enriched combustion atmosphere to improve combustion characteristics in terms of stability and temperature. In standard combustion process, it is used to eliminate the issue related to NOx emission because pure oxygen, possibly diluted with the exhaust gases from combustion, containing only CO_2 (after the separation of water) is used as oxidant; thus, no nitrogen is present in the combustion chamber and, consequently, there is no possibility to form NOx. It is also responsible of higher temperature that can be reached in these conditions. As a general view, oxyfuel combustion covers the upper part of the map of combustion technologies of Figure 4.6. In case of ammonia, as already stated, the use of oxyfuel combustion enlarges the limit of stability of the combustion process. This is clearly shown by laminar flame velocities reported by Murai *et al.* [21] in a 10 kW oxyfuel furnace. The temperature profiles collected in these conditions show that the 30 vol.% of O_2 enriched air concentration could result in higher temperature. Moreover, the presence of higher water concentration at higher oxygen concentration makes the total heat flux increase 1.4 times of the total heat flux of methane/air non-premixed condition, as shown in Figure 4.11.

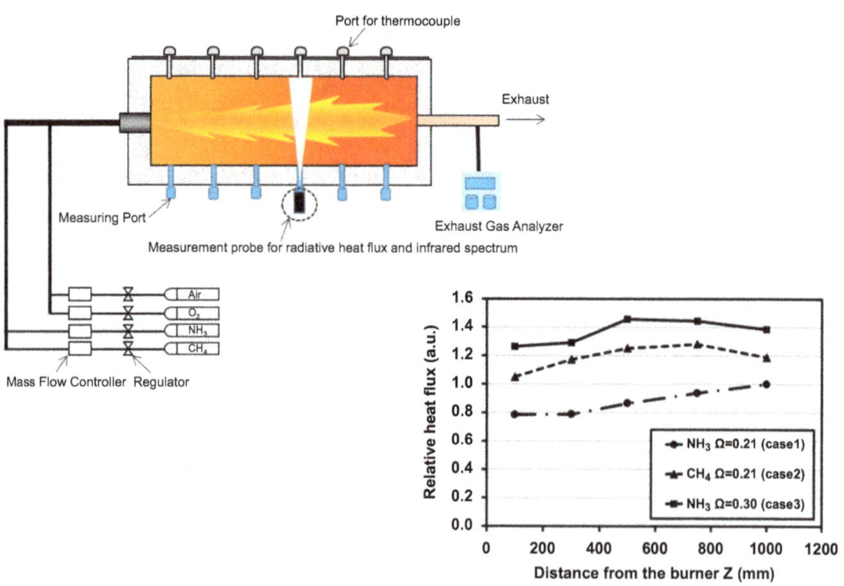

Figure 4.11 Oxy-fuel combustion of ammonia in 10 kW furnace: radiative heat flux

4.3 Boilers

Combustion boilers are the backbone of several industrial processes and domestic heating as they are designed to use the chemical energy to increase the energy content of steam and hot water for heating and power applications. To understand what are the opportunities that these systems have in facing the energy transitions, it is necessary to look at their main characteristics. Basically, the combustion of fuels with air occurs in burners of different typologies and produces hot exhaust gases that are subsequently cooled down as the heat is transferred to the water through the heat exchanger. The first boilers date back more to the industrial revolution and over the years they have undergone outstanding transformations, evolving from rudimental large kettles to high-efficiency and flexible systems.

Boilers application extends across numerous sectors, including agriculture, manufacturing, hospital, marine, and nuclear sectors, and they are produced in a variety of sizes and setups, tailored to the desired heating output (i.e. production of hot water or vapor), end use, and specific physical and chemical properties of the fuels.

They can be categorized in numerous ways, to mention some of them:

- *Based on the type of firing*, boilers can be classified into: mass/pile burning boilers; stoker-fired boilers; burner-fired boilers; bubbling fluidized bed combustion (BFBC) boilers; circulating fluidized bed combustion (CFBC) boilers; pulverized fuel (PF) boilers; liquor-fired boilers; waste heat boilers. This classification is in some extent related to the next typology of classification, based on fuel type.
- *Based on the type of fuel*, they can be classified into coal-fired, solid fuel-fired, oil-fired, and gas-fired. Typically, fossil fuels are fired in boilers, but also biomass, liquor, and waste in general can be used as fuels. The coal-fired boiler can be further subclassified into pulverized-coal (PC) fired, stoker-fired, and fluidized-bed combustion (FBC) boilers. All these typologies of boilers can be adapted for ammonia utilization. Indeed, as will be widely explained in the following, in boilers, ammonia is thought to be fed along with whatever other fuels that are generally used in such systems.
- *Based on the operating pressure*, boilers can be classified in subcritical or supercritical when the pressure is lower or higher than about 200 atm, respectively. Ammonia co-firing in a supercritical coal circulating fluidized bed has been analyzed by process simulation to evaluate the thermal efficiency with different co-firing ratio to identify the optimal operating conditions [16].
- *Based on the fluid* that flows in the tubes, boilers are commonly subdivided into water-tube or fire-tube units. In the first case, water is inside the tubes and hot gases outside; since the pressure is confined inside the tubes, such configuration is suitable for higher-pressure application and are generally fabricated in large sizes. In the fire-tube units, the exhaust hot gases pass through the tubes located inside a water-filled outer shell. The fire-tube boilers are commonly designed for lower-pressure applications and are suitable for cleaner fuels, since tubes are susceptible to fouling and erosion caused by

combustion products. Schematic representations of the water-tube and fire-tube boilers are reported in Figures 4.12 and 4.13, respectively. The utilization of a carrier as ammonia, but also as hydrogen, has to be taken into account in the choice of the configuration to be used. Indeed, the effect of combustion atmosphere deriving from these carriers and the carriers themselves can impact

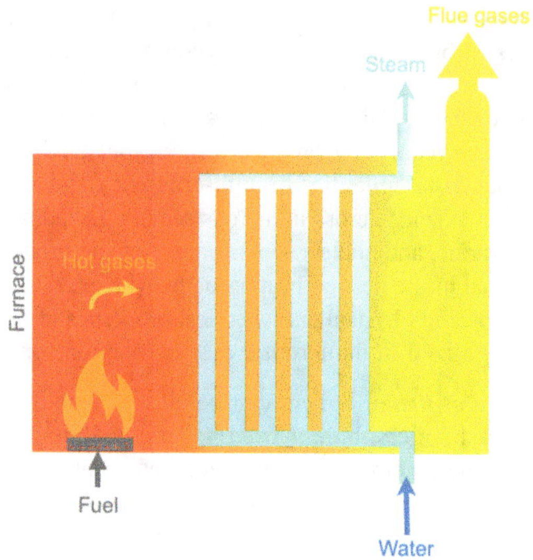

Figure 4.12 Schematic representation of a water-tube boiler

Figure 4.13 Configuration of a horizontal return tubular fire-tube boiler

on boilers tubes and the effect can be stronger in the case of fire-tube units where the confinement of flames is higher.

In addition to the main fuel, other fuel categories are necessary for the proper operation of a boiler, including [22–24]:

- *Start-up fuel*: used initially to ignite and load the boiler before switching to the main fuel. Examples include oil, gas, or charcoal depending on the boiler type.
- *Alternate fuel*: acts as a backup to the main fuel, ensuring uninterrupted operation in case of fuel disruption. It may require adjustments in equipment and typically incurs higher costs
- *Auxiliary fuel*: supplementary fuel burned with the main fuel, for disposal purposes or in co-firing mode. In case of small quantities (i.e. for disposal purposes), there are no specific requirements for the use of the auxiliary fuel, while co-firing is useful for auxiliary fuels with poor combustion properties or high content of moisture.

Undoubtfully, all these steps affect the impact of the process on emissions and have to be taken into account looking at the decrease in CO_2 and pollutant emissions. More specifically, also the substitution of these fuels with no carbon carrier can help in a decarbonation strategy for boilers. On the other hand, these fuels can be used as fuel enhancers in case of the low combustion intensity fuel as ammonia is used as main fuel.

The characteristics of fuels are crucial for the definition of the final size and shape of the boilers, and in particular on the typology of the burner to be used. Indeed, it must ensure the stability of the combustion process, a proper mixing between fuel and air and the maximum heat release from the fuel. Therefore, the utilization of new energy carriers as ammonia and hydrogen can strongly impact on the burner characteristic. Evidently, combustion systems for gaseous fuels are simpler in design compared to those for liquid or solid fuels, as the process of mixing gas is inherently less intricate than atomizing liquid fuel or dispersing solid fuel particles. Nevertheless, combustion strategies to stabilize combustion of ammonia require different solutions that impact on burner layout. This has to be also taken into account in relation to pollutant emissions. Indeed, burners have to be designed in order to comply with imposed emission limitations of pollutants as CO, NO_x, SO_x, and PM. For instance, low-NOx burners represent an effective strategy for NO_x emissions control, through the techniques of staged combustion and recirculation of flue gas with fresh reactants as in MILD and flameless combustion [22–24], as discussed in the introduction. This strategy has been proven to be effective also in case of ammonia where the NOx emission represents one of the main limitations in its utilization.

4.3.1 Energy conversion and efficiency

The maximization of the overall efficiency of boilers and the adoption of energy saving strategies can substantially reduce the amount of energy needed for boiler operations. The enhancement of the efficiency by just a small fraction would allow to reduce a vast amount of fuel consumption.

Depending on the end-use and the fuel type, the overall boiler efficiency can vary significantly from 20% up to 92%. For instance, the thermal efficiency of coal fired boilers varies between 81% and 85%, for oil ranges between 78% and 81%, and finally for gaseous fuel changes between 76% and 81% [25]. The boiler efficiency refers to the proportion of heat absorbed by the generated steam compared to the total heat supplied to the boiler. Therefore, to enhance boiler efficiency, it is imperative to reduce the heat wastage from the boiler by optimizing various parameters such as excess air, fuel flow rate, and steam demand. The main heat loss is related to the outgoing flue gas, as reported in Figure 4.14.

The exhaust gases produced in the burner transfer heat to water inside a heat exchanger and, as a result of this process, leave the boiler and are released into the environment at temperatures between 150 °C and 250 °C, dissipating a significant amount of thermal energy (about 10%–30%). Additionally, other sources of heat loss include radiations, blow-down, fly and bottom ashes losses.

In summary, the losses can be categorized as follows [25]:

- Heat loss due to stack gas;
- Heat loss due to unburned fuel, radiation, and convection losses;
- Heat loss due to the blow-down.

The energy loss through the gas stack is the major source of heat loss accounting for about 70%–80% of the total loss and depends on the temperature and volume of the gas leaving the boiler [25]. Ideally, the stack loss is null if the temperature of the flue gas is reduced to surrounding temperature. Nonetheless, due to limitation in heat transfer principles, these losses are unavoidable but they can be minimized by optimizing the excess air, ensuring that the heat exchange area remains clean and installing heat recovery systems. Excess air is required to ensure complete combustion and avoid carbon monoxide or smoke formation. On the other hand, low excess air would reduce the volume of stack gas, thus the gas velocity, extending the gas residence time in the heat exchanger to ensure a more efficient heat recovery.

As further factor, the evaluation of the ambient conditions, namely temperature, humidity, and altitude, has a significant role in determining the boiler

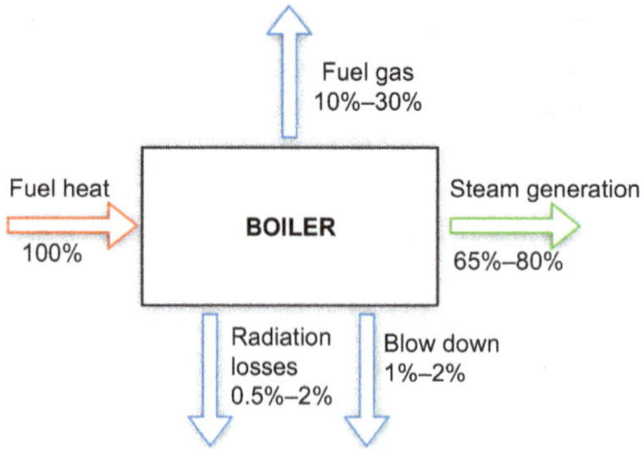

Figure 4.14 Typical heat balance of a boiler (adapted from [25])

efficiency and sizing the fans and volumetric devices of the boiler. The ambient temperature can change by more than 50 °C during the year, depending on the latitude, with consequences on the specific volume of the air.

4.3.2 Fuels for boilers

Gas-fired boilers are forecasted to be widely used for industrial applications in the next years to assist the energy transition. Among them, premixed boilers allow to obtain higher energy efficiency and lower pollutant emission with respect to non-premixed boilers and, when fueled with alternative fuels (mainly biogas, hydrogen, biomethane, and ammonia) would help to reduce the overall CO_2 emissions and ensure the security of the energy supply.

The main properties of gaseous fuel to consider are the composition, the calorific value, and specific gravity. The natural gas is considered to date the most suitable gaseous fuel for gas-fired boilers, as it does not require storage facilities and can be burnt in very compact burner with extremely simple and sensitive control systems [26]. Compared with coal and oil, natural gas has the highest H/C ratio and emits low CO_2.

Other common fuels for gas-fired boilers are the syngas (mainly composed of H_2 and CO) and commercial gases like propane and butane, the latter commonly used for domestic boilers.

Nevertheless, the gaseous fuels composition is expected to significantly vary in the next years due to the integration of non-fossil fuels, like biogas and hydrogen, on the gas grid [26]. Generally, gas boilers are tested with a reference gas, considering specifically extreme conditions (incomplete combustion and soot limit gas, light back limit gas, flame lift limit gas, overheating limit gas) to guarantee efficient and safe operation at various fuel compositions. Such expedients, however, are not sufficient to grant safe operation and full functionality in case of alternative fuels, as the different combustion features would affect the overall process. Indeed, the diverse combustion chemistry and resulting variations in flame speed also impact pollutant emissions, burner temperature, and subsequently burner longevity.

Fuel oils used for steam generation are generally distillation fractions of crude oil. Crude petroleum is burnt only infrequently, as the lighter fractions depress the flash point and present fire hazard, thus by only removing these components the oil becomes suitable for boiler applications. The properties to consider for fuel oils are viscosity, flash point and ignition point, sulfur content, and ash content. Viscosity influences both the transport and the atomizing quality of crude oil and depends on temperature and pressure. Different types of oils require different heating temperatures to attain the proper viscosity. Sulfur content has to be considered, since for sulfur greater than 0.3% can cause important metal corrosion on the low temperature heating surface. Depending on the sulfur content, oil can be divided into low sulfur ($S < 0.5\%$), middle sulfur ($S = 0.5\%–2.0\%$), and high sulfur ($S > 2\%$) grades [22].

Coal has traditionally served as the primary fuel source for power plants across many regions worldwide, except in areas abundant in oil and gas resources. Coal comprises complex compounds that originate geologically from ancient plant matter, undergoing transformation over millennia due to the earth's pressure, time, and heat. The resulting coal properties vary depending on the duration of exposure to these conditions. Older coals subjected to higher pressure tend to have converted the majority of the plant material into fixed carbon (FC), while younger coals experiencing lower pressure retain a significant proportion of volatile matter (VM). There are several classifications for coals. In the United States, they are classified by rank as per ASTM D388, depending on the content of FC and VC.

Coal contains also various amounts of oxygen, nitrogen, sulfur, moisture, and ash. The oxygen content can vary from 2% for anthracite up to 20% for lignite, while the nitrogen content in coal is smaller (0.5%–2%), even though it can significantly contribute to the formation of fuel-NOx during combustion. Sulfur is contained in different forms, like organic sulfur, FeS, and sulfates. Only organic sulfur and FeS can be oxidized. Ash comprises the inorganic solid residues left after the fuel is completely burned and the primary components are iron, calcium, silicon, and aluminum.

4.3.3 Ammonia as alternative fuel for co-fired/fired boilers

Fuel flexibility is certainly a desirable condition for burners and represents the greatest challenge as it cannot currently be warranted except at the expense of global efficiency and general deterioration in boiler performance. Nevertheless, ensuring flexibility while at the same time high combustion efficiency and low emissions become imperative to mitigate the fossil fuel consumption and minimize the environmental impact of energy production.

This aspect becomes especially urgent when it comes to industrial heat. Industrial heat is frequently produced on-site, making regulation more complex compared to centralized sectors like large thermal power generation. Moreover, it encompasses a broad range of temperature levels to accommodate diverse processes and end-uses.

There are few ongoing projects in the world with the ambitious goal to realize totally ammonia-fueled boilers. For instance, Mitsubishi Power announced the development of 100% ammonia-fed boilers for industrial heat applications and planned the demonstration of the reliability of the system by 2028 using a full-sized combustion system. Mitsubishi Power conducted combustion tests at the MHI Research & Innovation Center using basic combustion test furnaces that can simulate the combustion conditions of coal-fired boilers in power stations.

According to the Ammonia Energy Association [27], Mitsubishi also announced that optimal conditions for ammonia combustion have been successfully identified during the testing, including acceptable level of nitrogen oxide emissions, no residual ammonia, and stable flame during combustion [28].

Nevertheless, to date co-firing to coal power plants appears to be the most feasible solution for boilers and it is expect to be the first commercial application using ammonia as fuel in these systems.

In 2023, the Chugoku Electric Power Company conducted the ammonia mixed-combustion at coal-fire Mizushima Power Station. A 156 MW unit was used for the test campaign. Ammonia was pre-vaporized and accumulated before accessing the boiler under the required flowrates. Up to 450 kg/h of ammonia (equivalent to 400 kg/h of coal) was used to produce 1 MW power at a mixing rate of 0.6%. No corrosion in the system was observed. The results showed that recombination of ammonia in post-combustion sections reduced NO_x fumes, leaving the final emissions level similar to those of coal combustion [6].

The Ammonia Energy Association reports ongoing efforts and advancements toward an ammonia-based economy will be made in Asia. Japan, China, Singapore, and South Korea are making significant strides in developing these technologies to facilitate decarbonization across various processes, including coal power generation. India has also entered this framework, as it represents 8% of the global production of ammonia [6].

In South Korea, KEPCO and various public power generation companies funded a $12 M USD project, started in 2022, consisting of four sub-projects, namely ammonia co-firing to pulverized coal boilers, circulating fluidized bed boilers and gas turbines, and the development of the ammonia supply chains. The project engaged boiler makers and chemical companies, universities, and research centers.

Successive to the project, in 2023, Korea's Ministry of Trade, Industry and Energy (MOTIE) has funded a 5-year demonstration project aiming at 20% ammonia co-firing to four commercial-scale power plants: two pulverized coal boilers and two circulating fluidized bed boilers. Following the demonstration project, the 20% co-firing initiative will expand to 24 coal-powered generation units to comprehensively evaluate the progression of utilizing ammonia as a fuel source for industrial-scale power generation [29]. Ammonia co-firing trials are also going to commence in Philippines, Thailand, and Chile.

In order to assist the transition, it is crucial to possess a comprehensive understanding of the co-firing features of NH_3 and coal, as many issues still remain controversial. First, there is limited understanding regarding the impact of NH_3 on the formation and deposition of mineral ash, which is a crucial factor influencing the safety of heating surfaces within the boiler. Second, the influence of NH_3 addition on the fraction of unburned carbon in fly ashes and on the NOx emissions has to be evaluated.

Recent studies have been focused on the NOx emission of co-firing of NH_3 and coal [30], addressing particularly the impact of NH_3 injection position and co-firing ratio.

Yamamoto *et al.* [31] examined the co-firing dynamics of NH_3 and coal in a horizontal test furnace employing a single burner. Their findings indicate that the injection position of NH_3 has minimal impact on the concentration of NO emissions and there is no significant increase in NO emission in comparison with pure coal firing. In a separate study, exploring the co-firing of ammonia and coal in a single burner, Kimoto *et al.* [32] observed that NO_x concentration can be reduced by injecting ammonia into the pulverized coal combustion flame through the side of the furnace wall at an appropriate position.

Ishihara *et al.* [33,34] studied the effect of NH_3 nozzles position on NO_x emissions. In particular, they set the nozzles at three different locations in the mixing zone, main combustion zone, and burnout zone and concluded that the NO_x emissions are lower than that of the fired-coal conditions when NH_3 is injected into the main combustion zone. Further, they changed the NH_3 co-firing ratio ranging from 0% up to 80% and observed the lowest NO_x emissions at 20%.

Such experimental studies suggest that, although the presence of nitrogen chemically bound to the fuel could potentially increase the NO_x emissions, the presence of ammonia in appropriate reaction zones (in other words, under proper conditions of temperature and concentration) may instead positively affect NO_x reduction.

The effectiveness of NH_3 as reducing agent is well documented in literature; indeed several abatement technologies rely on the Selective Non-Catalytic Reduction (SNCR) that use NH_3 injection within a specific temperature range (850 °C–1100 °C) to operate a homogeneous gas-phase reduction of NOx to N_2 [35–38], under an O_2 atmosphere. An alternative process for NOx reduction is the Advanced Reburning (AR) that uses natural gas or coal as reburning fuel at temperatures above 1200 °C up to 1600 °C–1700 °C, in the absence of oxygen that can be detrimental for the reduction. The coupling of SNCR and AR, that is the injection of ammonia in the NO high temperature reduction zone in the furnace, is the key concept of the Advanced Air-Staged Combustion technology (AASC). Fan *et al.* [39] carried out experimental tests in a high-temperature fixed bed system with ammonia injection in the reduction zone of AASC. They report that the injection of ammonia into the pulverized coal combustion effectively reduced the NOx at very low oxygen or oxygen-free concentration and high temperature conditions favor the reduction process. In the reduction zone, the simultaneous presence of ammonia and char is beneficial for the NOx reduction, since their coupled reduction ability is higher than their respective reduction ability. Coal ash can further enhance the reduction efficiency of ammonia, especially at high temperatures.

Studying the AASC could give important suggestions for the NH_3/coal co-firing.

Indeed, recent works show that co-firing NH_3 with coal appears to synergistically promote the homogeneous reduction through NO heterogeneous reduction promoted by the char in a reduction zone at high temperatures and lean oxygen conditions, created by air-staged combustion [30,40]. Chen *et al.* [30] performed theoretical calculation to investigate the heterogeneous reduction mechanism of NO by NH_3 in the presence of NH and NH_2 free radicals. They showed that the energy barrier values for the rate-determining step in the heterogeneous reduction of NH-NO on char are even lower than that of the homogeneous reduction of NO by NH_3 and reveal that the formation of OH on the char surface is a key step to strengthen the reduction of NO by NH. On the down side, the energy barrier values for the rate-determining step in the heterogeneous reduction of NH_2-NO are higher than that in the homogeneous reduction of NO by NH_3. In particular, the rate constant for the rate-determining step in the NH-NO heterogeneous reaction is 20 times greater than that in the NH_2-NO heterogeneous reactions; thus, the NH has a dominant role in ammonia denitrification. The temperature has a key role in the reduction process, as the denitrification rate increases with the temperature.

The synergistic coupling of air-staging, SNCR, and injection of ammonia in the high-temperature reduction zone can lead to an effective reduction, with denitrification efficiency up to 90% [41].

Tamura *et al.* [42] carried out experimental tests for ammonia-pulverized coal co-firing in a bench-scale 1.2 MW-thermal coal-fired furnace, with the primary purpose to confirm stable combustion and control/reduction of NOx emissions by staged combustion. Different locations and various methods of ammonia injection were considered, in particular mixing ammonia with coal before supply to furnace, separate injection by insertion of ammonia gun into coal burner and side wall injection. The latter strategy was found to increase NOx emission, especially at high co-firing ratio, due to less residence time of NOx reduction process. On the other hand, NOx emission can be limited up to 30% co-firing when ammonia is injected from the gun of mixed in the coal nozzles. In particular, NOx can be reduced under staged combustion in the sub-stoichiometric zone. Further, the authors demonstrated that the designed gun for ammonia can be used for pure ammonia firing. In this case, air staging ratio is fundamental to obtain strong NOx reduction, achieving NOx below the level achieved in co-firing for air staging of 32% and 42%. It should be noted that pure ammonia combustion envisages lower combustion temperatures, given its lower adiabatic temperature compared with coal, that can be beneficial for less thermal NOx formation.

Besides the concern for the NOx emissions, systematic studies of the combustion features of NH_3-coal are fundamental to guarantee stable ignition and combustion, thus stable operation of the boilers. In a coal-fired boiler, flame speed is one of the most important parameters of flame stability in the burner. Therefore, exploring the flame propagation properties of ammonia-coal co-firing is essential for ensuring safe combustion, enhancing process control and efficiency, and optimizing the fuel blending to achieve peak performance and emission reduction [43].

The fundamental mechanism of the flame propagation phenomenon for pulverized coal/ammonia co-combustion was studied by Xia *et al.* [44], by systematically changing the equivalence ratio of the ammonia-oxidizer mixture in a spherical chamber with constant volume. They reported that the flame propagation velocity of pulverized coal/ammonia co-combustion is higher than that of pure ammonia combustion for ammonia lean condition, while it is almost the same at stoichiometric condition. Conversely, for ammonia-rich conditions, the flame propagation velocity of pulverized coal/ammonia co-combustion is lower than that of pure ammonia. To explain this trend, three different factors have been proposed to contribute differently depending on the equivalence ratio:

- Radiation from the luminous flame (positive effect);
- Increment of the local equivalence ratio by the addition of the volatile matter generated by pyrolysis;
- Heat absorption by coal particles in the preheat zone (negative effect).

Generally, the flame propagation velocity in gaseous fuel surpasses that of coal particles. Consequently, increasing the pyrolysis rate of coal particles could be achieved through enhanced heat transfer promoted by the flame propagation of ammonia, along with the introduction of volatile matter in ammonia/coal particle cloud co-combustion [45]. Nevertheless, the effect of the increment of the local equivalence ratio is not always positive for the flame propagation velocity. Xia *et al.* [44] reported that for ammonia lean conditions, both the effect of radiation and the increment of the local equivalence ratio contribute positively and are stronger than the heat absorption by coal particles, thus generating an increase in the flame propagation velocity. At stoichiometric conditions, the effects are comparable and the flame velocity of pulverized coal/ammonia co-combustion is the same as pure ammonia. Finally, for ammonia rich conditions, the increment of the local equivalence ratio affects negatively the process; as a consequence, the negative effects overcome the positive effect of the radiation resulting in the lower flame propagation velocity compared with that of pure ammonia.

Hadi *et al.* [45] performed experimental studies on the effect of turbulence intensity and different coal types on the turbulent flame propagation under co-firing conditions. In particular, they considered coal types with different fuel ratios, defined as the content of the fixed carbon divided by the content of volatile matter (from bituminous C5 to high fuel ratio coal as TW, KK, UL). The measurements of the flame speeds are reported in Figure 4.15. In pure coal combustion, only the C5 sustained flame propagation at all the turbulence intensities is considered, while at high-fuel ratio, KK and UL coals did not sustain the flame propagation for the same turbulence intensities.

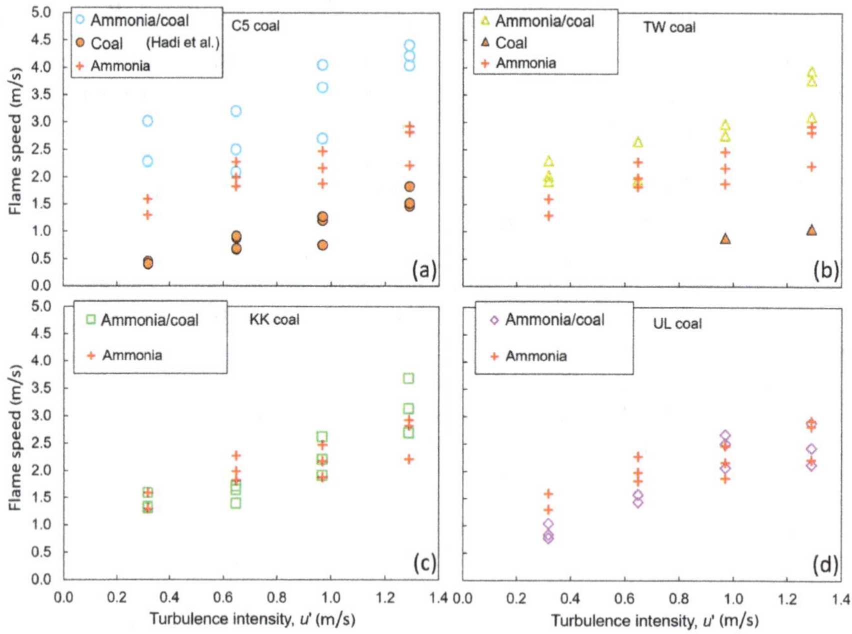

Figure 4.15 Flame velocity as a function of turbulence intensity for pure ammonia, pure coal, and co-firing conditions, for fuel-lean conditions. Reprinted, with permission, from [45].

In general, the turbulence intensity yields an increase in the flame speed, as it increases both the flame front area and the turbulent heat transfer rate at the flame front.

Consistently with the findings of Xia *et al.* [44], they found that, at fixed turbulence intensity and ammonia fuel-lean conditions, the velocity of bituminous coal combustion is three times greater that of pure coal combustion and twice the result obtained for pure ammonia combustion. On the other hand, for high-fuel-ratio coal cases, the difference in flame speed between ammonia/coal co-combustion and pure ammonia combustion is smaller, suggesting that the different tendencies depend on the difference of the volatile matter amount in the coal particles. Also, the soot tendency of the different coal types can affect the flame velocity. Basically, the soot formation on the post-reaction front is faster for bituminous soot, because of the sufficient amount of the volatile matter in the reaction front. Consequently, the distance between the reaction front and the luminous flame front is short and the coal particles in the preheat zone receive larger radiation heat flux, increasing the pyrolysis rate of coal particles. The mechanism of flame propagation in coal/ammonia cofiring is schematically represented in Figure 4.16.

Conversely, for high fuel-ratio coal, the volatile release rate is lower resulting in a slower soot formation rate. The luminous flame front is relatively far from the reaction front, resulting in lower radiation heat flux to coal particles. In this case, the heat sink effect of coal particles is overwhelming and the flame velocity can be even lower than that of pure ammonia, as reported for the UL coal/ammonia co-firing in Figure 4.16.

In summary, when co-fired with coal, ammonia ignites before the coal and the heat released accelerates the coal thermal decomposition and release of volatile matters, thus increasing the local gas-phase equivalence ratio, with effect on NOx formation and soot particles.

Another aspect that could significantly impact the boiler design and combustion efficiency is the radiative heat transfer features of the flame, since the radiative heat transfer in boilers is closely related to the used fuel and its combustion products. Traditional fuels (from gas to solid fuels) produce CO_2, H_2O, fly ash, and

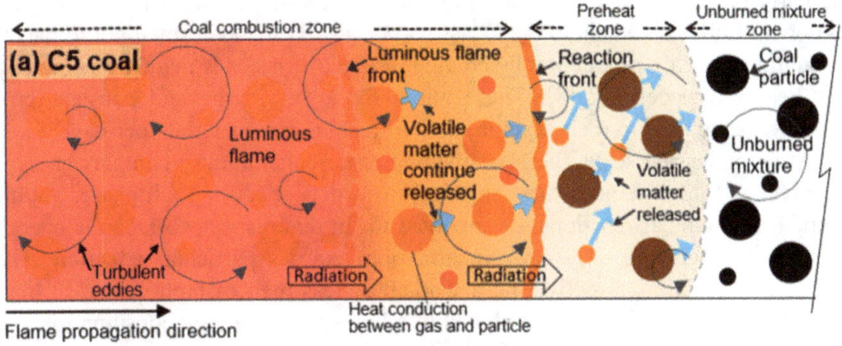

Figure 4.16 Mechanism of the flame propagation of coal/ammonia co-firing. Reprinted, with permission, from [45].

soot particles, while pure ammonia combustion forms only N_2 and H_2O, with different temperature distributions and radiative heat transfer properties. The different composition of the exhaust could also alter the convective heat transfer process within the heat exchanger, given the different heat capacity of the flue gases, richer in H_2O than CO_2. These aspects cannot be ignored also for ammonia-coal co-firing.

A comprehensive evaluation of the exergy efficiency was carried out by Xu *et al.* [46]. The authors performed simulations of a boiler system of a 600 MW coal-fired power plant unit in both coal-fired condition and ammonia-coal co-firing conditions, and evaluated the impacts of ammonia co-firing on the emission and thermal properties of flue gas, exergy loss, and exergy efficiency. Two different scenarios were compared, namely keeping the same AFT and the same excess air coefficient (EAC). The results showed that CO_2 emission was effectively reduced by 24.08 t/h for every 5% increase in the ammonia co-firing ratio, while the H_2O content increased with ammonia co-firing increasing the specific heat capacity of the flue gas. The variation of the thermal properties of the flue gases drastically decreased the temperature of the flue gases at the same position for the EAC scenario, while only a slight variation was observed for the AFT configuration. Compared with pure coal combustion, the exergy of the fuel in the furnace is higher but the exergy of the flue gases is lower; thus, the global exergy loss for both the ammonia co-fired scenarios increases with the ammonia co-firing ratio. On the other hand, the exergy loss of heat exchangers and fans decreased for ammonia co-firing; therefore, the resulting exergy efficiency of the boiler was only slightly reduced under ammonia co-firing conditions (53.82% vs. 53.27%–53.22%).

Such preliminary studies provide important information and lay the foundation for the understanding of ammonia-coal co-firing, but they are certainly not exhaustive and a lot more systematic explorations are further demanded in order to reveal the fundamentals of ammonia–coal co-firing process.

4.4 Glass melting furnaces

The early development of glassmaking and the specific furnaces used is largely unknown, similar to metalmaking. Glass was originally made in small pot furnaces, although large-scale production by continuous tank furnaces is mainly obtained by open-hearth furnaces [47,48]. These allow for a very large and constant productivity, being able to melt several hundred tons of raw materials per day.

At the industrial scale, the glass melting represents the most complex energy-intensive and high-temperature process, in which specific multiphase, multi-component chemical reactions are involved in. In this respect, to face the energy transition to meet the CO_2 net-zero emission requires particular attention. The melting process aims at delivering a workable glass melt of high quality at high production rates, as much as possible with a low specific-energy consumption and abiding with environmental legislation. These aspects entailed over the years the development of technologies and furnaces design with peculiar characteristics and improvements for the specific utilization.

The glass production process brings together issues related to reactor design and technology, bulk solid melting, particle dissolution, redox, and acid–base chemistry of the melt. In particular, the melting process for the vast majority of glasses (large-scale production of bottles, jars, and flat glass) takes place in specifically designed and continuous glass furnaces, whose production capacities range from a few tons to 1000 tons per day. In this respect, the glass industry is highly diversified, in terms of products, processes and furnaces. This of course impacts on the possible choice of strategies for decarbonizing the sector.

Although furnace type and design vary from the small electric furnace used in the glass wool sector to the huge continuous furnaces, generally glass fusion process involves the same several steps reported in Figure 4.17 [49,50].

The first step represents the careful preparation of the batch from the appropriate raw materials. The second and third take place at high temperatures through the various reactions that lead to a melt through complete dissolution of even the most refractory starting materials. The fourth and fifth steps aim at producing a homogenous, bubble-free product by physical and chemical fining. Finally, the last step aims at cooling the molten glass to make it suitable for subsequent manufacturing.

As reported, all the glass melting process steps require very high energy input to reach the needed operating temperatures in the range 1200 °C–1600 °C [51–53]. In this respect, the glass melting process is classified as "energy intensive" and the industry is constantly working to reduce its energy consumption. It is characterized by reaching and emitting considerable amounts of pollutant species such as CO_2, NO_x, CO, and SO_x into the atmosphere. Generally, a typical glass furnace consumes about 70%–80% of the total thermal energy input required by the whole glass plant. This is characterized by very long service lives with continuous operation, which can even reach 20 years for flat glass plants. This is why it is not economically advantageous in the short and medium term to focus on new technologies but to adapt the existing systems for the utilization of the new energy carriers.

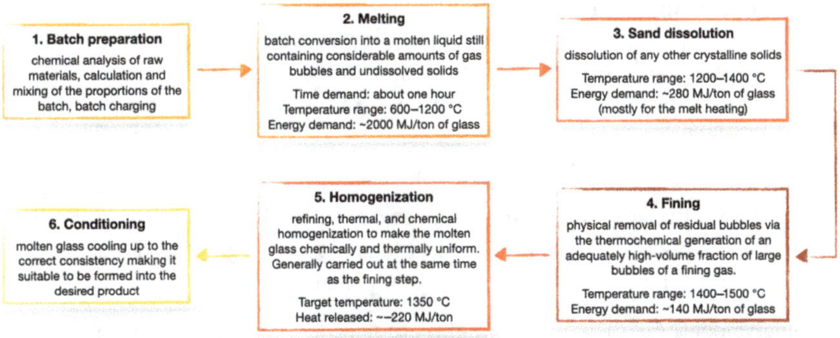

Figure 4.17 Main steps involved in the glass making process

In particular, for an efficiently operating furnace, the specific energy consumption is about 7–8 GJ/tons of glass [54]. Based on this calculation, the glass industry is estimated to consume approximately 1.2 TJ/year, thus resulting in a global glass production approximately equal to 150 million tons per year. To increase the glass melting process efficiency and sustainability, several solutions are being adopted and investigated. In particular, the melting furnace can be equipped with specific solutions aimed at recovering energy from combustion products (~1000–1500 kJ/kg), as well as using innovative fuels, such as hydrogen or ammonia, as it will be analyzed in the following.

Depending on the final product, raw materials, energy vector choice, size of operation, and economic factors, several types of melting furnaces can be identified as used in the glass industry [55]. In particular, melting furnaces can be classified according to:

- **Continuity of the melting process:**
  - Discontinuous or batch furnaces (pot furnaces): the various stages of glass melting are sequentially carried out in the same part of the furnace at different times, and the temperature system of the furnace is variable.
  - Periodical and continuous furnaces (tank furnaces): the glass melting stages are carried out at the same time and in different parts of the same furnace, and the temperature system in the furnace is constant.
- **Methodology for waste heat recovery equipment:**
  - Recuperative furnaces: flue gas waste heat recovered by heat exchange.
  - Regenerative furnaces: flue gas waste heat recovered by heat storage.
- **Heat source:**
  - Flame furnaces: fuel combustion is used as heat energy source. The fuel can be gaseous (natural gas), liquid (heavy oil), and, in a lesser extent, solid (petcoke).
  - Electric furnaces: electric energy is used as the heat source.
  - Hybrid furnace: combustion is the main heat source while electric energy the auxiliary one.
- **Flame development direction:**
  - Horizontal flame (side-port-fired) furnaces: the flame develops orthogonally with respect to the flow direction of the molten glass.
  - Horseshoe flame (end-port-fired) furnace: the flame develops with a horseshoe shape.
- **Furnace size:**
  - Large furnace: ~150–400 tons per day of produced glass.
  - Medium-sized furnace: ~50–150 tons per day of produced glass.
  - Small furnace: <50 tons per day of produced glass.

Independently of the considered glass furnace kind, key parameters to take into account to evaluate the glass furnace performance are the melting capacity (tons of glass produced per day), the thermal efficiency (η_{th} = how much of the supplied energy is transferred to the molten glass), and specific energy consumption (SEC=GJ/ton of produced glass).

4.4.1 Batch and continuous furnaces

In discontinuous or batch furnaces, a batch of raw materials is added to a single melting vessel and melted into glass [56]. Once the whole melting process is completed, the batch is removed from the melting tank and sent to the forming processes. Therefore, a new batch is introduced only once the previous batch has undergone the entire process. Moreover, the batch is melted, fined, homogenized, and conditioned within the same pot, before being extracted from the pot for forming.

Pot furnaces are structures constructed of refractory materials, in which several pots are located. The latter are made of resistant refractory materials to glass attack at high temperatures and contain the glass batch to be melt. These furnaces use the reverberatory principle. Basically, the fuel and the glass batch are kept apart, and the flames are made to strike back from the arched roof of the furnace upon the materials to be melted. In particular, by placing a cover over the pots, the purity of the molten glass batch is ensured. Therefore, in such a specific type of glass melting furnaces, there is no contact between the furnace structure and the glass itself. Indeed, several different types of glasses can be simultaneously melted in dedicated pots, over a number of hours and worked on 24 or 18 h for each cycle. In this respect, pot furnaces are intrinsically highly flexible, since enabling manufacturers to produce multiple formulations of different glasses at the same time and specifically depending on the particular customer demand, without the possibility of cross-contamination [57]. Such a characteristic covers a prominent role for pot furnaces, making them extremely useful when high purity of the produced glass is required. However, because of the melting process intermittency, this type of glass furnaces is characterized by limited productivity. In particular, a single average pot usually holds 600–700 kg of glass and can be used for about 30 melting cycles, thus producing ~20 tons of glass. However, because of their discontinuity, batch furnaces are typically used for limited production, often required for special glass types used in optics, electronics, glass formed by hand and mouth blowing.

Continuous glass melting furnaces, usually referred as **tank furnaces**, are the most widely used in large-scale glass production [57] where continuous flow of glass is needed to feed automatic glass forming machines. Generally, modern continuous glass furnaces are typically from 8 up to 40 m long, 5–15 m wide, and hold enough glass over typical depths of 0.9–1.4 m, with a daily production of glass ranging from 20 up to 1000 tons, for a total investment cost ranging from about 2 up to 15 M€. In tank furnaces, a mixture of raw materials for glass production is continuously fed at one end of the glass furnace, while melted glass is continuously extracted at the other. This characteristic ensures continuous and constant productivity. Indeed, continuous melting processes and tank furnaces are high-throughput and preferred wherever a high volume (100–500 tons per day) of the same type of glass is required such as for architectural glass, fiberglass, and food and beverage containers [58].

Modern industrial continuous glass lines can be typically divided into three main sections: the batch house, the hot end, and the cold end (Figure 4.18).

The batch house, upstream of the melting furnace, is the location in which raw materials are stored and the glass batch is prepared and delivered to the furnace. The hot end, in which the melting tank is located, corresponds to the part of the furnace where melting, forming, and annealing processes occur, and it represents the most energy intense part of the whole glass production process. Downstream of the forming and annealing processes the cold end is located, where postprocessing such as coating, cutting, and inspection takes place.

4.4.2 Regenerative and recuperative furnaces

In order to increase the glass furnace efficiency and fuel economy, simultaneously achieving high melting temperatures, various heat recovery strategies have been devised for both discontinuous (pot) and continuous (tank) furnaces, being currently integrated into the majority of industrial glass melting furnaces [59]. Indeed, the heat transfer process from the combustion atmosphere to the batch and melt is usually limited, thus resulting in the majority of energy input being dissipated through exhaust gases.

In this respect, regenerative and recuperative systems are the most widely used [60–62]. Indeed, whenever flue gases are produced, glass furnaces are equipped with either regenerators or recuperators of the same types as those used in other high-temperature processes [63,64]. Both these systems aim at recovering the waste heat of exhaust gases produced by the fuel oxidation, to preheat the combustion gases (the combustion air in particular) before they reach the glass furnace burners. These strategies allow for current modern furnaces to achieve high energy efficiency, with most furnaces showing energy consumptions ranging from 3.2 to 8 GJ/t of glass [59].

4.4.2.1 Regenerative furnaces

Regenerative glass furnaces recover heat from the exhaust gases to preheat the incoming combustion air by alternatively passing the hot exhaust gases and cold combustion air through large stacks refractory material that constitute the regenerators core. The regenerator consists of two or several chambers designed to store

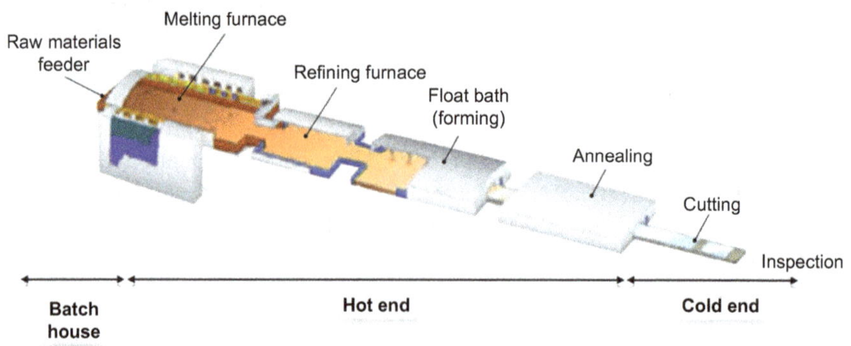

Figure 4.18 Schematic illustration of a float glass industrial production line. Reprinted, with permission, from [47].

as much heat as possible by a checkwork of staked refractory bricks, whose heat capacity is one of the key parameters determining the regenerative furnace performance.

The heat recovery and its successive transfer to the incoming fresh air occur discontinuously, by realizing alternating heat recuperation and heat restitution cycles. This occurs by alternatively operating two sets of regenerators for each glass furnace, whose general principle is shown in Figure 4.19. In particular, while one regenerator is being preheated by the exhaust gases, that typically cool down from about 1450 °C up to 600 °C–400 °C thus exchanging their sensible enthalpy with the regenerative system, the other one simultaneously transfers the previously stored heat to the incoming combustion air, increasing its temperature up to typically 1100 °C–1400 °C. After the regenerator has transferred the previously stored heat, the exhaust gases and combustion air flows are reversed. After reversing, the exhaust gases pass through and heat the checkers that had previously heated the combustion air. Such operating principle leads to very high thermal efficiencies and reduced fuel consumption owing to good heat recovery via the checker work regenerators.

In order to allow for efficient combustion air preheating and good thermal homogeneity in the furnace, the heat recuperation and heat restitution cycles alternate on a regular basis, approximately reversing every 20–30 min, depending on the furnace thermal load. In particular, the flows of both fluids alternate in less than one minute during which firing stops. This reversal period must be as short as possible to avoid excessive cooling of the glass furnace. The high air preheating temperatures achievable by regenerative furnaces entail very high flame temperatures and, thus, high NO emissions. In this respect, more recently, recuperative furnaces were operated at low excess of air and by using oxygen instead of air as oxidizer, in an attempt to reduce NO emissions from glass production.

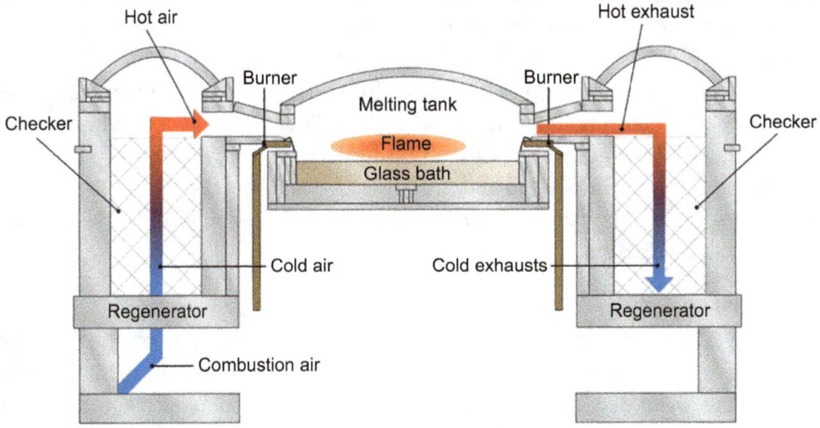

Figure 4.19 General principle of regenerative glass furnaces. Reprinted, with permission, from [47].

Concerning the furnace configuration, regenerators may be placed on the side or at the back of the glass furnace, thus identifying two main categories for regenerative glass furnaces: end-port-fired or side-port-fired (also identified as cross-fired furnaces), respectively. The former are characterized by burners and regenerators at the back of the furnace, whereas the latter show burners and regenerators on each side of the furnace itself.

With reference to end-port-fired furnaces (also identified as U-flame furnaces) [65], they are the most common because of their economically efficiency for several types of production, typically used for producing less than 230 tons of glass per day. These are characterized by two ports equipped with 2–4 burners each and located on a single side of the furnace, next to the regenerator. The flames originating from one port spread across the entire length of the furnace, then they turn 180° back and toward the opposite port from where the flue gases are extracted. This U shape ensures long average residence times, with generally longer flames than cross-fired furnaces.

The main advantages related to this type of furnaces are the high levels of preheating for combustion air (that make them very effective with respect to the use of low reactive fuels such as ammonia) and a relatively low energy consumption that entail high thermal and overall energy efficiency. Conversely, they allow for a limited energy distribution control along the longitudinal axis of the furnace, which results in some limitations with respect to high-quality glass production. Furthermore, their investments costs are relatively high, as their characteristic NO_x emissions requiring also particular maintenance work for the chamber cleaning. However, regenerative end-port-fired furnaces are typically used for container and tableware glass production.

With respect to side-port-fired [66] 4–8 burner ports are placed on two opposite sides of the glass furnace, next to the regenerators. Such a configuration allows to opportunely tailor the temperature profile along the longitudinal axis of the glass furnace, thus obtaining a localized hot spot of a given size and at the right place above the melting tank. This, in turn, allows to carefully control the achievable glass quality. When compared with side-port-fired regenerative furnaces, end-port-fired ones show lower energy efficiency. Furthermore, the arrangement of the regenerators increases the building costs, while a latent issue is related to the possible clogging of the regenerators of the first furnace section, due to the dust carried by the exhaust gases. Side-port-fired are typically used for producing greater than 230 tons of glass per day, commonly reaching about 460–630 tons of glass per day for float glass production.

4.4.2.2 Recuperative furnaces

Recuperative furnaces realize the incoming combustion air preheating continuously, by means of a heat exchanger (recuperator). In particular, the hot exhaust gases coming from the melting tank are directed toward one/two recuperators (depending on the furnace), in which the heat is transferred to the combustion air inlet flow, which is then sent to the burners to produce the flames above the tank [67]. Recuperators can operate in both co- and counter-current exchange. They are typically made of high-temperature-resistant alloys, while thin-walled materials are required to ensure efficient heat exchange, although severely limit the entry temperatures of the exhaust gases. The nature of heat transfer from the hot flue gas to the recuperator is mainly radiative; therefore, these devices are often termed radiative recuperators. On the other hand, the combustion air heat up by convection only.

Although recuperators are theoretically more efficient than regenerators, they do not reach the efficiencies of these latter, thus entailing a moderate air preheating, up to maximum 750 °C–800 °C. This results in lower NO_x emissions and investment costs, simultaneously entailing higher energy consumption with respect to regenerative furnaces. Furthermore, recuperative furnaces are relatively cheap and simpler in construction compared to regenerative ones, since not containing refractory checkers, thus also ensuring clean combustion air.

Recuperative furnaces are typically equipped with a large number or burners (5–15 per side, depending on the size of the furnace), independently controlled, that allow for fine tuning of the temperature profile along the melting tank. Furthermore, these furnaces are typically smaller than regenerative ones, with a maximum production capacity ranging between 20 and 400 tons per day, and mainly used for fiber-glass production, tableware, and container glass.

4.4.3 Oxy-fuel furnaces

Instead of regenerative and recuperative furnaces that run using fuel and air as oxidizer, oxy-fuel-fired furnaces use pure oxygen, thus significantly reducing the exhaust gas volume as the dilution by nitrogen is avoided. In these furnaces, the oxygen concentration in the exhaust gases is increased up to 91%–98% vol, thus minimizing energy losses due to the 78% N_2 and 1% Ar dilution of air. This allows to markedly increase the operating temperatures (>1700 °C), also reducing the NO_x emissions levels, the volume of flue gases and thus the energy required to reach a specific temperature [68,69].

Generally, oxyfuel furnaces have similar designs to recuperative furnaces, with multiple burners on each side of the melting tank, and usually not equipped with a heat recuperation system. Compared with the other typical glass-fired furnaces, oxy-fuel ones show much higher efficiency and low design costs, although usually requiring high-quality refractories [70]. These also show low environmental impact and reduced fuel consumption. In this respect, compared with a common furnace for float glass production, a 100% oxygen fired oxyfuel furnace may reduce the energy consumption by 25%, while CO_2, SO_2, and NO_x by 15%, 38%, and 90%, respectively. However, pure oxygen production and attainability entail very high operating costs that may exceed the gain provided by the reduction in fuel consumption.

Oxy-fuel furnaces are especially advantageous if higher melting temperatures are required, as for very high-quality glass production such as for continuous fibers, for which even microscale defects cause breakage of the thread in the drawing machine. In this respect, as reported in several studies, glass quality and production rate improve in most of the oxyfuel furnaces due to the enhanced fining reactions, in which water is involved since dissolved in the molten glass [71,72]. In fact, all commercial glasses contain water in the form of hydroxyls, which cause significant impact on the fining behavior. In some cases, manufacturers are able to opportunely control the amount of water in glass, to ensure more consistent product quality [72]. In this respect, oxyfuel melting typically increases the water in the molten glass by 30% or more, compared to air-fired furnaces.

4.4.4 Energy use and environmental impact associated with glass furnaces

As reported in the previous sections, glass production is a very energy-intensive process, for which the average theoretical minimum energy input required for glass melting is about 7–8 GJ/ton of glass [54]. Specifically, this required energy accounts for 15%–25% of the cost of glass products, depending on the sector and the country of manufacture. In this respect, the most energy-intensive sections of glass production and manufacture are melting and refining processes, which together account for 70%–80% of the total energy needs, while tin bath, forming/cutting processes, annealing, and other plant services account for the remaining energy.

Despite the different kinds of production and technologies adopted for glass-making, the required energy usually derives from natural gas combustion, electricity, and, to a lesser extent, heavy fuel oil and solid fuels.

Natural gas or heavy fuel oil are normally used for firing tank furnaces. Indeed, nowadays the energy use that consolidates the European and the US glass industries is dominated by natural gas, accounting for approximately 75%–85% of the used energy source [73], due to its higher purity with respect to heavy oil, that accounts for 5%–10% of used energy source, easiness of use, and not requiring specific on-site storage. On the other hand, solid fuels, in particular coke and petcoke (a residue from petrochemistry) [74], are mainly used in China. In this respect, only referring at the Chinese glass industry, its total energy consumption was about 9.6 million tons of coal by 2010 [75].

The particular fuel choice depends on several factors, including the type of used furnace, availability of fuel sources, costs, and required fuel purity. Furthermore, both air and pure oxygen can be used as oxidizer (as analyzed with respect to oxyfuel melting furnaces) depending on the specific glass type to be produced and the required operating temperatures.

On the other hand, since molten glass is an electrical conductor at high temperature, also electricity is usually used in smaller glass furnaces, such as fiberglass bushings, tin bath heating and annealing, or for electrical boosting techniques and hybrid melters. For this purpose, as previously analyzed, electrodes are directly immersed into the melt in order to generate Joule's heat [55]. However, due to its high cost, electricity is typically utilized to enhance the efficiency of gas or oil-fired furnaces. Nonetheless, advancements in electric melting technology have made it feasible to employ all-electric glass melting furnaces, despite the high electricity cost, entailing electricity accounting for approximately 10%–15% of the used energy source for glass melting industry [73].

Due to its energy requirement and the predominant use of fossil fuels (~90% of the total energy sources), the carbon intensity of the glass industry covers a prominent role with respect to the global CO_2 emissions, as proven by the over 60 million tons of CO_2 emissions per year [76] related to container and flat glass sectors (which combined account for 80% of the total glass production). Furthermore, to emphasize the environmental impact, glass production process also contributes to NO_x emissions, arising from the high-temperature combustion occurring in the glass furnaces, SO_x emissions, which mainly derive from fuel combustion and fining processes for sulfur-fined glass, and indirect emissions deriving from raw material degradation and the electricity originated in the grid.

Consequently, a key challenge to lower the energy intensity of the glass industry is the lack of technologies to operate large-scale float glass furnaces reducing the fossil fuels utilization and thus their environmental impact. In this respect, the vast majority of best-performing installations continue to be powered through a combination of fossil fuels and electrical boosts.

Among the several adopted strategies, the introduction of innovative and sustainable energy carriers, derived from processes aimed at boosting the integration of renewable sources, is receiving increasing attention. In particular, the sustainable use of fuels with high energy density has attracted attention from the viewpoints of economic efficiency and cost-effectiveness. In this respect, hydrogen (H_2), biofuels, organic hydrogen carriers (organic hydrides), and ammonia (NH_3) have been identified as potential energy carriers. However, for industrial applications with high heat demands, such as industrial furnaces and power generation, hydrogen and ammonia cover a prominent role [77,78].

In particular, hydrogen is characterized by well-known limiting aspects associated with the energy cost for liquefaction and transportation, as well as the need for CO_2 capture and storage during the fuel production, and combustion properties. Consequently, ammonia is receiving increasing attention toward its direct use as fuel in industrial furnaces such as glass melting ones. In this respect, instead of its scarce combustion properties and high NO_x emissions [79], several advanced thermochemical conversion technologies, such as MILD combustion [80,81] and Rich-Lean staging systems [82], are already able to directly use pure ammonia as fuel, ensuring high stability, reduced NO_x emissions, and operating flexibility [83,84], thus meeting the requirements of the glass industry in terms of operating temperature and reduced emissions.

4.4.5 Ammonia as alternative fuel in the glass industries

The use of ammonia in the glass sector still needs significant development.

Several studies were carried out aimed at evaluating the characteristics of oxynitride glasses, i.e. glasses produced by using ammonia as agent to add nitrogen in the glass structure. These include studies on silicate and aluminosilicate glasses vitrified by reaction with ammonia in the glass melt [85], or by bubbling gaseous ammonia directly into the molten glass [86] or melting glass in an ammonia atmosphere [87]. Specifically, these studied highlighted the transparency loss of the obtained glass if integrated nitrogen concentration increases over a critical value [88].

On the other hand, very few studies were devoted to use ammonia as direct fuel in glass furnaces to ensure its wide utilization on a large scale. In this respect, experimental campaigns were carried out to highlight the impact of using ammonia in lab-scale glass melting furnaces on the produced glass quality [89]. Specifically, glass characteristics such as color tone are affected by several operating parameters, among which the melting process, the melting temperature, and the combustion atmosphere are the most influent [90–93].

In this respect, using ammonia as source of heat in the melting furnace affects the glass-melting process because of the different composition of exhaust gases, especially high-temperature steam with no CO_2, and the lower achievable temperatures with respect to those typically derived from fossil fuels combustion such as natural gas. Indeed, the absence of CO_2 and the higher concentration of H_2O ($\sim$31% on volume basis) in the exhaust gases derived from ammonia combustion directly affects both the efficiency of the radiative heat exchange from the flame zone of the melting furnace toward the melting glass [21] and the glass melting reactions. Indeed, the water generated per unit calorific value during ammonia combustion is approximately twice that deriving from natural gas combustion. The higher amount of high temperature steam comes into contact and interacts with the molten glass, significantly affecting its physical properties and structure [94–96]. Specifically, the presence of water in the melting glass significantly lowers its viscosity and glass transition temperature [97], thus considerably affecting its characteristics and following processing.

This behavior covers a prominent role especially with respect to the optical properties of glasses containing transition metals, such as colored glasses, which depend on the oxidation state of the transition metal elements in the glass itself, directly affected by the water concentration [21,93–98].

In this respect, with reference to [89], glass melted by using ammonia as fuel was found exhibiting different color tones and lower glass transition temperatures compared with natural gas. In particular, the glass transition temperatures were found to be 9 °C lower than those obtained by electric melting.

The increasing interest with respect to introducing ammonia as fuel in the glass industry is also testified by international projects devoted to the development of glass production plants totally using ammonia as fuel. In this respect, AGC company [99,100] announced that it has recently succeeded in the world's first demonstration test of glass production using ammonia as fuel in an actual production furnace, in the framework of the project "Technology Development for the Utilization and Production of Ammonia as Fuel," commissioned by New Energy and Industrial Technology Development Organization (NEDO). For this project, a realistic examination of ammonia combustion technology in oxy-fuel burners (200 kW each) with low NO_x emissions was carried out, by considering a glass melting furnace dedicated to architectural glass production. Glass quality, required furnace materials, furnace control, as well as nitrogen oxide (NO_x) emissions were investigated to evaluate the system performance in order to comprehensively integrate ammonia into glass melting furnaces. Additionally, investigations were performed to extend the ammonia integration beyond glass industry, toward materials such as steel and aluminum. The test furnace and the ammonia flame in the furnace are reported in Figure 4.20.

Figure 4.20 *(a) Glass melting furnace and (b) inside the glass melting furnace at the AGC Yokohama Technical Center. Reprinted, with permission, from [99].*

Although the tangible and increasing interest related to the use of ammonia as fuel in the glass industry, huge efforts and research are still required for decreasing the environmental impact of glassmaking in the future. In this regard, the integration of cleaner energy sources, decarbonation, increased energy efficiency, and development of more flexible furnaces represent crucial areas for the future of the glass industry.

4.4.6 Ceramics

Looking at the hard-to-abate sector, as shown in Figure 4.21, the ceramic industry accounts for about 1Mt direct CO_2 emission [101]. Also in this case, electrification is not a feasible path for decarbonization because of the high temperature involved. The maximum temperature needed for the ceramic production depends on the typology of products.

Differently from glass that has an amorphous structure, ceramics have a crystalline structure that makes them exhibit high melting temperature and high thermal and chemical stability. Hardness and inertness are their main characteristics. For their characteristics, ceramics are used in very diverse sectors: for example, in the transport sector, ceramics are involved in brake disks, cylinder sleeves, and heat shields, whereas in the aerospace sector, a relevant application is as refractory insulation of thermal barrier and core blade in aircraft. In electrical and electronic applications, they are used as excellent electrical insulators also in high-voltage power lines or in semiconductor and piezoelectric devices. In bioengineering, very relevant utilizations are as bone grafts, dental restorations, and joint replacements. Other common fields of application are energy and power generation, built environment, as well as industry. In the energy and power generation sector, ceramics are becoming key enabler materials for energy transition because ceramics are involved in catalysis processes, solar energy elements, insulation,

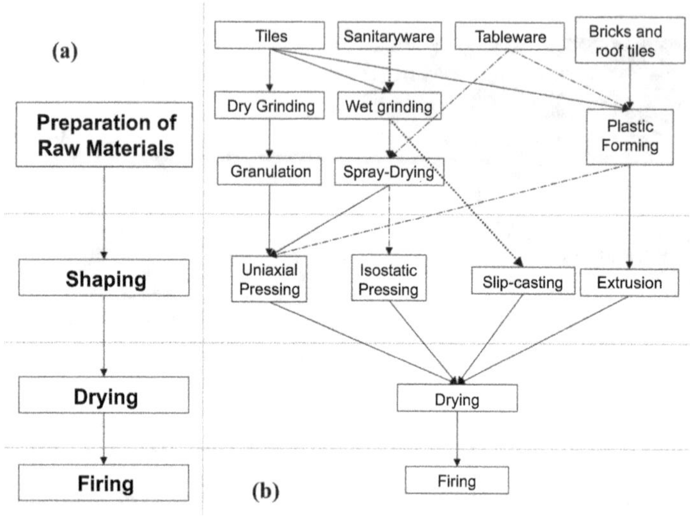

Figure 4.21 Main production steps for different ceramic products. Reprinted, with permission, from [143].

smart mobility, advanced sensors, and many more. In the building sector, tiles and bricks as well as electric storage heater components are the main applications. Overall, the global ceramics market was valued at USD 227.60 billion in 2021 and is expected to grow further beyond. Indeed, it is forecasted to reach about USD 360 billion by 2030 [102] in the market distributed all over the world.

The process chain of ceramic production is very complex. The main steps have been summarized in Figure 4.21 for different product chains and described in few words in the following. The preparation of raw material is focused on mixing together raw materials such as clay, feldspar, silica, talc, with additives and grinding them to reduce their size. Then, the shaping step follows downstream, where the grinded material is shaped in the so-called "green body." The technique used for the green body depends on the final use of ceramic product. Thus, dry pressing, extrusion, slip casting, or other alternative processes are used. Then, the drying of the green body is required to extract water and avoid problems in the following firing section. The firing of the "green body" is the main relevant part of the process because it makes the product reach its almost final characteristics in terms of consistency and properties. Some ceramic products are further processes with a second firing stage, as in the case of specific typology of tiles. Each step of the process requires energy. More specifically, grinding and shaping needs electrical energy, whereas firing and drying are carried out with thermal energy from fuel combustion. Therefore, though CO_2 emissions coming from ceramic production process is not only related to combustion process, this latter is responsible for about 64% of emission, while 19% is related to the process itself whereas the remaining part to indirect emissions. Overall, ceramic industry is responsible of emission of 1Mt CO_2 direct emission.

Indirect emission accounts for 19% of contribution to CO_2 emission and mainly refers to the electric energy needs. The further contribution to CO_2 is the process emissions generated by mineralogical transformation of the clay. The European Ceramic Industry Association elaborated the scenario for decarbonization of ceramic industry reported in Figure 4.21 showing that the reduction of CO_2 is possible but the main action to reduce CO_2 emission in ceramic industry is to decarbonize the thermal heating processes involved in the production line. As already highlighted, electrification is hardly feasible in ceramic firing process because of high temperature needed for the process. They strongly depend on ceramic typologies as summarized in Table 4.1. Nevertheless, it ranges from 800 °C to 1300 °C.

A combustion-based thermal process is then required. The potential strategies are those actually considered for whatever thermal conversion process in industry, ranging from reduction of process emission with innovative approach, looking at potential CCS/CCSU downstream processes and with fuel switching. The CO_2 reduction following the peak emission in 2000 was essentially related to the use of natural gas in place of coal. The switch to renewable fuels is the next step in the ceramic industry decarbonization, potentially reducing about 45% of CO_2 emission from the ceramic industry.

Hydrogen and ammonia are two no carbon fuels that are candidate to substitute.

4.4.7 Ammonia as alternative fuel in the ceramic industry

The utilization of ammonia in ceramic industry has the same limitations and issues encountered in other industrial sectors, as flame stability and NOx emissions. Advanced Zero-Carbon Combustion Technology Joint Innovation and Research Center, established by Foshan Xianhu Laboratory, DLT Technology, and Oceano Ceramics, successfully produced the world's first "green" ceramic tile using pure ammonia fuel [103,104]. For stabilizing the ammonia combustion process, a high swirl burner has been developed for ammonia/hydrogen combustion.

Table 4.1 Temperature requirement for different ceramics production [145]

Sector	Temperature requirements
Bricks and roof tiles	Bricks and roof tiles are heated at temperatures varying between 800 °C and 1300 °C.
Vitrified clay pipes	Temperatures range from 1150 °C to 1250 °C, while the firing time lasts between 30 and 80 h.
Refractory products	These materials are fired at temperatures ranging between 1250 °C and 1850 °C.
Expanded clay aggregates	These materials are subjected to a firing process of temperatures ranging between 1100 °C and 1300 °C.
Floor and wall tiles	The temperature required for floor and wall tiles varies between 1050 °C and 1300 °C.
Sanitaryware	Normally, the required temperatures for vitreous china ranges between 1200 °C and 1210 °C and is about 1220 °C for fireclay.

While the effective combustion of ammonia can be reached with appropriate strategies, one of the main critical points is to study the effect the atmosphere in the furnaces on the load. It has been shown that the effects on characteristics of produced material depend on working conditions and the typology of atmosphere. For example, in Pontikes *et al.* [105], the effect of atmosphere composition and the utilization of hydrogen was tested on the characteristics of ceramics made from Bayer process bauxite residue. It was shown that the utilization of N_2 in place of air in sintering process results in an increase in bulk density and water absorption with an average porous size higher in case of N_2 atmosphere. On the other hand, with the utilization of a small amount of hydrogen in the fuel, the shrinkage begins at lower temperature and the main crystalline phase composition changes while the final matrix shows a lower average porous dimeter and closed pores. The effect of oxidative atmosphere on ceramics is the topic of many research projects, as for example the one involving the British Ceramic Confederation [101] where different products, such as heavy clay, whitewares, and refractories, have been considered and fired in natural gas and hydrogen atmosphere. In general, no substantial differences were found between the material produced in these two different atmospheres for the initial firing trials, demonstrating the potential of firing ceramics using 100% hydrogen. Moreover, they are comparable with those obtained for 100% methane-fired products. The products meet the required technical specifications, with differences falling in those related to the test firing cycle variation. Within expected variations, products treated with hydrogen firing appear to show improvement with respect to water absorption, density, and compressive strength.

Though the impact on ceramic products of hydrogen fueled burner is useful also in case ammonia is used, further results are expected on the quality of material produced using ammonia.

4.5 Cement industry

Cement production is an energy-intensive process consuming thermal energy of the order of 3.3 GJ/t of clinker produced. In the full energy balance of the cement industry, the electric consumption also has to be considered, with an average impact of about 90–120 kWh/t of cement [106–109]. Higher-energy consumption are in those devices related to the kiln feed, the cooler, and the cement grinding and transport processes, which accounts for more than 50% of the total electrical energy consumption.

The other big issue related to the cement industry concerns the release of emissions, mainly in the form of carbon dioxide (CO_2) and nitrogen oxide (NO_x). Minor pollutant emissions are related to particulate matter (PM), ash, and metal oxides.

With respect to other sectors (agriculture, transport, building, others), the cement industry counts for CO_2 emissions equal to approximately 6%, depending on many factors, including the country economy, population growth, and manufacturing technologies, as shown in Figure 4.2 [110]. Referring to the industrial manufacturing processes, the cement industry is the second CO_2 larger emitter, after the iron and steel production sector [111–113].

4.5.1 Cement production technology

In order to address the contingency strategies that may be implemented to decarbonize the cement industry and reduce its environmental impact, the cement production chain has to be analyzed to identify the energy demand steps (Figure 4.22), along with the concrete production sub-processes responsible for pollutant emissions [114]. It is a multistage process that involves six stages:

- **Mining**: The process starts from the raw materials mining (limestone, sand) from quarries.
- **Grinding**: Then the raw material is crushed to small particle size and grinded to fine powder, in a rotating, cylindrical ball, or tube mills containing a charge of steel grinding balls. This raw material is then conditioned with some additives, containing calcium (Ca), silica (SiO_2), aluminum (Al), and iron oxides (Fe_2O_3) (among others).
- **Pre-calciner**: In modern cement plants, this powder is fed to the pre-calciner section, where it is heated up to 800 °C–900 °C through the exhausted enthalpy of combustion products from the kiln and/or extra-fuel oxidation, to initiate the dissociation of carbonate $CaCO_3$ to calcium oxide (CaO) and CO_2. The pre-calciner stage can process approximately up to 90% of the raw material before the kiln calcination step, depending on the pre-calcinizer type (air-through, air-separated). The pre-calciner allows to increase the energy efficiency of the kiln operation, where the rest of the calcination and material sintering (to avoid the pre-calciner) is processed.
- **Clinker**: The powder is then processed in a kiln at about 1450 °C, where calcium silicates and aluminates are formed (clinker).

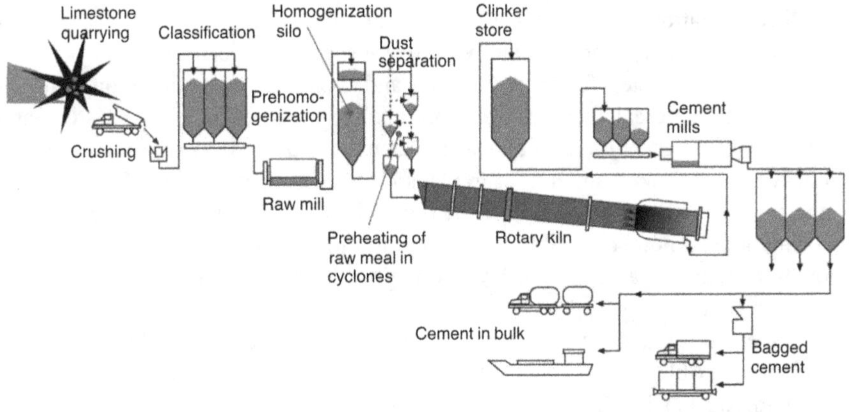

Figure 4.22 Cement process production chain. Reprinted, with permission, from [144].

- **Cooling and mixing**: Finally, the clinker is cooled and subsequently ground with gypsum (approximately 5%) or other materials such as slag, fly ash, pozzolans to form the cement.
- **Final grinding**: Cement is then grinded and re-grinded to powder in mills; then it is packaged.

Within this chain, it is very important to analyze the steps that lead to the release of pollutants, with a specific target to CO_2. This sector has a peculiar behavior as most of CO_2 ($\cong$50% of total emissions) is released from the calcination reactions (limestone->clinker+CO_2), and not from fuel oxidation ($CO_2\cong$28%) as in other sectors; thus, this emission results to be unavoidable for commercial portland cements. In addition, the other CO_2 sources are related to the electricity consumption needed for clinker cooling and raw material/cement grinding, and cement distribution and storage ($CO_2\cong$12% of the total emissions).

4.5.2 Alternatives to decarbonize the cement industry

Given this background, many interventions can be implemented for energy saving and CO_2 emissions mitigations as short-term strategies. First, all the actions aimed to increase the global process efficiency and fuel consumption, over all the cement production chain, might be undertaken [115–117].

The first step should be the upgrading of the cement industry from wet to dry technologies, as direct strategies to save the energy required from slurry (humid grinded limestone and addictive) drying, with a net energy recovery equal to 50% [116]. Second, any steps contemplated in the cement production chain can be ameliorated through less energy demanding technology. For instance, raw/cement grinding systems can be improved by replacing ball-type mills with vertical-type or high-pressure rollers [115–117]. Actions can be launched also for raw material transportation, raw material/additive blending, and separator/classifier units [115]. The same authors suggest also to act on the heat loss from the process, intervening on the kiln and on the cooling system, as the majority of the heat is lost from the kiln, thus insulation through specific refractory materials, while clinker cooling process can be ameliorated through reciprocating grate coolers.

A fundamental step for energy saving and CO_2 emission mitigation is to act directly on the pre-calciner sections, optimizing the design of the system and installing multistage burners to boost the calcination process to maximum degree, while avoiding clinker sintering for plausible clogging problems. This strategy could drastically allow fuel saving in the kiln burner section, with a net reduction of CO_2 emissions. Another possible action is to minimize the wastes from the cement production stages to a minimum level, utilizing them as raw material for cement production [118].

All these strategies have to be mandatorily supported by the implementation of robust and flexible monitoring/control strategies, through digital twin and soft sensing approaches [119].

These approaches can be considered as short-term updating actions to retrofit existing cement industries. The strict requirement to completely abate CO_2

emissions by 2050s from this sector implies the drastic switch to revolutionary actions through breakthrough technologies. To this aim, several strategies have a great potential to be exploited as mid/long-term solutions.

While there are many chances to decarbonize the cement industry for the kiln section changing the fuel and the combustion process to reach typical temperature of 1450 °C, the CO_2 emissions cannot be avoided for the calcination reactions, that, as previously reported, constitute more that 50% of the total CO_2 release. Here there are not many degrees of freedom. Only opportunities to mitigate the CO_2 emissions rely on the decrease of clinker-to-cement ratio through the substitution of limestone with calcined clays, fly ash, blast-furnace slag, limestone fillers, and natural pozzolans calcined clays, fly ash, blast-furnace slag, limestone fillers, and natural pozzolans additions, creating a class of alternative cements to the classical Portland ones, with concrete features comparable to traditional ones [117,120,121].

One of the most challenging technologies that can be implemented with huge potentials to mitigate the CO_2 emissions consists in carbon capture and storage strategies [122,123]. These would allow a CO_2 emission reduction of about 65%–75%. CO_2 is captured and compressed into liquid form to transport to geological storage facilities. Sequestrated CO_2 can be re-directed to several trade markets through the production of different chemicals (minerals, fuels, chemicals, and polymers), enabling carbon capture and utilization (CCU) strategies. CO_2 capture can be realized through absorption on chemical liquid (amine) or solid (metallic-organic framework, MOF) sorbents.

This concept is often also associated with a drastic change in the combustion mode, switching from conventional to oxy-fuel combustion, where a carbon-based fuel is combusted only in oxygen, producing exhausted gases composed of steam and CO_2 at high concentration (≅80%), easier to capture and store with respect to conventional combustion flue gas stream. As alternative to post-combustion methods, calcium looping technologies can be resorted for CO_2 capture. The spread-out of this technology is then hindered by the prohibitive costs and the use of extra energy for CO_2 liquefaction and transportation, with the necessity to find proper geological storage sites. In addition, the oxy-fuel concept can be applied to new kilns, opportunely designed for this combustion concept: retrofitting old kilns may result in costly investment. The CCS concept has been demonstrated in small plants; thus, scientific efforts should still be paid for its technological maturation, requiring proper investments.

Another alternative for the decarbonization of the cement industry comes through the electrification of the production chain. There are several companies in charge with this idea, as the Finland-based CoolBrook, in collaboration with CEMEX (https://www.cemex.com/w/cemex-and-coolbrook-electrify-cement-production-process), where they aim to decarbonize the cement industry by a partial electrification of both the calcination process and the Kiln. In their roadmaps, they are achieving a cut-off of CO_2 emissions equal to 45% in few years. The electrification of the calcination process would produce a pure CO_2 stream to destine to CCS processes, while in the electrified kiln, the CO_2 amount is drastically reduced as a lower amount CO_2 is released since less fuel is needed.

At the same time, the Swedish SaltX Technology firm (https://www.saltxtechnology.com/files/REPRINT-ICR-APRIL-2024-HR-SaltX-Electric-kilns-3p-MB.pdf) has developed a 300 kW arc electric calciner (EAC) that release heat through a thermal plasma system, to scale up to industrial sizes.

In China, AGICO cement has developed a silicon-carbide-rod electrified system to install into the kiln. All these approaches have to be handled with the support of renewable production/delivery electricity systems, and may be supported only in the case of cheap energy (https://www.cemnet.com/News/story/174030/the-electrified-commercial-cement-kiln.html).

Parra and Romano [5] have presented a thorough LCA analysis for the implementation of electric systems into the cement industries, considering several scenarios as (i) direct electrification and plasma technologies; (ii) H_2 combustion and oxy-combustion of alternative fuels; (iii) combining direct electrification, alternative fuels combustion, and post-combustion CO_2 capture; (iv) electrification of the hydraulic calcium hydro silicate production process. In general, they conclude that electrification is feasible only if the cost of electricity is kept down to unfeasible level for the moment, while CCS/CCUS systems remaining as more convenient solutions. The chance to electrify energy-intensive systems comes also through the material availability to construct electrical systems and difficulties to easily integrate them through retrofitting strategies for old-science/consolidated systems.

4.5.3 Fuel replacements

A large palette of fuels is purported to be used in the cement industry (both in the pre-calciner and the kiln) as alternative to the classical fuels (methane, coal, oils, petroleum coke, diesel, etc.). Three basic groups can be identified in relation to the fuel state of matter: gas (e.g. landfill gas, pyrolytic gas, and biogas), liquid (e.g. used oils and solvents), and solid (tyres, wood waste, plastics, municipal solid waste (MSW), sewage sludge (SS), and textiles, bone animal meal (MBM), and end-of-life tyres (ELT) [120].

The use of these alternative fuels has to be discussed with respect to cement final properties and pollutant emissions (specifically from chlorines and sulphur). In general, there are economic advantages from the use of waste materials, with large CO_2 mitigation potentialities in case of biomass and/or gasification/pyrolysis gases as carbon-neutral fuels. The use of biomasses may reduce also NO_x emissions as the nitrogen in biomass is released as ammonia that activates NO_x reducing chemistry. The use of waste materials comes also in compliance with European Union policies to reduce disposal sites and incineration plants. Operative conditions are also beneficial to avoid dioxins and furans formations.

On the other hand, given the alternative fuels lower calorific values, they can be used only under co-combustion conditions with traditional fuels, representing a partial solution for CO_2 mitigation. In addition, system operative conditions have to be tuned with respect to their physical/chemical properties and amount in co-combustion strategies.

4.5.4 Ammonia as alternative fuel in the cement industry

In parallel with other energy-intensive industrial sectors, ammonia has the chance to act as a valuable mean to decarbonize the cement production sector. Within this perspective, Mitsubishi UBE Cement Corporation, in the framework of Cross-Ministerial Strategic Innovation Promotion Program (SIP) [124] in Japan, has announced the first ammonia co-combustion in this cement manufacturing process in 2018. The driving idea came from a simple energy balance considering 1 year cement production in Japan and related CO_2 emissions from coal consumption: for ammonia-coal co-firing furnaces, the replacement of 1% of ammonia with coal would result in a CO_2 reduction of 160 kton/year (over 15921 kton per year), equal to 1% kton/year [125].

Starting from this idea, UBE Industries performed a series of experiments to value the chemical reaction rates and the cement properties in basic fuel-ammonia in co-combustion systems. First, an atmosphere-controlled electric furnace was used to produce 10 mm clinker pellets, invested, in a second stage, by synthetic flows mimicking coal and coal-NH_3 (30%) full conversion exhausted gases at several temperatures (1000 °C–1450 °C). UBE Industries demonstrated that the different environments did not change the clinker global reaction rate.

Second, experiments with heavy-oil and heavy-oil/ammonia (30%) in a 10 kW clinker furnace were run to value the cement characteristics in different environments [126]. The 10 kW clinker furnace burner consisted of a central oil injector, surrounded by an annular air injector, realizing a non-premixed configuration. A second annular injectors allowed to feed a NH_3-air premixed flow within the combustion chamber. Varying the heavy-oil/primary air ratio and the NH_3-air/secondary ratio, they demonstrated the possibility to control NO_x emissions, to value close to pure heavy-oil combustion.

The X-ray diffraction (XRD)/Rietveld analyses clearly showed the preservation of cement mineral composition and strength, albeit the combustion environment was different and that ammonia has no effect on cement features.

Fawzi [127] claimed ammonia added to cement in water solutions could also be beneficial to improve cement flexural strength, thus posing the basis for a further benefit to exploit in co-combustion systems.

UBE Industries performed then CFD analyses on the cement characteristics in different combustion environments derived from heavy-oil and heavy-oil/ammonia (30%) oxidation, through a dedicated software (KilnSimu) [126]. They claimed some issues related to cement composition at the exit because of slightly lower temperatures in the clinker for heavy-oil/ammonia (30%), due to different combusted gas emissivity's. They claimed CFD simulations should support the development of burner design with the definition of furnace boundary conditions, feeding flow rates and primary/secondary fuel/air equivalence ratio. With respect to this approach, they warned about some CFD issues related to chemistry/turbulence interaction, due to slow NH_3 oxidation chemistries and the need to use fine-chemistry approaches.

The development of the ammonia co-combustion in clinker furnaces comes through the progress of the state-of-art of ammonia combustion in furnaces; thus once it has been demonstrated, clinker reaction rates and concreate properties do not substantially change with the co-combustion environment; with some exceptions for radiative issues, improvements in co-combustion technologies will sustain the expected ammonia role in the cement industry. Given this background, UBE industries has launched several announcements to boost the coal/NH_3 co-firing level to a 20%–30% of ammonia for cement production systems, with the aim to solve all the energy conversion and emissions (NH_3 slip/fuel-NO_x release) issues within 2050, toward achieving carbon neutrality for this energy-intensive sector.

Beyond the potentialities to implement the "NH_3 economy" within the cement production sector, this fuel can also be exploited as an energy-dense cost-effective "hydrogen carrier" to crack into pure hydrogen for cement kiln. This approach is being developed by the Cranfield University[*], with the formulation of a new auto-thermal ammonia cracking process in collaboration with Stopford[†] and Heidelberg Materials UK[‡]. The same idea has been proposed by the Pennsylvania Aggregates and Concrete Association (Pennsylvania, USA)[§], where the NH_3 cracking is processed through photo-catalysis, induced by LED-powered devices. Conventionally, the endothermic ammonia cracking catalytic process requires 650 °C–1000 °C, being the bottleneck of the process. The researcher from the Rice University has developed a photo-catalysis cracker (close to commercialization) based upon copper-iron (less expensive than conventional copper-ruthenium) doped with "antenna" nanoparticles, able to absorb energy from light sources (LEDs) and release it as short-lived "hot electrons/photons," enabling ammonia cracking reactions at environmental temperatures.

Albeit with the potentialities of hydrogen as a carbon-free fuel (no CO_2 emissions), the need to redesign kiln burners and kiln geometry is mandatory to ensure an effective combustion and heat transfer to the feed material, along with issues related to NO_x emissions and the kiln materials resistance [128].

For both ammonia and hydrogen direct utilization in pre-calcination and clinker section, the chance to use novel combustion concepts as MILD combustion has to be thoroughly analyzed, given the potentialities of the process to ensure high stable oxidation conditions, uniform temperatures distribution within the combustion chamber (save fuel consumption, decrease combustion chamber), high radiative heat transfer mechanisms, and low NO_x emissions. The above-mentioned process has another intriguing feature: it is very flexible with respect to the fuel properties, thus able to tolerate fuel typology, fuel availability and fluctuation composition (biomasses, e-fuels), and fuel co-combustion.

[*]https://www.cranfield.ac.uk/press/news-2024/revolutionizing-the-cement-industry-with-ammonia
[†]https://www.stopford.co.uk/
[‡]https://www.heidelbergmaterials.co.uk/en
[§]https://www.specifyconcrete.org/blog/fuel-cement-kilns-with-green-hydrogen-from-ammonia

4.6 Furnaces for the steelmaking industry

The steel industry is a major consumer of fossil fuels and it is responsible for about 8% of global CO_2 emissions (the highest contribution from an industrial subsector) and it accounts for more than 25% of world carbon dioxide emissions in the major energy-intensive industries, primarily due to its reliance on carbon-based fuels such as coal and natural gas [129].

This is also largely due to the extreme temperatures required in furnaces to reduce iron ore.

The need to decarbonize the steel industry cannot be understated, as its climate performance is intertwined with supply chains that span across several high-emitting sectors including automotive and construction. By focusing emissions reduction efforts on the steel industry, supply chains can be decarbonized at the source.

The increase in energy efficiency is not always the right solution to obtain the desired reductions in terms of emissions and it can only partly solve the issues. There is, therefore, a strong interest in the steelmaking sector in implementing two main solutions in such processes: carbon capture and storage technologies and/or fuel switching solutions by substituting coal or natural gas with hydrogen or hydrogen vectors. Such decarbonization strategies based on the transitioning to low-carbon alternatives are crucial for meeting international climate targets.

Figure 4.23 reports a scheme for a typical steel manufacturing process, including steel production, energy conversion, and energy storage systems [130]. The main two routes for steel production are the BF-BOF (blast furnace-basic oxygen furnace) that uses iron ore and coal as raw materials and energy sources, whereas the second one is the EAF (electric arc furnace) where the raw materials

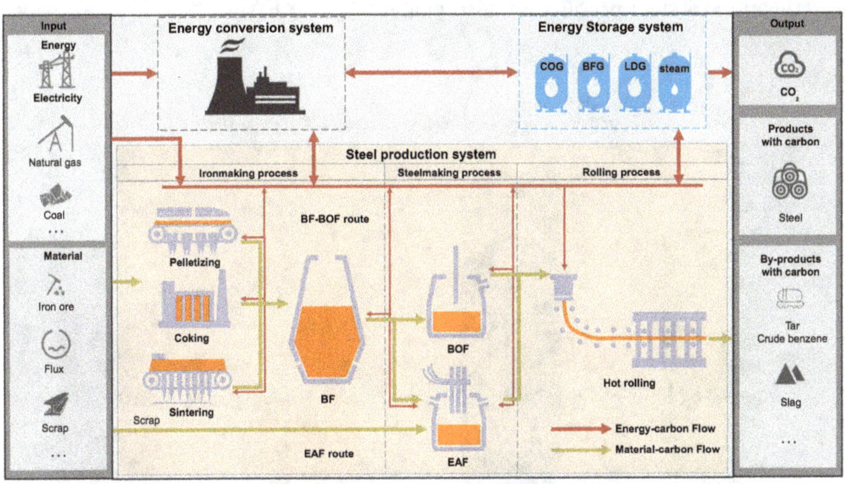

Figure 4.23 Schematic representation of typical steel manufacturing process. Reprinted, with permission, from [130].

and energy sources are scrap and electricity which usually employs part of the iron in the EAF.

The primary direct reduced iron (DRI) and electric arc furnace (DRI-EAF) process employs syngas, produced from natural gas or gasified coal, along with electricity to operate the EAF. In this method, iron ore is transformed into metallic iron using the syngas, and the resulting DRI is subsequently processed into steel in the EAF.

The secondary scrap steel and electric arc furnace (scrap-EAF) process utilizes electricity. Recycled scrap steel is heated and melted in an EAF to create new steel.

The plant energy conversion system mainly uses gas byproducts, waste heat, and energy from the onsite power plant to produce power for self-use. Furthermore, the company buys electricity and gas from the main grid and natural gas network to meet its energy requirements.

The steel industry worldwide is under mounting pressure to decrease its environmental impact because of the rising concerns about climate change and strict environmental laws. Conventional steel production methods, like the BF-BOF route, heavily use coal and natural gas, making them very carbon-intensive.

In 2022, the global average carbon dioxide emission intensity for steel production was 1.91 metric tons of CO_2 per metric ton of crude steel cast, as reported in Figure 4.24. The blast furnace-basic oxygen furnace (BF-BOF) route had the highest emissions intensity, producing 2.33 tCO_2/t-crude steel. In contrast, the scrap-based electric arc furnace process (scrap-EAF) emitted only 0.68 tCO_2 per metric ton of crude steel cast [131].

Such global average value for both intensities also incorporates a contribution from DRI-based EAF steel production, given its rising importance as a key steelmaking process.

A possible solution to decarbonize some of the steelmaking processes is the electrification of steel production, also enabled by the EAF.

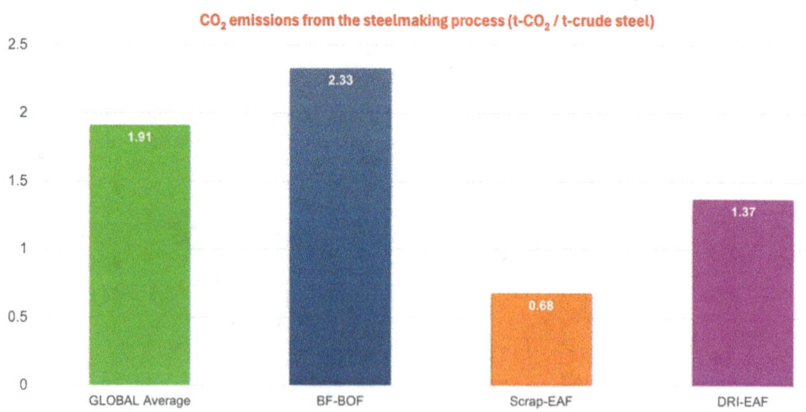

Figure 4.24 CO_2 emissions from steelmaking process in 2022

Despite this, several key factors contribute to the ongoing widespread use of the BF-BOF route. The established BOF infrastructure has historical roots that require substantial capital and time to achieve a complete conversion to EAF. Regions with plentiful and affordable coal and iron ore continue to find the BOF process attractive. Additionally, the integration of low-carbon fuels (such as hydrogen) and CCU technologies in BOF facilities reduces the urgency for complete electrification [132].

From previous analysis, it seems that the most emitting steelmaking processes are BF-BOF and direct reduction iron ones.

In this framework, switching to low-carbon solutions is crucial for meeting sustainability targets in steelmaking processes. The integration of hydrogen in such industry is possibly the most attractive solution to decarbonize the sector and achieve the net-zero emission targets [133]. Preliminary results obtained a replacement of about 30% of fossil fuels with pure hydrogen. However, significant research and development in process and product qualities is needed in this field to obtain hydrogen-integrated BF-BOF plants available by 2030.

However, to date, the direct reduction of iron with an integrated electric arc furnace (H2DRI-EAF) has the potential to be the most advanced technology for integrating the hydrogen use, potentially leading to an 80%–90% reduction in CO_2 emissions when compared with traditional methods. However, the biggest challenge lies in obtaining cost-effective green hydrogen and top-quality ore. Apart from the various technological challenges, the primary driver for the successful transition from fossil fuel to hydrogen-based steel manufacturing is the cost-competitive and sustainable hydrogen supply.

Among the possible alternatives to hydrogen, ammonia is being explored as a potential carbon-free fuel because of its high hydrogen content and the already available infrastructures for its transportation and storage. Some peculiar aspects related to the possibility of using ammonia in furnaces for steelmaking processes are reported in the next sub-sections.

4.6.1 Ammonia as a fuel for steel furnaces

Ammonia, with its high hydrogen content and zero carbon emissions when burned, emerges as a promising candidate for direct utilization as a fuel in steel furnaces.

When ammonia is used as an energy carrier in an industrial furnace, there are some issues that need to be taken into account, mainly related to very low flammability and flame stability, high levels of NOx emissions, and low radiative heat flux compared with fossil fuels-based furnaces [134].

Ammonia is usually considered as a by-product of several hard-to-abate processes, leading many facilities to burn off or release it into the atmosphere through flares. Consequently, there are multiple potential applications for recovering and utilizing such waste ammonia to produce heat for industrial processes or steam generation.

Hagihara *et al.* [135] conducted research on utilizing ammonia for degreasing steel sheets with an impinging jet burner. To address the difficulties of using ammonia in this process, a new burner was required to be designed exclusively for co-firing ammonia with natural gas. Various shapes and setups were examined to enhance the heating profile for degreasing galvanized steel sheets used in construction, appliance, and automotive industries. Results showed that an optimal design enables the effective use of ammonia for this applications, as reported in Figure 4.25.

Some applications of ammonia as a fuel for semi-industrial furnaces were reported in the literature by investigating the features of MILD combustion with ammonia as energy vector.

In the study by Xie *et al.* [136], MILD combustion is explored as a solution for addressing several drawbacks associated with NH_3 combustion. They reported a numerical study of ammonia/methane combustion in a semi-industrial furnace employing a conjugate heat transfer (CHT) model for furnace walls heat transfer. The study compares NO emissions, temperature distribution, flame structure, and furnace heat transfer behaviors between traditional swirling combustion and MILD combustion at various NH_3 blending ratios.

Figure 4.26 shows images of the flame captured from two observation windows of the combustion furnace, depicting swirling combustion and MILD

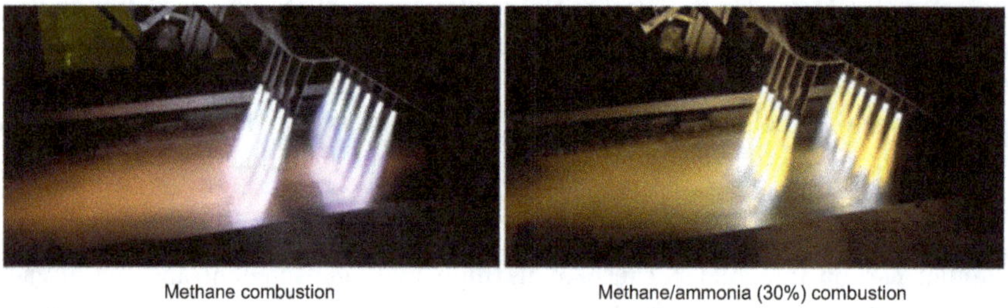

Figure 4.25 Methane and methane/ammonia combustion for degreasing steel sheets. Reprinted, with permission, from [135].

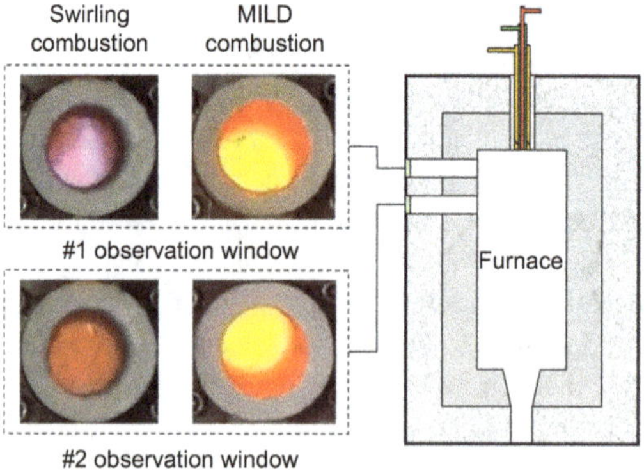

Figure 4.26 Images of the flame from swirling and MILD combustion modes. Reprinted, with permission, from [136].

combustion, respectively. A bluish conical "V"-shaped flame is clearly visible for the swirling combustion mode through window #1. In contrast, no visible flame is obtained for MILD combustion through the windows. Additionally, CO emissions are below 10 ppm for both combustion modes, whereas NO emissions significantly decrease from 42 to 19 ppm when passing from swirling to MILD combustion. These findings indicate the efficient stabilization of MILD combustion with the current burner.

The model shows that the heat flux on the furnace wall follows similar trends in both combustion modes, but the heat utilization efficiency of MILD combustion is slightly lower and that when ammonia replaces methane, heat utilization efficiency gradually decreases in both combustion modes due to lower peak temperatures and decreased soot radiation.

The diagram in Figure 4.29 displays the heat flux patterns along the side-wall for two combustion modes, with and without CHT models, in a pure CH_4-firing scenario. In the simulation result with a CHT model (solid lines), the heat flux profiles consistently show positive heat transfer, indicating that heat is being drawn out by the furnace side-wall in all locations. Conversely, in simulation results without a CHT model (represented by dashed lines), negative heat transfer is obtained in both upstream and downstream regions under swirling combustion, and mainly in the upstream region under MILD combustion. This negative heat transfer implies that heat is being transferred from the furnace side-wall to the combustion gas.

On the other hand, the heat flux profiles with the CHT model (solid lines in Figure 4.27) indicate that there is no significant difference between swirling combustion and MILD combustion, but slightly higher heat flux is obtained by swirling combustion along almost the entire furnace side-wall. Simulations showed that the total net heat flux absorbed by the furnace side-wall increases from 2.102 to

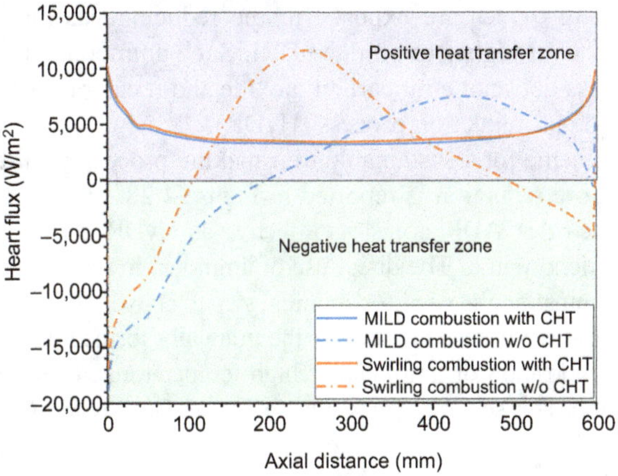

Figure 4.27 Heat flux profiles on furnace side-wall for swirling and MILD combustion modes with and without CHT model in pure CH4 firing case. Reprinted, with permission, from [136].

2.209 kW when passing from MILD to swirling combustion. This is mainly related to the decreased convection heat transfer amount on the furnace side-wall under MILD conditions.

In the same context, Mohammadpour *et al.* [137] performed CFD simulations of MILD combustion in a furnace with pure NH_3. The establishment of MILD combustion condition along with the temperature spatial distributions inside the furnace (temperature uniformity) was investigated. The results showed that when the furnace wall temperature is decreased, the reaction zone volume is raised and that the same effect has been obtained with an increase in the oxidizer dilution level. Such variation in the wall temperature and/or dilution levels alleviates the average and maximum temperatures of the furnace, by increasing the uniformity of the furnace temperatures, promoting an optimal distribution of heat fluxes.

Such results are very important in providing information on the possible use of ammonia in steel reheating furnaces and the issues to overcome before its deployment on large scales.

4.6.2 Ammonia as a reductant of iron ore and nitriding effects in steelmaking processes

Several steel industries are pursuing different approaches to decrease CO_2 emissions in the different steelmaking processes. Direct Iron ore reduction with hydrogen is usually identified as a promising solution to reduce the process carbon intensity, but hydrogen is currently not produced in nearly sufficient quantities to put steel production on a more or less climate-neutral route. A possible alternative relies on the use of ammonia for this direct reduction route ammonia.

Ma *et al.* [138] reported a sustainable method for the utilization of ammonia in the reduction of iron ore by adding directly the ammonia into a direct-reduction reactor where solid oxides are exposed to the reducing gas (ammonia) at temperatures of 700 °C or higher to produce DRI. Such ammonia-based direct reduction (ADR) process does not emit carbon dioxide and it can be an alternative to the hydrogen-based direct reduction process (HyDR).

A possible scheme for a sustainable iron-making process by directly deploying ammonia in iron ore reduction is reported in Figure 4.28.

Results showed that ADR is just as effective as HyDR at 700 °C in producing environmentally friendly iron. The direct use of ammonia in the reduction process creates a shortcut, eliminating the need for an initial step to convert ammonia into hydrogen and nitrogen. As the redox reaction occurs, the gradually formed porous iron serves to catalyze the decomposition of ammonia at high temperatures, releasing hydrogen for iron oxide reduction. This self-catalyzing reaction paves the way for increased efficiency and cost savings. The in situ nitriding process also shields the pure iron from environmental deterioration, which would otherwise require additional energy-intensive and logistically complex steps. This protective nitride phase can be completely dissolved and removed during a subsequent melting process. Therefore, ADR offers a new approach to using intermittent renewable energy for a revolutionary and sustainable shift in

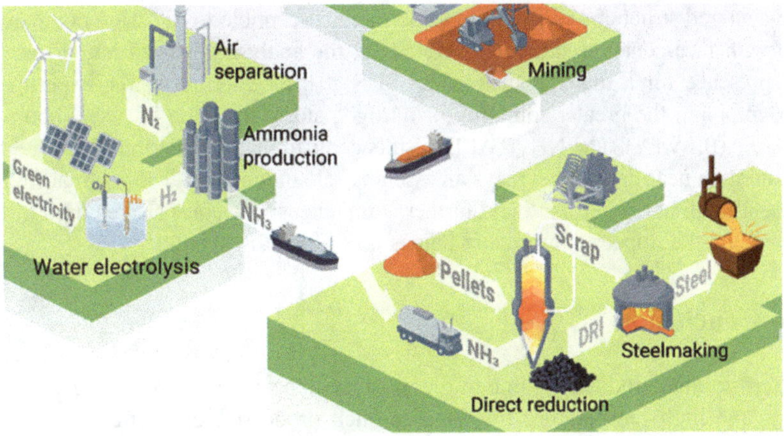

Figure 4.28 Future steel industry with deploying intermittent renewable energy mediated by green ammonia. Reprinted, with permission, from [138].

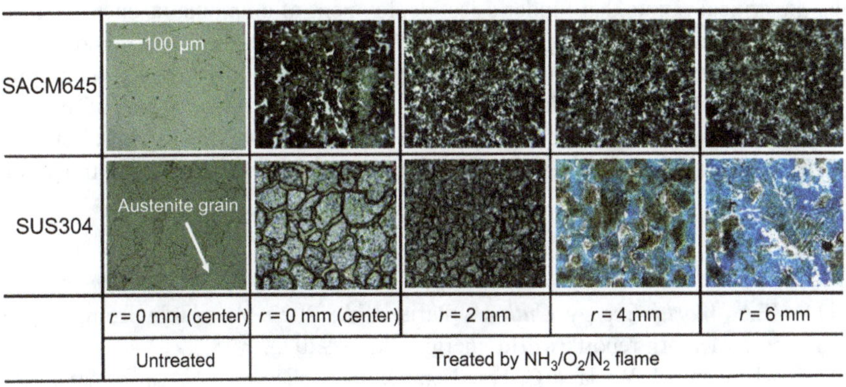

Figure 4.29 Optical micrographs of the SACM645 and SUS304 test plate surfaces after being exposed to the $NH_3/O_2/N_2$ flame at 550 °C for 5 h. Reprinted, with permission, from [139].

metallurgical processes. With these advantages, it bridges two of the most greenhouse gas-intensive industries (steel and ammonia production) and provides a means to make them more environmentally friendly and sustainable. Additionally, it can overcome logistical and energetic drawbacks associated with the transport of pure hydrogen.

In this context, Wang *et al.* [139] investigated the use of materials commonly used in steel nitriding (SACM645) and in structural/ducting components (SUS304 and SUS301S). Such materials were subjected to ammonia flames, resulting in deterioration due to nitriding effects (Figure 4.29). The findings indicated that the diffusion of nitrogen atoms occurred at varying depths (approximately 25–50 μm), influenced by the crystallographic structure of the respective material. Interestingly, materials with cubic-centered structures were found to be more susceptible to diffusion compared to those with face-centered structures. Moreover, the analysis revealed higher nitrogen atom presence for a material denoted as SUS at the centerline of the experimental setup, implying the greater solubility of nitrogen atoms in γ-Fe (4.5 wt%) compared with α-Fe (0.1 wt%). The NH_3-TALIF results confirm that the distribution of the NH_3 concentration near the surface plays an essential role in nitriding. These initial findings represent the first of their kind and further comprehensive studies are required to fully comprehend the impact of ammonia flames on different materials.

4.7 Conclusion

The hard-to-abate industry is one of the most challenging industries to decarbonize. Due to the high temperature involved in such processes, electrification is not a solution in the short/medium period. Therefore, the use of no carbon fuels is the viable path to make the sector meet the requirements of net-zero CO_2 emission. In this framework, ammonia can play a relevant role in many industries. It has been showed that there are several efforts and different approaches that are in play to face the critical points that hinder a fast deployment of ammonia as energy carrier on a large scale. Though many steps have to be still done, it has been demonstrated that ammonia can be burnt in an efficient way with relatively low pollutant emissions if appropriate strategies are used, meeting the requirement of energy-intensive industries. If the required amount of green ammonia will be available, it has great potential to be among the most relevant carriers beyond energy transition.

References

[1] IEA. *World Energy Outlook.* Paris: IEA; 2023. Available from: https://www.iea.org/reports/world-energy-outlook-2023.

[2] Kleinhans U, Wieland C, Frandsen FJ, and Spliethoff H. Ash formation and deposition in coal and biomass fired combustion systems: progress and challenges in the field of ash particle sticking and rebound behavior. *Progress in Energy and Combustion Science* 2018;68:65–168. https://doi.org/10.1016/j.pecs.2018.02.001.

[3] Sahu SG, Chakraborty N, and Sarkar P. Coal – biomass co-combustion: an overview. *Renewable and Sustainable Energy Reviews* 2014;39:575–86. https://doi.org/10.1016/j.rser.2014.07.106.

[4] Wei M, McMillan CA, and De La Rue Du Can S. Electrification of industry: potential, challenges and outlook. *Current Sustainable Renewable Energy Reports* 2019;6:140–8. https://doi.org/10.1007/s40518-019-00136-1.

[5] Quevedo Parra S, and Romano MC. Decarbonization of cement production by electrification. *Journal of Cleaner Production* 2023;425:138913. https://doi.org/10.1016/j.jclepro.2023.138913.

[6] Valera-medina A, Vigueras-zuniga MO, Shi H, Mashruk S, and Alnajideen M. Ammonia combustion in furnaces: a review. *International Journal of*

Hydrogen Energy 2024;49:1597–618. https://doi.org/10.1016/j.ijhydene.2023.10.241.

[7] Mei B, Zhang J, Shi X, Xi Z, and Li Y. Enhancement of ammonia combustion with partial fuel cracking strategy: laminar flame propagation and kinetic modeling investigation of $NH_3/H_2/N_2$/air mixtures up to 10 atm. *Combustion and Flame* 2021;231:111472. https://doi.org/10.1016/j.combustflame.2021.111472.

[8] An Z, Zhang W, Zhang M, *et al.* Experimental and numerical investigation on combustion characteristics of cracked ammonia flames. *Energy Fuels* 2024;38:7412–30. https://doi.org/10.1021/acs.energyfuels.4c00312.

[9] Tornatore C, Marchitto L, Sabia P, and De Joannon M. Ammonia as green fuel in internal combustion engines: state-of-the-art and future perspectives. *Frontiers of Mechanical Engineering* 2022;8:944201. https://doi.org/10.3389/fmech.2022.944201.

[10] Han X, Wang Z, Costa M, Sun Z, He Y, and Cen K. Experimental and kinetic modeling study of laminar burning velocities of NH_3/air, NH_3/H_2/air, NH_3/CO/air and NH_3/CH_4/air premixed flames. *Combustion and Flame* 2019;206:214–26. https://doi.org/10.1016/j.combustflame.2019.05.003.

[11] Wang Z, Han X, He Y, *et al.* Experimental and kinetic study on the laminar burning velocities of NH_3 mixing with CH_3OH and C_2H_5OH in premixed flames. *Combustion and Flame* 2021;229:111392. https://doi.org/10.1016/j.combustflame.2021.02.038.

[12] Mei B, Zhang X, Ma S, *et al.* Experimental and kinetic modeling investigation on the laminar flame propagation of ammonia under oxygen enrichment and elevated pressure conditions. *Combustion and Flame* 2019;210:236–46. https://doi.org/10.1016/j.combustflame.2019.08.033.

[13] Jian J, Hashemi H, Wu H, and Glarborg P. Study of ammonia oxidation with ozone addition. *Applications in Energy and Combustion Science* 2023;14:100137. https://doi.org/10.1016/j.jaecs.2023.100137.

[14] Yang W, Al Khateeb AN, and Kyritsis DC. The effect of hydrogen peroxide on NH_3/O_2 counterflow diffusion flames. *Energies* 2022;15:2216. https://doi.org/10.3390/en15062216.

[15] Lhuillier C, Brequigny P, Lamoureux N, Contino F, and Mounaïm-Rousselle C. Experimental investigation on laminar burning velocities of ammonia/hydrogen/air mixtures at elevated temperatures. *Fuel* 2020;263:116653. https://doi.org/10.1016/j.fuel.2019.116653.

[16] Kim S, Lim M, Lee Y, Lee J, and Yang W. Evaluation of effects of ammonia co-firing on the thermal performances of supercritical pulverized coal and circulating fluidized bed boilers. *Energy Conversion and Management* 2023;276:116528. https://doi.org/10.1016/j.enconman.2022.116528.

[17] Wang Z, Jia Z, Zhao C, Xiao Y, Feng S, and Chu H. Effect of ammonia on the characteristics of coal combustion. *Journal of the Energy Institute* 2023;109:101275. https://doi.org/10.1016/j.joei.2023.101275.

[18] Cavaliere A, and de Joannon M. Mild combustion. *Progress in Energy and Combustion Science* 2004;30(4):329–366.

[19] Ariemma GB, Sorrentino G, Ragucci R, de Joannon M, and Sabia P. Ammonia/methane combustion: stability and NOx emissions. *Combustion and Flame* 2022;241:112071. https://doi.org/10.1016/j.combustflame.2022.112071.

[20] Ariemma GB, Bozza P, De Joannon M, Sabia P, Sorrentino G, and Ragucci R. Alcohols as energy carriers in MILD combustion. *Energy Fuels* 2021; 35:7253–64. https://doi.org/10.1021/acs.energyfuels.0c03862.

[21] Murai R, Omori R, Kano R, *et al.* The radiative characteristics of $NH_3/N_2/O_2$ non-premixed flame on a 10 kW test furnace. *Energy Procedia* 2017; 120:325–32. https://doi.org/10.1016/J.EGYPRO.2017.07.232.

[22] Basu P, Kefa C, and Jestin L. *Boilers and Burners: Design and Theory.* Berlin: Springer Science & Business Media; 2012.

[23] Tawil E. *Boiler Classification and Application*. Continuing Education and Development, Inc., https://www.cedengineering.com/userfiles/Boiler%20Classification%20and%20Application.pdf.

[24] Rayaprolu K. *Boilers for Power and Processes*. Boca Raton, FL: CRC Press; 2009.

[25] Barma MC, Saidur R, Rahman SMA, Allouhi A, Akash BA, and Sait SM. A review on boilers energy use, energy savings, and emissions reductions. *Renewable and Sustainable Energy Reviews* 2017;79:970–83. https://doi.org/10.1016/j.rser.2017.05.187.

[26] Schiro F, Stoppato A, and Benato A. Gas fired boilers: perspective for near future fuel composition and impact on burner design process. *E3S Web of Conferences-EDP Sciences* 2017;22:00154. https://doi.org/10.1051/e3sconf/20172200154.

[27] Ammonia Energy Association n.d. https://www.ammoniaenergy.org/.

[28] Mitsubishi Power to develop ammonia combustion systems for thermal power plant boilers. 2021. https://power.mhi.com/news/20210826.html (accessed February 28, 2024).

[29] Atchison J. Kepco agrees to ammonia offtake from Gladstone. 2022. https://ammoniaenergy.org/articles/kepco-agrees-to-ammonia-offtake-from-gladstone/.

[30] Chen P, Fang Y, Wang P, Gu M, Luo K, and Fan J. The effect of ammonia co-firing on NO heterogeneous reduction in the high-temperature reduction zone of coal air-staging combustion: experimental and quantum chemistry study. *Combustion and Flame* 2022;237:111857. https://doi.org/10.1016/j.combustflame.2021.111857.

[31] Yamamoto A, Kimoto M, Ozawa Y, and Hara S. Basic co-firing characteristics of ammonia with pulverized coal in a single burner test furnace. In 2018 *AIChE Annual Meeting. AIChE. AIChE Annual Meeting*, 2018.

[32] Kimoto M, Yamamoto A, Ozawa Y, and Hara S. Ammonia co-firing technology applicable to pulverized coal fired boilers. *Combustion Society of Japan* 2019;61:299–303.

[33] Ishihara S, Zhang J, and Ito T. Numerical calculation with detailed chemistry of effect of ammonia co-firing on NO emissions in a coal-fired

boiler. *Fuel* 2020;266:116924. https://doi.org/10.1016/j.fuel.2019.116924.

[34] Ishihara S, Zhang J, and Ito T. Numerical calculation with detailed chemistry on ammonia co-firing in a coal-fired boiler: effect of ammonia co-firing ratio on NO emissions. *Fuel* 2020;274:117742. https://doi.org/10.1016/j.fuel.2020.117742.

[35] Tayyeb Javed M, Irfan N, and Gibbs BM. Control of combustion-generated nitrogen oxides by selective non-catalytic reduction. *Journal of Environmental Management* 2007;83:251–89. https://doi.org/10.1016/j.jenvman.2006.03.006.

[36] Lyon RK, and Hardy JE. Discovery and development of the thermal DeNOx process. *Industrial and Engineering Chemistry Fundamentals* 1986;25:19–24.

[37] Kasuya F, Glarborg P, Johnsson JE, and Dam-johansen K. The thermal DeNOx process: influence of partial pressures and temperature. *Chemical Engineering Science* 1995;50:1455–66.

[38] Glarborg P, Dam-Johansen K, Miller JA, Kee RJ, and Coltrin ME. Modeling the thermal DENOx process in flow reactors. Surface effects and nitrous oxide formation. *International Journal of Chemical Kinetics* 1994;26:421–36. https://doi.org/10.1002/kin.550260405.

[39] Fan W, Wu X, Guo H, *et al.* Experimental study on the impact of adding NH_3 on NO production in coal combustion and the effects of charcoal, ash, and additives on NH_3 reducing NO under high temperature. *Energy* 2019;173:109–20. https://doi.org/10.1016/j.energy.2019.02.065.

[40] Ma P, Huang Q, Si T, Yang Y, and Li S. Experimental investigation of NOx emission and ash-related issues in ammonia/coal/biomass co-combustion in a 25-kW down-fired furnace. *Proceedings of the Combustion Institute* 2023;39:3467–77. https://doi.org/10.1016/j.proci.2022.07.223.

[41] Cremer MA, Montgomery CJ, Wang DH, Heap MP, and Chen JY. Development and implementation of reduced chemistry for computational fluid dynamics modeling of selective non-catalytic reduction. *Proceedings of the Combustion Institute* 2000;28:2427–34. https://doi.org/10.1016/S0082-0784(00)80656-6.

[42] Tamura M, Gotou T, Ishii H, and Riechelmann D. Experimental investigation of ammonia combustion in a bench scale 1.2 MW-thermal pulverised coal firing furnace. *Applied Energy* 2020;277:115580. https://doi.org/10.1016/j.apenergy.2020.115580.

[43] Wang Q, Hu Z, Shao W, Wang Z, and Liu H. The present situation, challenges, and prospects of the application of ammonia – coal co-firing technology in power plant boilers. *Journal of the Energy Institute* 2024;113:101531. https://doi.org/10.1016/j.joei.2024.101531.

[44] Xia Y, Hadi K, Hashimoto G, and Hashimoto N. Effect of ammonia / oxygen / nitrogen equivalence ratio on spherical turbulent flame propagation of pulverized coal / ammonia co-combustion. *Proceedings of the*

Combustion Institute 2021;38:4043–52. https://doi.org/10.1016/j.proci.2020.06.102.

[45] Hadi K, Ichimura R, Hashimoto G, and Xia Y. Effect of fuel ratio of coal on the turbulent flame speed of ammonia / coal particle cloud co-combustion at atmospheric pressure. *Proceedings of the Combustion Institute* 2021; 38:4131–9. https://doi.org/10.1016/j.proci.2020.06.358.

[46] Xu Y, Wang H, Liu X, Zhu J, Xu J, and Xu M. Mitigating CO_2 emission in pulverized coal-fired power plant via co-firing ammonia: a simulation study of flue gas streams and exergy efficiency. *Energy Conversion and Management* 2022;256:115328. https://doi.org/10.1016/j.enconman.2022.115328.

[47] Musgraves JD, Hu J, and Calvez L (eds). *Handbook Glass*. Berlin: Springer; 2019.

[48] Conradt R. Prospects and physical limits of processes and technologies in glass melting. *Journal of Asian Ceramic Societies* 2019;7:377–96. https://doi.org/10.1080/21870764.2019.1656360.

[49] Sinton CW. Glass and energy. *Encyclopedia of Energy*. 2004:1–10. https://doi.org/10.1016/B0-12-176480-X/00374-0.

[50] CABLE M. The development of flat glass manufacturing processes. *Transactions of the Newcomen Society* 2004;74:19–43. https://doi.org/10.1179/TNS.2004.002.

[51] Zier M, Pflugradt N, Stenzel P, Kotzur L, and Stolten D. Industrial decarbonization pathways: the example of the German glass industry. *Energy Conversion and Management: X* 2023;17:100336. https://doi.org/10.1016/J.ECMX.2022.100336.

[52] Hartwell R, Coult G, and Overend M. Mapping the flat glass value-chain: a material flow analysis and energy balance of UK production. *Glass Structures and Engineering* 2023;8:167–92. https://doi.org/10.1007/S40940-022-00195-9/TABLES/7.

[53] Rio DDFD, Sovacool BK, Foley AM, *et al.* Decarbonizing the glass industry: a critical and systematic review of developments, sociotechnical systems and policy options. *Renewable and Sustainable Energy Reviews* 2022;155:111885. https://doi.org/10.1016/J.RSER.2021.111885.

[54] Westbroek CD, Bitting J, Craglia M, Azevedo JMC, and Cullen JM. Global material flow analysis of glass: from raw materials to end of life. *Journal of Industrial Ecology* 2021;25:333–43. https://doi.org/10.1111/JIEC.13112.

[55] Schaeffer HA, and Langfeld AR. *Mit Großer Zukunft W. Werkstoff Glas*. Berlin: Springer; 2014.

[56] Bourhis EL. *Glass: Mechanics and Technology*. Weinheim: Wiley-VCH Verlag GmbH & Co. KGaA; 2014.

[57] Hubert M. *Industrial Glass Processing and Fabrication*. Berlin: Springer; 2019. pp. 1195–231. https://doi.org/10.1007/978-3-319-93728-1_34/FIGURES/30.

[58] Trier W, and Loewenstein KL. *Glass Furnaces: Design, Construction and Operation*. Sheffield: Society of Glass Technology; 1984.

[59] Joint Research Centre: Institute for Prospective Technological Studies, Sissa AQ, Sancho LD, Roudier S, Scalet BM, and Muñoz MG. *Best Available*

Techniques (BAT) Reference Document for the Manufacture of Glass. Luxembourg: Publications Office of the European Union; 2013. Available from: https://data.europa.eu/doi/10.2791/70161.

[60] *Regenerative and Recuperative Furnace*. Vicenza: BDF Industries; n.d. https://www.bdfindustriesgroup.com/melting/melting-furnace/ (accessed May 25, 2024).

[61] Karem Y. A new review in glass furnaces energy saving field by pairing between recuperative and regenerative systems. *International Journal of Advanced Research and Development* 2018;3:17–20.

[62] Kuzyak VA. Recuperative glass tank furnaces. *Glass and Ceramics* 1956; 13:497–501. https://doi.org/10.1007/BF00669740/METRICS.

[63] Nicholson R. Recuperative and regenerative techniques at high temperature. *Journal of Heat Recovery* 1983;3:385–404.

[64] Jouhara H, Khordehgah N, Almahmoud S, Delpech B, Chauhan A, and Tassou SA. Waste heat recovery technologies and applications. *Thermal Science and Engineering Progress* 2018;6:268–89. https://doi.org/10.1016/J.TSEP.2018.04.017.

[65] Horn Glass Industries. *End Fired Furnaces. n.d.* https://www.hornglass.com/products/melting-furnaces-and-equipment/end-fired-furnaces (accessed 21 November 2024).

[66] Horn Glass Industries. *Float Glass Furnaces. n.d.* https://www.hornglass.com/products/melting-furnaces-and-equipment/float-glass-furnaces (accessed 21 November 2024).

[67] Horn Glass Industries. *Recuperative Furnaces. n.d.* https://www.hornglass.com/products/melting-furnaces-and-equipment/recuperative-furnaces (accessed 21 November 2024).

[68] Brown JT. 100% Oxygen-fuel combustion for glass furnaces. *Proceedings of the 51st Conference on Glass Problems*. 2008. pp. 594–609. https://doi.org/10.1002/9780470313237.CH18.

[69] Gärtner S, Rank D, Heberl M, *et al.* Simulation and techno-economic analysis of a power-to-hydrogen process for oxyfuel glass melting. *Energies* 2021;14:8603. https://doi.org/10.3390/EN14248603.

[70] Yao Y, He JY, Chen Q, Li T, Li B, and Wei XL. Analysis of energy, exergy and CO_2 emissions in a fiberglass furnace with oxy-fuel combustion. *Fuel* 2023;348:128484. https://doi.org/10.1016/J.FUEL.2023.128484.

[71] Brown JT, and Kobayashi H. How the water content of glass varies. *The Glass Industry* 1996;77:25.

[72] Kobayashi H. Advances in oxy-fuel fired glass melting technology. *Proceedings of XX International Congress on Glass*. 2004.

[73] U.S. Energy Information Administration (EIA). Glass Manufacturing Is an Energy-Intensive Industry Mainly Fueled by Natural Gas. 2013. https://www.eia.gov/todayinenergy/detail.php?id=12631.

[74] Drummond WH (ed.). *67th Conference on Glass Problems: Ceramic Engineering and Science Proceedings*. Hoboken, NJ: John Wiley & Sons, Inc., The American Ceramic Society; 2007.

[75] Li Z, He X, Wang Y, and Zhang B. Design of a flat glass furnace waste heat power generation system. *Applied Thermal Engineering* 2014;63(1):290–296.

[76] Springer C, and Hasanbeigi A. *Emerging Energy Efficiency and Carbon Dioxide Emissions-Reduction Technologies for the Glass Industry.* Berkeley, CA: Energy Analysis and Environmental Impacts Division, Lawrence Berkeley National Laboratory; 2017.

[77] Ministry of Economy, Trade and Industry (METI). *Technologies Supporting Zero-Emission Thermal Power Plants. n.d.* https://www.enecho.meti.go.jp/category/others/for_energy_technology/pdf/pamphlet_0201.pdf (accessed 21 November 2024).

[78] Valera-Medina A, Xiao H, Owen-Jones M, David WIF, and Bowen PJ. Ammonia for power. *Progress in Energy and Combustion Science* 2018; 69:63–102. https://doi.org/10.1016/j.pecs.2018.07.001.

[79] Kobayashi H, Hayakawa A, Somarathne KDKA, and Okafor EC. Science and technology of ammonia combustion. *Proceedings of the Combustion Institute* 2019;37:109–33. https://doi.org/10.1016/j.proci.2018.09.029.

[80] Cavaliere A, and de Joannon M. Mild combustion. *Progress in Energy and Combustion Science* 2004;30:329–66. https://doi.org/10.1016/j.pecs.2004.02.003.

[81] Sabia P, Sorrentino G, Ariemma GB, Manna MV, Ragucci R, and de Joannon M. MILD combustion and biofuels: a minireview. *Energy Fuels* 2021;35:19901–19. https://doi.org/10.1021/acs.energyfuels.1c02973.

[82] Kurata O, Iki N, Inoue T, *et al.* Development of a wide range-operable, rich-lean low-NOx combustor for NH_3 fuel gas-turbine power generation. *Proceedings of the Combustion Institute* 2019;37:4587–95. https://doi.org/10.1016/j.proci.2018.09.012.

[83] Sorrentino G, Sabia P, Bozza P, Ragucci R, and de Joannon M. Low-NOx conversion of pure ammonia in a cyclonic burner under locally diluted and preheated conditions. *Applied Energy* 2019;254:113676. https://doi.org/10.1016/j.apenergy.2019.113676.

[84] Ariemma GB, Sabia P, Sorrentino G, Bozza P, de Joannon M, and Ragucci R. Influence of water addition on MILD ammonia combustion performances and emissions. *Proceedings of the Combustion Institute* 2021; 38:5147–54. https://doi.org/10.1016/j.proci.2020.06.143.

[85] Garcia ÀR, Clausell C, and Barba A. Oxynitride glasses: a review. *Boletín de La Sociedad Española de Cerámica y Vidrio* 2016;55:209–18. https://doi.org/10.1016/J.BSECV.2016.09.004.

[86] Mulfinger H-O. Physical and chemical solubility of nitrogen in glass melts. *Journal of the American Ceramic Society* 1966;49:462–7. https://doi.org/10.1111/J.1151-2916.1966.TB13300.X.

[87] Elmer TH, and Nordberg ME. Effect of nitriding on electrolysis and devitrification of high-silica glasses. *Journal of the American Ceramic Society* 1967;50:275–9. https://doi.org/10.1111/J.1151-2916.1967.TB15108.X.

[88] Duval A, Houizot P, and Rouxel T. Review: elaboration, structure, and mechanical properties of oxynitride glasses. *Journal of the American Ceramic Society* 2023;106:1611–37. https://doi.org/10.1111/JACE.18824.

[89] Yamamoto H, Nakatsuka N, Hori S, *et al.* Effect of ammonia-air combustion melting on the color and physical properties of soda lime silicate. *Ceramics International* 2024;50:13237–49.

[90] Li M, Duan Q, Zhao H, Tao H, and Jiang H. Microstructure and performances of glasses melt under oxy-fuel combustion. *Journal of Wuhan University of Technology, Materials Science Edition* 2017;32:19–22. https://doi.org/10.1007/S11595-017-1550-9/METRICS.

[91] Yanık MCÖ, Sarıgüzel M, Öztürk Y, *et al.* Influence of different process conditions on mechanical, optical and surface properties of silver ion exchanged soda-lime silicate glass. *Journal of Non-Crystalline Solids* 2018; 493:1–10. https://doi.org/10.1016/J.JNONCRYSOL.2018.04.024.

[92] Deng W, Backhouse DJ, Kazi FK, *et al.* Alternative raw material research for decarbonization of UK glass manufacture. *International Journal of Applied Glass Science* 2023;14:341–65. https://doi.org/10.1111/IJAG.16637.

[93] Rossano S, Khomenko V, Bedidi A, *et al.* Glass colourations caused by Mn-Fe redox pair: application to ancient glass technology. *Journal of Non-Crystalline Solids* 2022;594:121710. https://doi.org/10.1016/J.JNONCRYSOL.2022.121710.

[94] Bartholomew RF. High-water containing glasses. *Journal of Non-Crystalline Solids* 1983;56:331–42. https://doi.org/10.1016/0022-3093(83)90490-8.

[95] Davis KM, and Tomozawa M. Water diffusion into silica glass: structural changes in silica glass and their effect on water solubility and diffusivity. *Journal of Non-Crystalline Solids* 1995;185:203–20. https://doi.org/10.1016/0022-3093(95)00015-1.

[96] Cardoso AV, and Abreu WMD. Water and the glass transition temperature of organic (caramel) glasses. *Journal of Non-Crystalline Solids* 2004; 348:51–8. https://doi.org/10.1016/J.JNONCRYSOL.2004.08.125.

[97] Tomozawa M. Water in glass. *Journal of Non-Crystalline Solids* 1985; 73:197–204. https://doi.org/10.1016/0022-3093(85)90346-1.

[98] Volotinen TT, and Pehlivan IB. Optical studies of mutual redox and partitioning of coordination sites between Ce3+/Ce4+ and Fe2+/Fe3+ ions in Na2O–CaO–SiO2 glass. *Journal of Non-Crystalline Solids* 2022; 593:121749. https://doi.org/10.1016/J.JNONCRYSOL.2022.121749.

[99] World's First Demonstration Test of Glass Production Using Ammonia as Fuel in Actual Production Furnace | News | AGC. AGC n.d. https://www.agc.com/en/news/detail/1203960_2814.html (accessed June 13, 2024).

[100] AGC Glass Europe, a European leader in flat glass. *n.d.* https://www.agc-yourglass.com/en-IT/homepage/homeowners (accessed 21 November 2024).

[101] British Ceramic Confederation: Hydrogen for the ceramics sector. 2023. https://assets.publishing.service.gov.uk/media/649aa7eede86820013bc8e8f/British_Ceramic_Confederation_-_Hydrogen_for_the_Ceramics_Sector_-_IFS_Feasibility_Report.pdf.

[102] *Ceramics Market Size Report*, 2024 – 2030. n.d. https://www.industryarc.com/Research/ceramics-market-research-800587 (accessed May 19, 2024).

[103] *Ammonia Hydrogen Combustion Technology Video | Facebook n.d.* https://www.facebook.com/watch/?v=530362889090475 (accessed June 13, 2024).

[104] KEDA Machinery Green development with ammonia-hydrogen low-carbon combustion technology in China's building ceramics and sanitaryware industry. n.d. https://kedagroup.com/news/index124.html (accessed May 19, 2024).

[105] Pontikes Y, Rathossi C, Nikolopoulos P, Angelopoulos GN, Jayaseelan DD, and Lee WE. Effect of firing temperature and atmosphere on sintering of ceramics made from Bayer process bauxite residue. *Ceramics International* 2009;35:401–7. https://doi.org/10.1016/j.ceramint.2007.11.013.

[106] Szabó L, Hidalgo I, Císcar JC, Soria A, and Russ P. CO_2 emission trading within the European Union and Annex B countries: the cement industry case. *Energy Policy* 2006;34(1):72–87.

[107] Chinyama MPM. Alternative fuels in cement manufacturing. In: Manzanera M (ed.). *Alternative Fuel*. London: InTech; 2011. https://doi.org/10.5772/22319.

[108] Calvo VLV, Giner-Santonja G, Alonso-Fariñas B, and Aguado JM. The effect of the European Industrial Emissions Directive on the air emission limit values set by competent authorities in the permitting procedure: the case of the Spanish cement industry. *Science of The Total Environment* 2021;773:145491. https://doi.org/10.1016/j.scitotenv.2021.145491.

[109] Supino S, Malandrino O, Testa M, and Sica D. Sustainability in the EU cement industry: the Italian and German experiences. *Journal of Cleaner Production* 2016;112:430–42. https://doi.org/10.1016/j.jclepro.2015.09.022.

[110] Meshram A. Optimizing and predicting compressive strength of one-part geopolymer concrete. *IJRASET* 2023;11:7457–72. https://doi.org/10.22214/ijraset.2023.53436.

[111] Feng X-Z, Lugovoy O, and Qin H. Co-controlling CO_2 and NOx emission in China's cement industry: an optimal development pathway study. *Advances in Climate Change Research* 2018;9:34–42. https://doi.org/10.1016/j.accre.2018.02.004.

[112] Mikulčić H, Cabezas H, Vujanović M, and Duić N. Environmental assessment of different cement manufacturing processes based on Energy and Ecological Footprint analysis. *Journal of Cleaner Production* 2016;130:213–21. https://doi.org/10.1016/j.jclepro.2016.01.087.

[113] Kajaste R, and Hurme M. Cement industry greenhouse gas emissions – management options and abatement cost. *Journal of Cleaner Production* 2016;112:4041–52. https://doi.org/10.1016/j.jclepro.2015.07.055.

[114] Lea FM, and Mason TO. Cement. *Encyclopedia Britannica*. 2024. https://www.britannica.com/technology/cement-building-material (accessed 6 November 2024).

[115] Sahoo N, Kumar A, and Samsher S. Review on energy conservation and emission reduction approaches for cement industry. *Environmental Development* 2022;44:100767. https://doi.org/10.1016/j.envdev.2022.100767.

[116] Madlool NA, Saidur R, Hossain MS, and Rahim NA. A critical review on energy use and savings in the cement industries. *Renewable and Sustainable Energy Reviews* 2011;15:2042–60. https://doi.org/10.1016/j.rser.2011.01.005.

[117] Habert G, Miller SA, John VM, *et al.* Environmental impacts and decarbonization strategies in the cement and concrete industries. *Nature Reviews Earth and Environment* 2020;1:559–73. https://doi.org/10.1038/s43017-020-0093-3.

[118] Oh D-Y, Noguchi T, Kitagaki R, and Park W-J. CO_2 emission reduction by reuse of building material waste in the Japanese cement industry. *Renewable and Sustainable Energy Reviews* 2014;38:796–810. https://doi.org/10.1016/j.rser.2014.07.036.

[119] Ige OE, Von Kallon DV, and Desai D. Carbon emissions mitigation methods for cement industry using a systems dynamics model. *Clean Technologies and Environmental Policy* 2024;26:579–97. https://doi.org/10.1007/s10098-023-02683-0.

[120] Aranda Usón A, López-Sabirón AM, Ferreira G, and Llera Sastresa E. Uses of alternative fuels and raw materials in the cement industry as sustainable waste management options. *Renewable and Sustainable Energy Reviews* 2013;23:242–60. https://doi.org/10.1016/j.rser.2013.02.024.

[121] Gabrielli P, Rosa L, Gazzani M, *et al.* Net-zero emissions chemical industry in a world of limited resources. *One Earth* 2023;6:682–704. https://doi.org/10.1016/j.oneear.2023.05.006.

[122] Plaza MG, Martínez S, and Rubiera F. CO_2 capture, use, and storage in the cement industry: state of the art and expectations. *Energies* 2020;13:5692. https://doi.org/10.3390/en13215692.

[123] Monteiro J, and Roussanaly S. CCUS scenarios for the cement industry: is CO_2 utilization feasible? *Journal of CO_2 Utilization* 2022;61:102015. https://doi.org/10.1016/j.jcou.2022.102015.

[124] Japanese cross-ministerial strategic innovation promotion program "Innovative AI Hospital System"; how will the 4th industrial revolution affect our health and medical care system? *JMA J* 2022;5:1–8. https://doi.org/10.31662/jmaj.2021-0133.

[125] Kujiraoka H, Izumi T, Yoshizuru Y, *et al.* Evaluation of the cement clinker fired in the combustion furnace of heavy-oil and NH_3 2018. https://ammoniaenergy.org/presentations/evaluation-of-the-cement-clinker-fired-in-the-combustion-furnace-of-heavy-oil-and-nh3/.

[126] Yoshizuru Y, Suemasu T, Nishio M, Murai R, and Akamatsu F. Optimization of the NOx reduction condition in the combustion furnace for the combustion of "heavy-oil – NH3 system" using CFD. 2018. https://ammoniaenergy.org/presentations/optimization-of-the-nox-reduction-condition-in-the-combustion-furnace-for-the-combustion-of-heavy-oil-nh3-system-using-cfd/.

[127] Fawzi NM, and Kareem SL. Effect of incorporating ammonia solution on some mechanical properties of concrete, *International Culture, Art & Islamic Architecture Conference (ICAIAC-16)*, 2016.

[128] Pisciotta M, Pilorgé H, Davids J, and Psarras P. Opportunities for cement decarbonization. *Cleaner Engineering and Technology* 2023;15:100667. https://doi.org/10.1016/j.clet.2023.100667.

[129] Bararzadeh Ledari M, Khajehpour H, Akbarnavasi H, and Edalati S. Greening steel industry by hydrogen: lessons learned for the developing world. *International Journal of Hydrogen Energy* 2023;48:36623–49. https://doi.org/10.1016/j.ijhydene.2023.06.058.

[130] Gu Y, Pan C, Sui Y, *et al.* CO_2 emission accounting and emission reduction analysis of the steel production process based on the material-energy-carbon correlation effect. *Environmental Science and Pollution Research* 2023;30:124010–27. https://doi.org/10.1007/s11356-023-30830-z.

[131] Lopez G, Galimova T, Fasihi M, Bogdanov D, and Breyer C. Towards defossilised steel: supply chain options for a green European steel industry. *Energy* 2023;273:127236. https://doi.org/10.1016/j.energy.2023.127236.

[132] Hasanbeigi A, Springer C, and Global E. How clean is the US steel industry. An international benchmarking of energy and CO_2 intensities. 2019. https://www.globalefficiencyintel.com/us-steel-industry-benchmarking-energy-co2-intensities.

[133] Shahabuddin M, Brooks G, and Rhamdhani MA. Decarbonisation and hydrogen integration of steel industries: recent development, challenges and technoeconomic analysis. *Journal of Cleaner Production* 2023;395:136391. https://doi.org/10.1016/j.jclepro.2023.136391.

[134] Murai R, Nakatsuka N, Higashino H, and Akamatsu F. Review of fundamental study on ammonia direct combustion in industrial furnaces. In: Aika, Ki., and Kobayashi, H. (eds.), *CO_2 Free Ammonia as an Energy Carrier* Singapore: Springer; 2023. pp. 627–40.

[135] Hagihara Y, Yamamoto Y, Numata M, and Matsuda T. Development of impinging jet burner using ammonia fuel for degreasing steel sheets. In: Aika, Ki., and Kobayashi, H. (eds.), *CO_2 Free Ammonia as an Energy Carrier*. Singapore: Springer; 2023. p. 641–51.

[136] Xie M, Tu Y, and Peng Q. Numerical study of NH_3/CH_4 MILD combustion with conjugate heat transfer model in a down-fired lab-scale furnace. *Applications in Energy and Combustion Science* 2023;14:100144. https://doi.org/10.1016/j.jaecs.2023.100144.

[137] Mohammadpour A, Mazaheri K, and Alipoor A. Reaction zone characteristics, thermal performance and NO_x/N_2O emissions analyses of ammonia MILD combustion. *International Journal of Hydrogen Energy* 2022;47:21013–31.

[138] Ma Y, Bae JW, Kim S, *et al.* Reducing iron oxide with ammonia: a sustainable path to green steel. *Advanced Science* 2023;10:2300111.

[139] Wang D, Lee M, and Suzuki Y. Nitriding effects of ammonia flames on iron-based metal walls. *Journal of Ammonia Energy* 2023;1. https://doi.org/10.18573/jae.8.

[140] Thiel GP, and Stark AK. 2021 To decarbonize industry, we must decarbonize heat. *Joule* 2021;5(3):531–550.

[141] NUTEC Bickley. Heat treatment in industrial furnaces – all there is to know. 2023. https://nutecbickley.com/newsroom/articles/heat-treatment-in-industrial-furnaces/ (accessed March 9, 2024).

[142] Valera-Medina A, Amer-Hatem F, Azad AK, *et al.* Review on ammonia as a potential fuel: from synthesis to economics. *Energy & Fuels* 2021;35(9):6964–7029. https://pubs.acs.org/doi/full/10.1021/acs.energyfuels.0c03685.

[143] Agrafiotis C, and Tsoutsos T. Energy saving technologies in the European ceramic sector: a systematic review. *Applied Thermal Engineering* 2001; 21:1231–49. https://doi.org/10.1016/S1359-4311(01)00006-0.

[144] Kääntee U, Zevenhoven R, Backman R, and Hupa M. Cement manufacturing using alternative fuels and the advantages of process modelling. *Fuel Processing Technology* 2004;85:293–301. https://doi.org/10.1016/S0378-3820(03)00203-0.

[145] Furszyfer Del Rio DD, Sovacool BK, Foley AM, *et al.* Decarbonizing the ceramics industry: a systematic and critical review of policy options, developments and sociotechnical systems. *Renewable and Sustainable Energy Reviews* 2022;157:112081. https://doi.org/10.1016/j.rser.2022.112081.

Chapter 5
Ammonia-fueled internal combustion engines

H. Shi[1,2], *E. Boulet*[1], *D. Dong*[3], *M. Alnajideen*[1], *S. Mashruk*[1], *Z. Zhang*[3] *and A. Valera-Medina*[1]

This chapter provides a comprehensive review of the potential of green ammonia as an alternative fuel for internal combustion engines (ICEs). The study focuses on the latest advancements in ammonia combustion characteristics, engine performance, and system integration. Key challenges, such as ammonia's low flammability and high auto-ignition temperature, are addressed, along with potential mitigation strategies including fuel blending and engine modifications. The chapter analyzes experimental findings from both spark ignition and compression ignition engines, highlighting the progress made and remaining obstacles. By examining the state-of-the-art technologies and novel approaches, this review contributes to understanding the prospects and challenges of transitioning to ammonia-fueled ICEs for a more sustainable transportation sector.

5.1 Ammonia as a renewable fuel

Global climate changes, primarily driven by anthropogenic greenhouse gas (GHG) emissions, are imposing critical challenges on the environment and biodiversity in recent years. The transportation sector is a major contributor to these emissions, with conventional fossil-fuel-based fuels exacerbating the issue. In Europe, transportation is the largest GHG emitter, accounting for over a quarter of total emissions and demonstrating a persistent upward trend since 1990 [1]. The imperative to transition to a low-carbon economy is intensified by this escalating climate crisis. The utilization of carbon-neutral or low-carbon chemicals offers a potential pathway to harness renewable, intermittent energy sources.

[1]School of Engineering, Cardiff University, UK
[2]Reactive Flows and Diagnostics, Department of Mechanical Engineering, Technical University of Darmstadt, Germany
[3]School of Naval Architecture, Ocean and Energy Power Engineering, Wuhan University of Technology, China

Ammonia emerges as a promising carbon-free fuel candidate due to its potential for flexible infrastructure and cost-effective power generation [2]. The existing infrastructure and the feasibility of ammonia production from renewable sources have accelerated its exploration as a sustainable fuel alternative [3]. The Haber–Bosch process, developed in 1909 by Fritz Haber and Carl Bosch, enabled ammonia synthesis from nitrogen and hydrogen, marking a pivotal moment in its utilization as a fuel. This process remains the primary method for ammonia production, with large-scale plants capable of producing thousands of tons daily, a capacity projected to expand significantly [2,4]. Recent interest in ammonia has been driven by its potential as a hydrogen carrier and renewable energy storage medium. Possessing a high hydrogen density of 1.5 mol per mole, ammonia can be directly combusted, particularly in internal combustion engines (ICEs) [5]. Given its suitability for power, heat, and propulsion generation, ammonia has garnered significant industrial acceptance. The transportation sector, a primary contributor to carbon emissions, presents a potential application for ammonia-fueled ICEs.

5.2 Ammonia properties compared with other engine fuels

Ammonia, as a carbon-free fuel, holds substantial promise for a low-carbon future, thereby driving intensified research into its combustion characteristics. Numerous studies have delved into critical aspects of ammonia combustion, including ignition delay times [6], flammability limits [7], flame speed [8], and combustion instabilities [9]. To facilitate comparative analysis, Table 5.1 presents key properties of ammonia (e.g. volumetric energy density, auto-ignition temperature, minimum ignition energy) alongside those of conventional hydrocarbon fuels like hydrogen, methanol, and gasoline.

Ammonia, as a carbon-free fuel, holds substantial promise for a low-carbon future, thereby driving intensified research into its combustion characteristics. Numerous studies have delved into critical aspects of ammonia combustion, including ignition delay times [10], flammability limits [11], flame speeds [12], and combustion instabilities [13]. To facilitate comparative analysis, Table 5.1 presents key properties of ammonia (e.g. volumetric energy density, auto-ignition temperature, minimum ignition energy) alongside those of conventional hydrocarbon fuels like hydrogen, methanol, and gasoline.

Ammonia presents several advantages as a renewable fuel for ICEs, including high energy density (liquid ammonia contains more than 1.5 times the energy of liquid hydrogen per unit volume), ease of liquefaction and storage (e.g., −33 °C at 1 bar, 20 °C at 9 bar), and a robust production infrastructure via the Haber–Bosch process [2]. As illustrated in Figure 5.1, these properties render ammonia a promising alternative to conventional fuels. Despite recent advancements in ammonia-fueled ICE research, significant challenges persist. Notably, ammonia's high heat

Table 5.1 Key properties of ammonia and common fuels

	Ammonia	**Hydrogen**	**Methane**	**Propane**	**Methanol**	**Ethanol**	**Gasoline**	**Diesel**
Density (g/L, 25 °C, 1 atm)	0.703	0.082	0.657	493	787	789	740	862
Volumetric energy density (MJ/m^3)	13.72	10.7	33.0	25.3×10^3	15.8×10^3	21×10^3	33×10^3	35×10^3
Boiling point (K, 1 atm)	239.8	20.28	111.2	231	337.9	351.5	310–477	436–630
Freezing point (K, 1 atm)	195.5	13.99	91.15	85	175.55	159.05	215	243
Latent heat of vaporization (kJ/kg)	1370	461	510	428	1165	919	348.7	232.4
Specific heat capacity (C_p, $kJ/(kg \cdot K)$)	2.19	14.30	2.483	1.63	2.53	2.57	2.22	1.75
Octane number	130	130	120	120	108.7	108.6	86–94	5–10
Auto-ignition temperature (°C)	651	571	537	450	470	365	230	250
Minimum ignition energy (MJ)	8	0.011	0.28	0.25	0.14	0.28	0.8	–
Adiabatic flame temperature (°C)	1800	2110	1950	1980	1949	2082	1997	2027
Lower heating value (MJ/kg)	18.8	120.1	50	46.4	19.7	26.8	44.5	45
Maximum laminar burning velocity (m/s) at ambient temperature and pressure	0.07	2.91	0.37	0.43	0.50	0.47	0.47	0.87
Stoichiometric air/fuel ratio (mass basis)	6:1	34.3:1	17.24:1	15.5:1	6.4:1	9:1	14.7:1	14.6:1
Flammability limit (equivalence ratio)	0.63–1.4	0.10–7.1	0.50–1.7	0.51–2.5	0.55–2.9	0.66–2.4	0.7–4	–

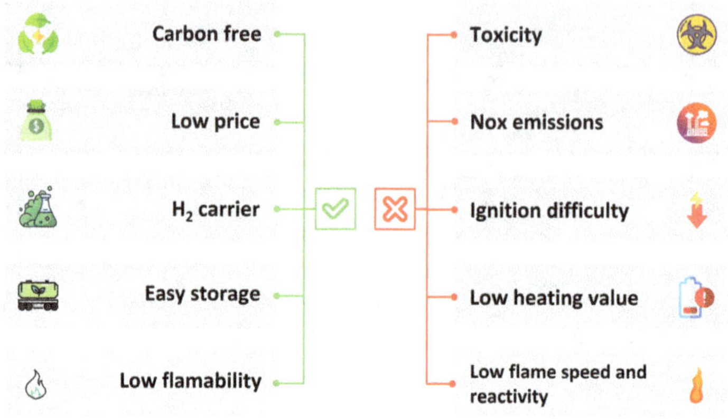

Figure 5.1 Advantages and disadvantages of ammonia as a fuel

of vaporization, low calorific value, high minimum ignition energy, and slow burning velocity hinder its direct application in ICEs [14].

Nevertheless, recent studies have demonstrated the feasibility of ammonia combustion in ICEs with relatively minor engine modifications, suggesting potential for retrofitting existing engines [10,11,15]. A critical issue is mitigating NO_x emissions while maintaining engine performance. Comprehensive assessments of ammonia as a sustainable ICE fuel have considered thermodynamic performance, system efficiency, driving range, fuel tank compactness, and economic factors [5,12]. These analyses consistently indicate the need for blending ammonia with other fuels to overcome its combustion limitations, as evidenced by the prevalence of dual-fuel approaches in the literature [13,16].

5.3 Ammonia as a fuel for internal combustion engines

ICEs exhibit remarkable versatility in terms of both fuel compatibility and application. They are prevalent in various transportation systems, including automobiles, locomotives, ships, and light aircraft, as well as in stationary applications such as power generators. Ammonia has a long history as a proposed vehicular fuel, dating back to 1920 when Wilhelm Pflaum patented the first ammonia combustion engine. Early successful attempts to use ammonia as a fuel for cars and buses occurred on a small scale during the 1930s and 1940s [17]. Subsequent investigations in the 1960s by U.S. institutions using cooperative fuel research (CFR) engines established the feasibility, albeit with challenges, of operating ICEs on ammonia-based fuels [18–22]. Despite these early successes, the focus shifted to conventional hydrocarbon fuels until the late 2000s when tightening carbon emission regulations renewed interest in ammonia. As illustrated in Figure 5.2, recent projects such as the University of Michigan's NH_3 car [23] and the ACTIVATE tractor [24] exemplify efforts to convert ICEs to ammonia-based operation. Major engine manufacturers, including Wärtsilä and MAN Energy Solutions, are actively

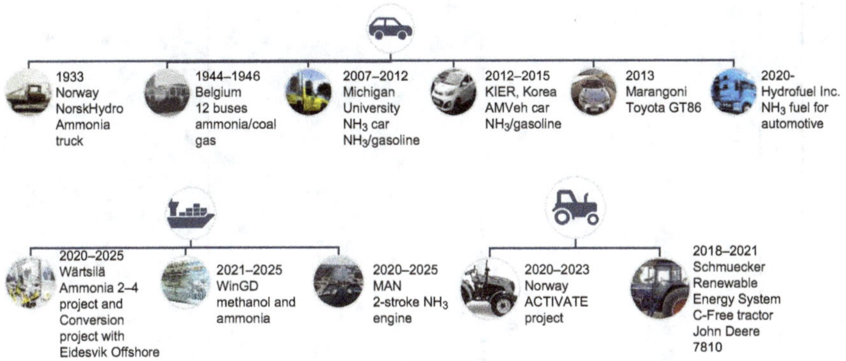

Figure 5.2 Non-exhaustive list of projects using ammonia-based fuel in internal combustion engine

developing ammonia engines, with commercial availability anticipated by 2024 [25,26].

Research into ammonia-fueled ICEs has primarily focused on two conventional engine architectures: spark ignition (SI) and compression ignition (CI). While current projects often lean toward CI engines [27], ammonia's low auto-ignition propensity renders SI combustion a more promising approach. Ammonia's narrow flammability limits and low flame speed can lead to incomplete combustion in SI engines. However, these challenges can be mitigated through the addition of hydrogen or gasoline to enhance ignition, or by increasing in-cylinder temperatures via higher compression ratios (CRs) or supercharging without inducing knock.

5.4 Ammonia-fueled SI engines

The application of ammonia as a fuel for SI engines dates back to the late 1960s when it was considered for military vehicles [18]. Lhuillier *et al.* [28] reported an indicated efficiency of approximately 36% for pure ammonia at full load, comparable to methane. However, ammonia's inherent low flame speed and combustion instability hinder its performance in SI engines. To address these limitations, recent research has predominantly focused on blending ammonia with secondary fuels, such as hydrogen [29] and methane [30], to improve combustion characteristics. A broader range of fuels, including gasoline, alcohol, natural gas, syngas, and methanol, have been explored as blending components. Additionally, increasing CRs and engine boost have been investigated as potential strategies to enhance ammonia combustion performance. Given the growing interest in ammonia-fueled SI engines, Table 5.2 provides a comprehensive overview of recent experimental studies, summarizing key findings and recommendations (Figure 5.3).

Table 5.2 Summary of the literature with experimental results related to spark ignition engine with ammonia in the fuel blend

Engine tested	Combustion enhancer	Main conclusions and recommendations from authors	Reference
CFR Single cyl CR: 9.4 Fuel: NH_3 and H_2	H_2	*Main conclusions*: • Poor performance with conventional SI engine • Improvement methods tested: Spark energy ignition, CR increase, Supercharging, H_2 addition • NH_3 dissociation investigated to produce H_2 • High NO_x emissions and NH_3 in the exhaust • 2% H_2 by mass showed promising results • H_2 necessary for part load • The addition of a supercharger to a conventional engine would develop the engine power • The other method would involve also the addition of a supercharger but only moderate supercharging of the engine, an increase in the compression ratio, and the addition of a small amount of hydrogen to the ammonia *Recommendations*: • Improvement method 1: supercharger added to conventional ICE • Improvement method 2: moderate supercharging + CR increase + H_2 • Challenge will be auxiliary equipment and controls for ammonia dissociation (also bulkier than for conventional) • Supercharger added to conventional ICE • Moderate supercharging + CR increase + H_2	Cornelius *et al.* [87]
CFR CR: 6, 8, 10 Fuel: NH_3 and H_2	H_2 by catalytic dissociation	*Main conclusions*: • Little engine modification required including ammonia flow control and spark timing • Indicated output is 70% of that with HC fuel • H_2 concentration critical: minimum 4%–5% by weight @1800 rpm for successful operation • Performance impacted by engine speed, spark timing, and pressure • Calculations showed engine output higher than with HC if liquid NH_3 injected into the cylinder *Emissions*: NO	Starkman *et al.* [88]

(Continues)

CFR CR: 16	H_2, acetylene, butane…	*Main conclusions*: ● Initial power with gasoline: 65 hp. 41 hp achieved on 100% ammonia, 53 hp with 1.5% H_2 (i.e. 80% of gasoline power) ● Speed above 3000 rpm: performance decreases due to low flame speed. Addition of H_2 (1.5%) allows increasing high speed operation output ● Operation in cold climate expected to be very difficult ● Higher firing pressure recorded: need to consider structural limits *Recommendations:* ● Main modifications required: increase CR, increase ignition energy and locate spark plug near center of combustion chamber ● Develop and test spark plugs with long electrode and suitable for high pressure ● Design and test compact combustion chambers (close to spherical) ● Use exhaust gas to vaporize ammonia ● Develop NH_3 cracker to produce 1.5% H_2 by weight (for speed above 3200 rpm)	Pearsall and Fryer [89]
CFR CR: 4.5, 6, 8, 10, 12.3, 16 Fuel: NH_3 and H_2	H_2, acetylene, butane, …	*Main conclusions*: ● Test additives tendency to increase oxidation rate of ammonia and/or ignite the mixture ● Ten additives were tested ● H_2 was best additive (1% weight needed), followed by acetylene (6% weight needed) ● Good operation between 900 and 1800 rpm with H_2 at less than 1%, with CR at 12:1 and SA at more than 50° ● Reasonable CO and NO_x	Gray *et al.* [90]

(Continues)

Table 5.2 (Continued)

Engine tested	Combustion enhancer	Main conclusions and recommendations from authors	Reference
CFR	H_2 by catalytic dissociation	*Main conclusions*: • Expected lower NO_x but higher NO_x than with HC (in some cases, one order of magnitude) • Different NO_x formation mechanism as for HC, possibly pyrolytic	Sawyer *et al.* [20]
CFR Bore: 83.4 CR: 7–16	Gasoline	*Main conclusions*: • CFR using a supercharger • Focus on operation at stoichiometry • @Idle, only gasoline was possible • @Normally aspirated + wide open throttle: 70% NH_3 and 30% gasoline possible • @Supercharger: >70% NH_3 possible • Knock-free operations when using ammonia with MBT spark timing and high CR and higher loads • Less gasoline needed as load increases *Recommendation:* • CR 10:1 recommended for gasoline/NH_3 operations, possibly 12:1 with better chamber geometry	Grannell *et al.* [36]
CFR Bore: 83.4 CR: 9,10 Fuel: Gasoline and NH_3	Gasoline	*Main conclusions*: • Standard three-way catalytic converter can be used • Standard oxygen sensors can be used • Clean-up region identified: stoichiometry to 0.2% rich • Adiabatic flame temperature lower when NH_3 used • @Engine-out: N_2O when NH_3 used ~10–40 ppm	Grannell *et al.* [39]

(Continues)

		• @Post-catalyst: N_2O when NH_3 used ~20–50 ppm • N_2O and NO formed in the catalytic converter due to over-oxidation of NH_3 when slightly lean *Recommendation:* • Lean operation must be avoided *Emissions*: NH_3, HC, NO, CO, N_2O	
CFR Bore: 82.6 CR: 6.23-13.58 Fuel: NH_3 and H_2	H_2	*Main conclusions*: • Analyze storage of NH_3 in metal ammine • Vary excess air and NH_3/H_2 ratio best found was 10% (vol.)H_2 • Propose to use exhaust heat to desorb NH_3 from metal amine • Best efficiency for excess air slightly >1 • Highest mean effective pressure for excess air slightly <1 • Exhaust heat not high enough to desorb ammonia at required rate but covers for most of the need • Suggested that fuel NO_x dominates over thermal NO_x and influenced by ignition timing variation • Higher CR compensates for reduced air flow *Recommendation:* • Reduce NO_x by using SCR after-treatment *Emissions*: NO_X	Mørch *et al.* [91]
Toyota Single-cylinder Bore: 86, 135, 137 CR: 14, 16, 16.5 Fuel: Gasoline, NH_3, and hydrogen-nitrogen mixture	H_2 Gasoline H_2 by catalytic reformer	*Main conclusions*: • Implement autothermal cracker to provide H_2 • Autothermal cracker proved to be successful and control the H_2 ratio by changing the air-NH_3 ratio • Tested under charged condition • No knocks observed *Recommendations:* • Reduce emissions @ cold start	Koike *et al.* [37]

(Continues)

Table 5.2 (Continued)

Engine tested	Combustion enhancer	Main conclusions and recommendations from authors	Reference
		• Improve combustion efficiency with less or no H_2 *Emissions*: NO, NO_2, NH_3, H_2, O_2	
Lombardini LGW 523 MPI (2 cyl.) Bore: 72 CR: 10.7 Fuel: NH_3 and H_2	H_2 by catalytic reformer	*Main conclusions*: • Modified injection system to inject NH_3 and H_2 in gaseous phase • Confirm that H_2 needed for improving ignition and combustion • H_2/NH_3 ratio needed depends mainly on load and less on engine speed (7% @full load, 11% @ half load) • Tested H_2 obtained from NH_3 by on-board catalytic reforming • Higher H_2/NH_3 ratio needed @ cold start • Lower performance obtained as for gasoline • NO_x and NH_3 in the exhaust • Electric heater needed for cold start to ensure reformer operability *Recommendations:* • Verify durability of the injection system • Develop on-board catalytic reformer for vehicle	Frigo and Gentili [38]
CFR Single cyl CR: 7–15 Fuel: NH_3 and H_2 Ammonia volume ratio: 80%	H_2	*Main conclusions*: • SCR after treatment investigated • N_2O sensitive to spark ignition timing (SIT) and excess air • NH_3 emission not influenced by SIT • Higher excess air: more NH_2 radicals and higher NO_2 and N_2O level	Westlye *et al.* [40]

(Continues)

		• NH_3 level above allowance of 10–25 ppm • Crevice mechanism proposed to explain NH_3 presence in the exhaust *Recommendation:* • NH_3 slip after cold start: ammonia trap needed (too low temperature for SCR to work @ cold start) *Emissions*: NH_3, HC, NO, NO_2, N_2O, CO	
CFR Bore: 82.5 CR: 10 Fuel: Gasoline and NH_3 and H_2 Enhanced: GDI injection	Gasoline	*Main conclusions*: • Effect of direct injection of ammonia on performance • Gasoline used to provide baseline power of 0.6 kW • Peak pressures were lower when using NH_3 compared to pure gasoline • NH_3 injection timing was advanced and injection duration was increased • Brake specific energy consumption was similar • When using ammonia: Increased NO_x emissions and NH_3 slip • Study the effect of NH_3 dissociation catalyst into H_2 on performance and exhaust • Catalyst used: NH_3 dissociation catalyst coated with 2% ruthenium on 3.175 mm aluminum pellets • Strong reduction of NO_x, NH_3, CO, and HC emissions at low loads *Recommendation:* • Developing catalyst to improve the conversion rate at high flows of NH_3 *Emissions*: CO, HC, NH_3, NO_x	Ryu *et al.* [41]
CFR	Gasoline + H_2 by catalytic reformer	*Main conclusions*: • Study effect of NH_3 dissociation catalyst into H_2 on performance and exhaust • Catalyst used: NH_3 dissociation catalyst coated with 2% ruthenium on 3.175 mm aluminum pellets	Ryu *et al.* [92]

(Continues)

Table 5.2 (Continued)

Engine tested	Combustion enhancer	Main conclusions and recommendations from authors	Reference
		• Strong reduction of NO_x, NH_3, CO and HC emissions at low loads *Recommendation:* • Develop catalyst to improve conversion rate at high flow of NH_3 is needed	
Hyundai 3 cylinders CR: 10.5 Fuel: NH_3 Enhanced: various spark timing	Gasoline	*Main conclusions*: • NH_3 burns slower than gasoline and thus spark timing must be advanced • NO_x and ammonia slip occur • Strong reduction of CO_2 emissions: CO_2, HC, NH_3, NO_x	Woo *et al.* [93]
Lombardini LGW 523 MPI (2 cyl.) Bore: 72 CR: 10.7 Fuel: NH_3 and H_2	H_2 with on board catalytic reformer	*Main conclusions*: • Develop H_2 production system from ammonia, can supply 1.4 Nm^3/h of H_2 • Cracking reactor with ruthenium-based catalyst • More H_2 needed at cold start (supplied via extra H_2 tank) • Flow limitation due to ammonia injector • Minimum H_2 flow rate and exhaust gases temperature more affected by engine rotational speed (rpm) than by load • Ammonia cracking reaction could be auto-thermally operated by only using the engine exhaust gases in the majority of the studied experimental conditions • A lower hydrogen flow rate could be produced in order to achieve satisfactory engine operation, however, this choice led to higher combustion temperatures and consequently to higher NO_x emissions *Recommendation:* • To reduce H_2 needed at cold start: Increase CR or modified ignition system *Emissions*: NO_x	Comotti and Frigo [44]

(Continues)

In-line 4 cylinders Bore: 88 CR: 10.4 Fuel: Gasoline NH_3 Ammonia volume ratio: 0%–17.2%	Gasoline + ethanol (0%–30%)	*Main conclusions*: • Study Gasoline/NH_3 blend using ethanol as emulsifier to improve NH_3 solubility in gasoline • Gasoline with 30% ethanol can retain 17.35% (vol.) NH_3 in the liquid phase • Optimum blend identified: gasoline with 20% (vol.) ethanol and 12.9% (vol.) ammonia • Engine performance similar to 100% gasoline for the optimum blend *Recommendations:* • Improve capability of engine control unit (ECU) as not able to maintain proper AFR with studied blend • It is possible to dissolve higher amount of NH_3 in E85 with 85% ethanol	Haputhanthri *et al.* [42]
Single cylinder test engine Bore: 86 CR: 13.8 Fuel: NH_3 Enhanced: In-line adsorption system	H_2	*Main conclusions*: • Propose in-line adsorption system at exhaust to reduce NH_3 emission • Adsorption system: redox converter and ammonia adsorber • Tested at cold start and during warm-up period • Adsorb NH_3 can be oxidized later by catalytic reaction. No slip observed • NH_3 free emissions achieved by the system • Tested 3 different adsorbers: H-ZSM-5, Cu-ZSM-5, and Pt-ZSM-5 • Cu-ZSM-5 and Pt-ZSM-5 regenerated after adsorption without NH_3 slip *Recommendation:* • Adsorber selection depends on amount of NH_3 emissions *Emissions*: NH_3, NO_x, N_2O	Koike and Suzuoki [94]

(Continues)

Table 5.2 (Continued)

Engine tested	Combustion enhancer	Main conclusions and recommendations from authors	Reference
PSA 1 cyl. Operation CR: 10.5 Fuel: NH_3 and H_2	H_2	*Main conclusions*: ● Pure NH_3 at various ER and intake pressure ● Compared with CH_4/air and NH_3/H_2/air ● Combustion analysis shows that H_2 acts as combustion promoter ● Partial dissociation of NH_3 could explain successful operation with pure ammonia ● Poor combustion efficiency below 95% for pure ammonia *Recommendations:* ● Use higher CR but with too early SIT otherwise lose higher CR benefit ● User supercharger to increase intake pressure and therefore power output ● Near-stoichiometric operation to limit emissions ● Rich operation with EGR ● Further investigation on LBV and turbulent behavior ● Implementation of SCR as after-treatment *Emissions*: NH_3, NO_x, HC, CO	Lhuillier *et al.* [47]
PSA 1 cyl. Operation CR: 10.5 Fuel: NH_3 and H_2	H_2	*Main conclusions*: ● Test of various H_2 fractions, ER, intake pressure ● Confirm NH_3 suitable for SI engine ● Exhaust: High NO_x @lean, High NH_3 at rich ● H_2 promotes NO_x formation ● Indicated efficiencies >37% were reached (using supercharger and 10% vol. H_2) ● Spark ignition timing closer to top dead center (TDC) when H_2 increase *Recommendations:* ● Supercharger to improve power output and indicated efficiency	Lhuillier *et al.* [95]

(Continues)

		• Slightly rich operation with EGR interesting and to be investigated • After-treatment should take advantage of NH_3 in the exhaust and of the exhaust heat itself *Emissions*: NH_3, NO, NO_2, N_2O	
PSA Single cyl Bore: 77 CR: 10.5 Fuel: NH_3 and H_2	H_2	*Main conclusions*: • Influence of laminar burning velocity (LBV) of the mixtures (includes non-stoichiometric and highly H_2 enriched blend) • Moderate H_2 addition: Highest indicated pressure @ slightly rich condition and highest efficiency @ slightly lean • H_2 allows performance and stability improvements. Wall heat loss could be a key player at higher H_2 fractions • LBV does not fully explain combustion behavior *Recommendation:* • Further study to understand ammonia combustion in turbulent conditions *Emissions*: NH_3, NO_X	Lhuillier *et al.* [33]
PSA Single cyl CR: 10.5 Fuel: NH_3, CH_4, and H_2	H_2 CH_4	*Main conclusions*: • NH_3 blend combustion characteristic in turbulent conditions • Laminar and turbulent flame experiment in constant volume vessel • Enrichment with H_2 or CH_4 could not be explained by laminar burning velocity measurement • Kinetic models used: (Nakamura *et al.* 2017; Okafor *et al.* 2018; Otomo *et al.* 2018) • Unusual combustion regime observed *Recommendation:* • Investigate further ammonia turbulent combustion with experiment and numerical simulation	Lhuillier *et al.* [34]

(Continues)

Table 5.2 (Continued)

Engine tested	Combustion enhancer	Main conclusions and recommendations from authors	Reference
PSA 1 cyl. Operation GDI EP6 PSA 4 cylinders Bore: 77 CR: 10.5 Fuel: NH_3 and H_2 Hydrogen addition ratio: 10% Enhanced: GDI injection	H_2	*Main conclusions*: ● Determine lowest load limit when fueled with NH_3/H_2 ● Provide data for cold start condition ● Evaluate dilution to reduce emissions (turbo-like): NO_x reduced in lean conditions with equivalent EGR ratio of less than 10% ● Higher engine speed: pure ammonia combustion is harder ● 10% H_2 (vol) allows running in most conditions ● Highest level of NO_x when lean and decrease with ER increasing ● Ammonia slip in rich condition ● H_2 at the exhaust, local NH_3 decomposition ● NH_3 at the exhaust decreases with an engine speed increase, with highest levels reached for rich mixture ● Dilution with N_2 in turbocharged liked conditions is interesting in decreasing NO_x at the exhaust, especially for lean and stoichiometric mixture *Recommendation:* ● Investigate dilution opportunities further, EGR *Emissions*: NO_X, NO_2, NO, N_2O	Mounaïm-Rousselle [35]
Toyota 4 cylinders Bore: 101 CR: 9 Fuel: NH_3 and H_2 Enhanced: on-board fuel reformer	H_2 via autothermal reforming	*Main conclusions*: ● Focus on cold start ● Use of autothermal reformer (H_2) to start combustion ● Use of ammonia adsorber to capture NH_3 (using three-way catalyst) ● Adsorber regeneration feasible without ammonia slip during lean regime *Recommendations:* ● N_2O and NO_x should be lowered	Koike *et al.* [46]

(Continues)

		• Focus on warm-up stage when catalyst not fully activated • Autothermal reforming (ATR) catalyst improvements *Emissions*: NH_3, NO_X, CO	
Heavy duty 6 cylinders CR: 10.5 Fuel: NH_3 and Natural gas Ammonia volume ratio: 0%–50%	Natural gas (NG)	*Main conclusions*: • Low speed regime to replicate marine engine behavior • Vary λ and NG/NH_3 ratio • CO_2 reduced by 28% at 50% NH_3 with deterioration of brake torque • AFR high limit at λ=1.5 due to slow flame speed • Initial burned duration decreased with AFR and split that increased • CA50 (50% burned mass timing) was retarded with ammonia • Lower thermal NO_x *Emissions*: NO_2, CO_2	Oh *et al.* [43]
Kia Cerato 4 cylinders Bore: 86 CR: 10.5 Fuel: Gasoline and Ethanol and NH_3	Gasoline/ethanol	*Main conclusions*: • The effect of the port injection of ammonia in a gasoline/ethanol dual-fuel engine was investigated • NO_x emissions significantly decreased by 50% across the engine speed range • BSFC, CO, and HC emissions were worsened with added ammonia(10% ammonia, BSFC, CO, and HC increased by 3%, 30%, and 21%) • Ammonia can positively affect the engine knock *Recommendation:* • The proposed combustion mode could be studied for heavy-duty diesel engines, with possible positive impacts on efficiencies and GHG (due to higher efficiency) *Emissions*: NO_X, CO, HC	Salek *et al.* [45]

(Continues)

Table 5.2 (Continued)

Engine tested	Combustion enhancer	Main conclusions and recommendations from authors	Reference
EP6 LC single cylinder Bore: 77 CR: 11.75 Fuel: NH_3 and H_2	H_2 via upstream NH_3 dissociation	*Main conclusions*: • Impact of upstream NH_3 dissociation on performance and emissions • Experimental + Chemkin model • 10% upstream NH_3 dissociation improve combustion stability • NH_3 cracking reduce mixture energy content at intake • Efficiency reduced because of lower LHV (and additional energy needed to crack ammonia) • H_2 exhaust emission $<2\%$ • N_2O reached 60 ppm @ 15%NH_3 dissociation rate • NO_x emission reduction by adapting equivalence ratio • NH_3 slip reduction: adjust dissociation rate, ER has no significant impact *Recommendations:* • Rich operation should be favored (less NO_x and NH_3 slip due to crevice mechanism and not to additional NH_3) • Optimizing engine geometry reduce NH_3 slip *Emissions*: NH_3, NO, NO_2, N_2O	Mercier *et al.* [96]
Single-cylinder four-stroke water-cooled SI engine CR: 12–15 Fuel: NH_3 and H_2	H_2	*Main conclusions*: • The high-octane number of NH_3/H_2 allows high CR, increasing efficiency without knock • NO_x emissions are found to increase as the CR and H_2 energy fractions increase *Recommendations:* • The use of ammonia/hydrogen blends can be a promising alternative to conventional fuels • Further research is needed to optimize the combustion process and reduce NO_x emissions	Dinesh *et al.* [31]

(Continues)

		• The effect of hydrogen addition on the combustion characteristics of ammonia needs to be studied in detail • The impact of ammonia/hydrogen blends on engine durability and reliability should be investigated *Emissions*: CO, CO_2, O_2, HC, NO_x	
Heavy-duty turbocharged spark ignited engine six cylinders Bore: 123 CR: 10.5 Fuel: CH_4 and NH_3	CH_4	*Main conclusions*: • The addition of ammonia as a fuel resulted in lower brake power due to its lower LHV compared to natural gas • The net indicated efficiency was slightly decreased due to increased pumping and incomplete combustion loss • The dominant emission was fuel NO_X with the introduction of ammonia *Emissions*: HC, CH_4, CO_2, CO, NH_3, NO_X	Oh *et al.* [97]
CFR Single cyl Bore: 77 CR: 11.7 Fuel: Syngas and NH_3 and CH_4 Ammonia addition ratio: 0%–15%	Syngas	*Main conclusions*: • Study syngas combustion enriched with NH_3 • Decrease of CO and CO_2 proportional to NH_3 amount and slight increase of NO_x • Performance lower with syngas than with CH_4 • Boosting intake pressure by 20% allowed to obtain similar performance as with CH_4 • NH_3 increases output work (higher energy content) but slower flame speed leading to lower indicated efficiency • Compared to CH_4: NO_x emissions with 15%NH_3/85% syngas is similar and lower HC and CO • Lower combustion efficiency for concentration >10% NH_3 possibly due to ammonia slip *Recommendations:* • Quantify ammonia in the exhaust • Fundamental study of ammonia/syngas combustion *Emissions*: CO, CO_2, NO_X, HC	Brequigny *et al.* [98]

(Continues)

Table 5.2 (Continued)

Engine tested	Combustion enhancer	Main conclusions and recommendations from authors	Reference
GDI 4 cylinders Bore: 77 CR: 10.5 Fuel: NH_3 and H_2 ammonia addition ratio: 5.2%, 7.96%, 10.68%		*Main conclusions*: ● Adding ammonia changes the combustion characteristics of hydrogen ● As the volume fraction of added ammonia increases, the rate of peak pressure increases decreases ● Adding ammonia increases the power density and thermal efficiency of the hydrogen fuel engine ● The addition of ammonia increases the NO_x emissions of hydrogen fuel engines *Emissions*: NO_X, NO, HC	Xin *et al.* [86]
GDI 4 cylinders Bore: 77 CR: 9.5 Fuel: NH_3 and H_2 ammonia addition ratio: 0%, 2.2%		*Main conclusions*: ● NH_3 is confirmed as a very suitable additive and the addition of ammonia reduced the PRRmax of the engine ● Under lean-burn conditions, the addition of ammonia improved the engine performance and thermal efficiency, and the peak IMEP and ITE of the engine were increased ● The addition of ammonia extended the CA0-10 and CA10-90 ● Compared to pure hydrogen fuel, ammonia addition ● increased engine NO_x emissions and did not decrease with delayed ignition timing *Emissions*: NO_X, NO	Xin *et al.* [50]
DV6F PSA SI ICE Single cyl Bore: 77		*Main conclusions*: ● The increase of the energy share of ammonia improves the working capacity of the engine	Xin *et al.* [48]

(Continues)

CR: 10.7 Fuel: NH_3 and H_2 ammonia volume ratio: 80%	• After the excess air ratio exceeds 1.4, the BTE of the engine first decreases slightly with the increase of the ammonia percentage, and then gradually increases • Under the working conditions of this study, when the proportion of ammonia exceeds 80%, the engine begins to misfire • The exhaust temperature of the ammonia hydrogen engine is low, and it is necessary to consider matching a small inertia turbocharger to increase power • The NO_x emission of the ammonia hydrogen engine is relatively high, and it is necessary to consider using a selective catalytic reduction device for treatment *Recommendations:* • Consider matching a small inertia turbocharger due to the relatively low exhaust temperature of the ammonia–hydrogen engine • Consider the production rate of hydrogen to ensure stable operation when using the online catalytic reforming system *Emissions*: NO_X	
DV6F PSA 4 cylinders Bore: 77 CR: 10.7 Fuel: NH_3 and H_2 Enhanced: variable valve timing	*Main conclusions*: • When the intake valve opening time is advanced, that is, the valve overlap angle is increased, the combustion of the mixture in the cylinder is unstable • Increasing the valve overlap angle will significantly reduce the combustion rate of the mixture in the cylinder and prolong the combustion duration • The ammonia–hydrogen fuel was used on the Miller cycle engine, increasing the valve overlap led to deterioration of engine performance	Xin *et al.* [49]

(Continues)

Table 5.2 (Continued)

Engine tested	Combustion enhancer	Main conclusions and recommendations from authors	Reference
Single- cylinder SI engine CR: 13 Fuel: CH_4 and NH_3		*Main conclusions*: • Ammonia combustion presents a lower peak value of heat release and longer combustion duration compared to methane, resulting in large cycle-to-cycle variations and lower thermal efficiency • The difference in the main combustion duration between ammonia and methane is small compared to laminar burning velocity • There is little difference in flame propagation speed between ammonia and methane	Zhang *et al.* [51]
DV6F PSA Single cyl Bore: 87.5 CR: 10 Fuel: NH_3 and CH_4 Ammonia volume ratio: 0%–60%		*Main conclusions*: • The lower heating value of the fuel mixture was reduced with increasing the share of ammonia in the NH_3/CH_4 fuel blend • Cycle-to-cycle combustion variations increased with the increasing share of ammonia in the fuel mixture • The lower flame speed of ammonia reduced the flame propagation speed of the fuel mixture	Kurien *et al.* [52]
DV6F PSA Ricardo E6 engine Single cyl Bore: 80 CR: 14 Fuel: NH_3 Enhanced: RCTJI		*Main conclusions*: • The concept of reactivity-controlled turbulent jet ignition (RCTJI) for ammonia engines was proposed and experimentally verified in a single-cylinder engine • The RCTJI approach achieved stable combustion and improved engine performance compared to conventional spark ignition and compression ignition combustion modes • The results suggest that RCTJI could be a promising approach for ammonia combustion in transportation applications	Liu *et al.* [99]

(Continues)

	Recommendations: • Further research is needed to optimize the RCTJI approach for ammonia engines and to evaluate its performance under different operating conditions • The variations in chemical reactivity of the pre-chamber mixture can also be achieved by using different fuels, and further investigation is needed to explore the potential of different fuel blends for RCTJI • The chapter provides insight into the use of ammonia as a zero-carbonization energy source in transportation applications, and further research is needed to evaluate its feasibility and potential benefits	
Single cyl Bore:80 CR:14 Fuel: NH_3 and H_2 and gasoline and methane Enhanced: Pre-chamber turbulent jet ignition (TJI)	*Main conclusions*: • A novel self-designed air-assisted pre-chamber TJI system • TJI ammonia engine fueled with high-reactivity fuels in the pre-chamber, improved combustion stability, higher IMEP, and lower ISFC • Reduced NO_x and NH_3 emissions compared with the SI mode *Emissions*: NO_x, NH_3, NO, NO_2, N_2O, CO_2, CO	Liu *et al.* [100]
Single cyl Bore: 80 CR: 14 Fuel: NH_3 and H_2 Enhanced: Reactivity controlled turbulent jet ignition (RCTJI)	*Main conclusions*: • Compared with Mode 1, it can be found that Mode 2 has a faster combustion rate and a better performance in ignition delay, combustion phase, and combustion durations • The formation of the faster jet flame benefits the ignition of the in-cylinder mixture • The two-stage combustion of RCTJI, including the prechamber jet flame-dominant stage and the main combustion highest reactivity *Recommendation:* • The hydrogen is a promising candidate for RCTJI-adopted ammonia engines to solve the combustion issues of ammonia engines	Liu *et al.* [101]

(Continues)

Table 5.2 (Continued)

Engine tested	Combustion enhancer	Main conclusions and recommendations from authors	Reference
CFR CR: 16–24 Fuel: NH_3		*Main conclusions*: ● The addition of a spark plug to a diesel engine allows for efficient ammonia combustion, even at lean equivalence ratios ● Machine learning models can be used to analyze engine performance and establish the relationship between engine control variables and IMEP ● The artificial neural network algorithm performed the best in learning this relationship ● The conversion of compression ignition engines to ammonia spark ignition operation is promising *Recommendations:* ● Further research is needed to optimize the engine performance and reduce noise ● Future studies should focus on the effects of different engine parameters on ammonia combustion, such as compression ratio and spark timing ● The use of machine learning models can be expanded to other engine types and fuels to improve engine performance and reduce emissions	Liu *et al.* [102]

HC = Hydrocarbons, ICE = Internal Combustion Engine, CR = Compression Ratio, CFR = Cooperative Fuel Research engine, MBT = Maximum Brake Torque, SIT = Spark Ignition Timing, SCR = Selective Catalytic Reduction, ER = Equivalence Ratio, TDC = Top Dead Center, CA = Crank Angle, EGR = Exhaust Gas Recirculation, BSFC = brake specific fuel consumption, LHV = Lower Heating Value.

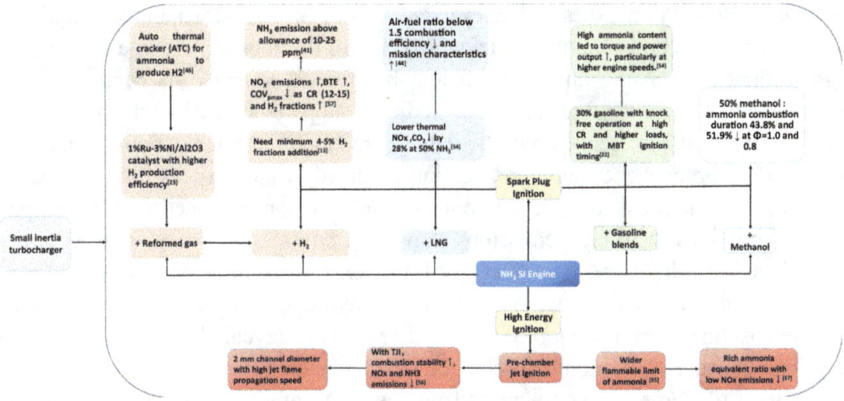

Figure 5.3 Strategies for enhancing ammonia combustion in SI engines and the responses of engine performance and emission characteristics.

5.4.1 Ammonia/H_2 blends

For SI ammonia engine, ammonia is typically introduced in the gas phase through valve injection at the intake pipe, which can be defined as gas engine. However, the pure ammonia's ignition energy is so large (14 mJ) that the mixture is not easy to ignite. For hydrogen with lower minimum ignition energy (0.011 mJ) can be easily ignited by traditional spark plug. Thus, the hydrogen is a good choice for assisted ignition of ammonia/air mixtures. Moreover, the high laminar flame velocity of hydrogen, recorded at 291 cm/s, significantly exceeds that of ammonia, which is below 8 cm/s in stoichiometric conditions. Consequently, the propagation of hydrogen combustion flames in an ammonia/air mixture leads to an expanded flame region, resulting in the acceleration of ammonia combustion. Dinesh *et al.* [31] conducted a study on the performance, combustion, and NO_x emission characteristics of a single cylinder four stroke water-cooled engine fueled with ammonia/hydrogen mixed fuel under CRs. The study found that the higher the CR, the faster the flame development, flame propagation speed, and combustion cycle. Increasing the hydrogen content also improved combustion, leading to a rapid increase in pressure and temperature. Xin *et al.* [49] investigated the effect of variable valve timing on a hydrogen rich ammonia engine on a water-cooled four cylinder four stroke SI engine. The results showed that as the intake valve advanced and the valve overlap angle increased, the combustion stability deteriorated, the flame development period (CA 0-10) extended, and cycle fluctuations increased, The COV_{pmax} will exceed 25%, and the braking thermal efficiency will rapidly decrease from 32.8% to 30%. When the intake valve timing is delayed by 25°, the maximum average intake volume and minimum intake volume increase by 8% and 16%, respectively.

Xin *et al.* [50] focused on studying the effects of stoichiometric and lean burn conditions on the performance of a Miller cycle SI engine using ammonia with hydrogen substitution fuel. The findings indicated that augmenting the ammonia energy fraction under stoichiometric conditions enhanced the operational efficiency of the engine. Beyond an excess air coefficient of 1.4, in both stoichiometric and lean burn scenarios, the brake thermal efficiency (BTE) of the engine initially experienced a slight decline with rising ammonia content before subsequently showing a gradual improvement. Otherwise, the ammonia addition suppressed the knock in hydrogen engines [34]. With the escalation in the volume fraction of ammonia, both the duration of flame development and flame propagation extended, while the peak heat release rate decreased. The augmentation in ammonia volume fraction exhibited minimal impact on NO_x emissions, with NO_x emissions showing a progressive rise as ignition timing was delayed. Furthermore, Xin *et al.* [86] also examined the suppressive impact of ammonia addition on the heat release rate of hydrogen-fueled engines in a four-cylinder gasoline direct injection SI engine. The findings revealed that the introduction of a small quantity of ammonia decelerated the combustion rate of the blend, extended the duration of flame development and propagation within the engine, and lowered both the peak cylinder pressure and peak heat release rate during lean burn configurations. This enhancement resulted in an increase in the peak indicated average effective pressure and peak indicated thermal efficiency (ITE) of the engine, subsequently leading to elevated NO_x emissions. Frigo *et al.* [36] investigated the influence of hydrogenation on the combustion attributes of ammonia in a four-stroke, two-cylinder SI engine. The empirical findings demonstrated that the introduction of hydrogen into the ammonia-air blend is an essential tactic to enhance ignition and boost combustion velocity. The proportion of hydrogen added predominantly relies on the engine load, exhibiting minimal influence on the engine speed. Mørch *et al.* [91] investigated the combustion and emission characteristics of a CFR engine under different excess air coefficients and ammonia hydrogen ratios. The results showed that efficiency and power were improved due to the possibility of higher CRs. Westlye *et al.* [40] studied the nitrogen-based emissions from hydrogen-rich ammonia-fueled engines using a single-cylinder cycle engine. The ignition timing impacted the temperature progression during the expansion and exhaust strokes, with delayed ignition leading to elevated temperatures and consequently facilitating N_2O production. A higher excess air coefficient accelerated the formation of NH_2 radicals and increased NO_2 levels, thereby fostering N_2O generation. Conversely, N_2O production was lower under the Maximum Brake Torque (MBT) timing condition. Mounaïm-Rousselle *et al.* [35] gave a set of experimental investigations on a four-stroke, four-cylinder direct injection ammonia engine. The findings indicated that the engine could maintain stable operation across a wide range of operating conditions by incorporating less than 10% hydrogen (constituting 10% of the total fuel) into NH_3. Lhuillier *et al.* [33] investigated the combustion and emission characteristics of hydrogen ammonia engines under different equivalence ratios and intake pressures in a four-cylinder four stroke gasoline direct injection SI engine. The results showed that ammonia is a very suitable fuel for ignition engine operation, as high-power output and satisfactory efficiency can be achieved through hydrogen concentration, increased intake pressure, or a combination of both. However, high concentrations of NO_x and unburned NH_3 emissions were also observed, requiring additional mitigation measures. Lhuillier *et al.* [41] conducted experimental research on the performance, combustion characteristics, and pollutant emissions of a new type of SI engine using a premixed ammonia/hydrogen/air mixture. The results showed that when appropriate promotion strategies were adopted, its performance could be comparable to conventional fuel operations. Specifically, adding up to 20% hydrogen by volume to the fuel can improve cycle stability, avoid misfire, and provide the best power output and indicated efficiency close to the stoichiometric ratio. The higher the hydrogen fraction, the lower the efficiency.

5.4.2 Ammonia/LNG blends

Natural gas engine has become popular engine type for power system and electric system. The storage of natural gas is not a problem due to its wide use condition. For ammonia engines, the NH_3/CH_4 SI engine is a solution. Oh *et al.* [43] conducted natural gas-ammonia dual-fuel combustion tests on a six-cylinder SI engine with different air-fuel ratios and ammonia split ratios. The study found that the air-fuel ratio was limited to below 1.5, leading to a deterioration in combustion efficiency and emission characteristics. In terms of nitrogen oxide emissions, with the increase of ammonia content, the combustion phase is delayed, and the combined fuel emissions account for a large proportion of the total nitrogen oxide emissions. The combustion speed is slow, and there is a large amount of minimum ignition energy. Kurien *et al.* [52] investigated the effects of ammonia fuel ratio and engine load on the combustion stability and engine characteristics (combustion and performance) of gaseous (NH_3/CH_4) gasoline engines. The results showed that at higher engine operating loads, ammonia can be maximized to replace methane, thereby achieving the benefits of clean fuel utilization. Lhuillier *et al.* [33] conducted laminar and turbulent flame experiments related to engines under 445 K and 0.54 MPa conditions in a constant volume container to evaluate the combustion behavior of ammonia/hydrogen/air, ammonia/methane/air, and methane/hydrogen/air mixtures in all metal single cylinder SI engines, in order to elucidate the combustion characteristics of ammonia mixtures under engine related turbulent conditions. The results indicate that the acceleration or deceleration effects of hydrogen or methane enrichment observed in SI engines cannot be explained by the measured laminar combustion velocity, and the flame of ammonia mixed fuel is more affected by turbulence.

5.4.3 Ammonia/gasoline blends

Liquid ammonia is another type for ammonia application. The direct mixing of liquid ammonia with other liquid fuel to form a mixed fuel is a simple and effective method. SI gasoline engine is widely used in vehicles, and the gasoline is easily store and ignition. Grannell *et al.* [22] investigated the combustion and detonation characteristics of gasoline/ammonia mixtures at different levels in a CFR engine. The study found that a large portion of gasoline used in SI engines can be replaced by ammonia, requiring approximately 100% gasoline at idle. However, a fuel mixture containing 70% ammonia/30% gasoline on an energy basis can be used under normal suction and large valve opening conditions. Compared to using gasoline alone, ammonia and gasoline can achieve knock free operation at higher CRs and higher loads, and have MBT ignition timing. Haputhanthri *et al.* [42] used a small constant temperature vapor-liquid equilibrium high-pressure pool with ethanol as an emulsifier to enhance the ammonia's solubility in gasoline. The potential for creating ammonia-gasoline mixing liquid fuel was explored in a four-cylinder SI engine. Gasoline containing 30% ethanol can hold 17.35% of ammonia in the liquid phase on a volume basis. Results from engine tests indicate that fuels with high ammonia content lead to an increase in torque and power output, particularly at higher engine speeds.

5.4.4 Ammonia/reformed gas blends

However, the storage of hydrogen, especially liquid hydrogen, is expensive with high security risks which is not very suitable for engine vibration conditions. On-board reforming for hydrogen production is promising. Hydrogen catalyst preparation and reforming gas structure design are the key factors for ammonia on-board reforming to produce hydrogen. Zhang *et al.* [121] extensively studied advanced catalysts for ammonia reforming reactions using DFT calculations and experiments. Non-precious metal nickel (Ni), precious metal ruthenium (Ru), and Ru-Ni composite surfaces were evaluated. NH_3 showed stable adsorption on Ni and Ru surfaces. Ni exhibited superior NH_3 adsorption compared to Ru. Ru-based catalysts were more active at lower temperatures. The Ru-Ni bimetallic catalyst had the lowest energy barrier for N_2 synthesis due to Ni's adsorption capability and Ru's catalytic activity. Experimental data showed 1%Ru–3%Ni/Al_2O_3 had 6.2% higher H_2 production efficiency compared to 1% Ru/Al_2O_3, highlighting its stability. Ryu *et al.* [92] studied the effects of hydrogen combustion generated by ammonia decomposition catalysts on engine performance and emissions in gasoline engines. The results indicate that the combustion of hydrogen generated by ammonia decomposition catalysts leads to improved engine performance and reduced exhaust emissions. The conversion rate of ammonia to hydrogen is affected by the flow rate of ammonia, and the catalyst is very effective at low to medium flow rates, significantly increasing engine power and reducing fuel consumption. After using this catalyst, the emissions of CO, HC, NH_3, and NO_x were significantly reduced. H_2 seems the best choice to improve ammonia combustion speed, however, the storage and transportation of hydrogen, especially liquid hydrogen, are very expensive. The technology of on-board reforming for hydrogen production is a promising solution. Koike *et al.* [94] developed an auto thermal cracker (ATC) of ammonia to produce H_2 which was blended with ammonia to investigated to improve the combustion stability of an ammonia single cylinder engine. Predicted high brake engine torque and thermal efficiency were achieved than diesel engine. This method of ammonia on-board reforming for hydrogen solved the problem of hydrogen storage and transportation. Syngas, also referred to as producer gas or wood gas, is a gaseous fuel containing H_2 that can be generated through the gasification of biomass (Figure 5.4).

5.4.5 Pre-chamber jet ignition ammonia

Pre-chamber turbulent jet ignition (TJI), characterized by high ignition energy, offers several advantages, including enhanced combustion stability, reduced ignition delay, and improved control over combustion timing. This ignition approach

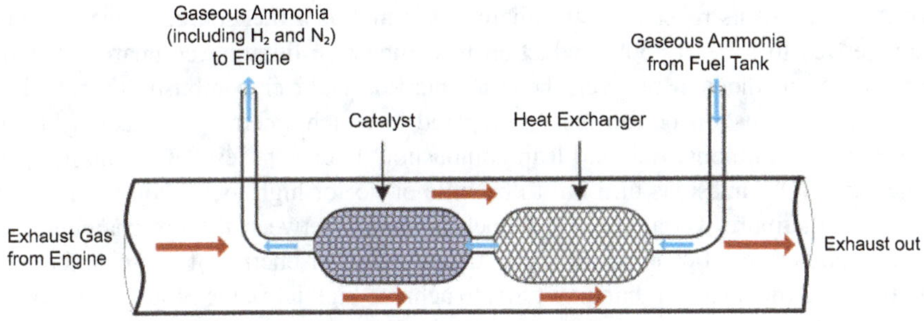

Figure 5.4 Schematic of ammonia dissociation catalyst assembly. Reprinted, with permission, from [23].

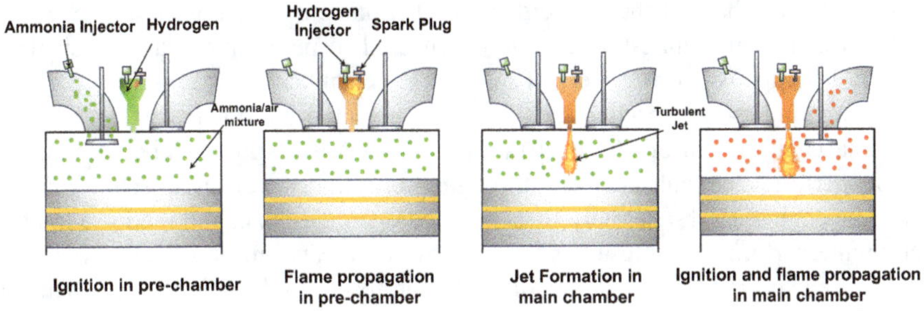

Figure 5.5 Pre-chamber turbulent jet ignition process description. Reprinted, with permission, from [50].

can also promote more efficient combustion of ammonia, leading to decreased NH_3 emissions (Figure 5.5).

The various fuels injected into the pre-chamber which induce different jet ignition ability. These fuels include ammonia, hydrogen, methanol, natural gas, and gasoline. Liu *et al.* [100] introduced a novel gas-assisted pre-chamber with scavenging functionality on a single-cylinder, four-stroke water-cooled engine. They conducted a preliminary study on the controlled turbulent jet ignition (CTJI) of ammonia engine reactivity. The study revealed that the CTJI combustion process comprises two distinct stages: pre-chamber jet combustion and main combustion reactivity. CTJI not only manages the chemical reactivity within the pre-combustion chamber but also has a minor regulatory impact on reactivity within the cylinder. Furthermore, Liu *et al.* outlined two TJI modes using self-designed scavenging pre-combustion chambers for testing ammonia engine performance on an engine test bench. Findings indicated that the TJI system enhances combustion stability, accelerates combustion rates, and reduces NO_x and NH_3 emissions in ammonia engines. Dong *et al.* [122] investigated the pre-chamber jet ignition performance to accelerate the ammonia combustion, employing visual experiments and instantaneous pressure measurements. The findings revealed that the methanol-fueled pre-chamber enhanced ignition and combustion of ammonia compared to ammonia-fueled pre-chamber. The optical experiments revealed that utilizing methanol as a fuel source in the ignition chamber led to a nearly 50% reduction in combustion duration compared to conventional SI methods. Moreover, the ammonia lean and rich combustion resulted in decreased combustion performance compared to stoichiometric combustion [123]. However, the ammonia rich and lean combustion induces to less NO_x emissions in which the NO_x emissions turn out to be little or no for high to 1.2 equivalent ratio [122]. Furthermore, variations in channel diameters between the pre-chamber and main chamber significantly influenced the jet ignition energy of pre-chamber jet ignition. A 2 mm channel diameter turns to achieve high jet flame propagation speed in main chamber. For ammonia lean combustion, a larger channel diameter of 4.5 mm can expand the limit to equivalent ratio of 0.55. In addition, Dong *et al.* [124] investigated the combustion characteristics of ammonia/methanol blends triggered by methanol jet ignition using pre-chamber. Introducing 50% methanol energy into the main chamber enhanced the active thermal environment, resulting in a 43.8% and 51.9% decrease in combustion duration compared to pure ammonia at equivalent ratio of 1.0 and 0.8, respectively. Thus, the pre-chamber TJI seems to accelerate ammonia combustion speed which has the potential to obtain high-efficiency ammonia engines. Liu *et al.* [100] introduces air-assisted pre-chamber design for engine performance enhancement through pre-chamber TJI in a spark-ignition ammonia engine. High-reactivity fuels like hydrogen, gasoline, and methane impact TJI engine performance. Results show improved indicated mean effective pressure and reduced specific fuel consumption with high-reactivity fuels in the pre-chamber under TJI mode. TJI mode optimizes combustion with shortened ignition delay and duration, leading to double-peak phenomena in pressure and heat release rate curves. Optimal engine performance is achieved with hydrogen in the pre-chamber, showing higher heat release rate peaks and decreased ignition delay and combustion duration. TJI mode exhibits lower NO_x and NH_3 emissions, especially with hydrogen. NO and N_2O emissions are below 1000 ppm and 50 ppm, respectively (Figure 5.6).

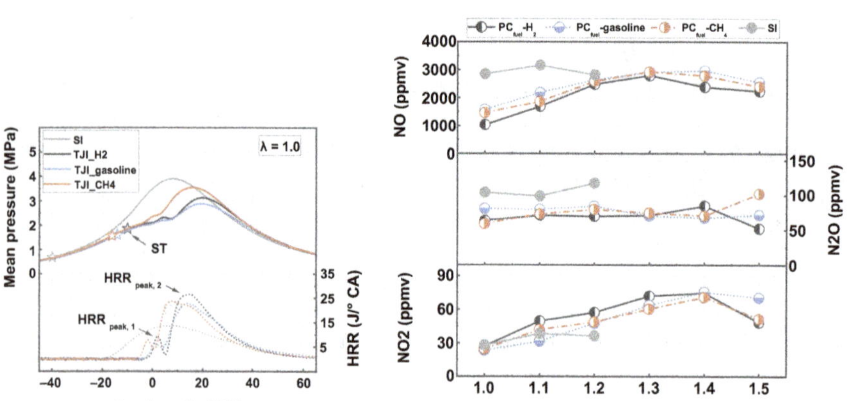

Figure 5.6 Left: Pressure and HRR at MBT with different fuels in the pre-chamber. Right: NO, N_2O, and NO_2 emissions with different fuels in the pre-chamber. Reprinted, with permission, from [86].

5.5 Ammonia-fueled CI engines

One of the main challenges of using ammonia as a fuel for CI engines is its low reactivity and high auto-ignition resistance, which make it difficult to achieve CI under stoichiometric conditions. Achieving CI with pure ammonia necessitates impractically high CRs. Van Blarigan [32] demonstrated that a CR of 40 or higher is needed for ammonia to ignite by compression without preheating the intake air. Even with preheated intake air at 148 °C, a CR of 35 is still necessary [53]. These extreme CRs lead to substantial heat and friction losses, potentially offsetting the efficiency gains from increased compression. To mitigate these issues, dual-fuel combustion strategies have been proposed to lower ammonia's auto-ignition temperature and reduce the necessary CR. For instance, Pearsall and Garabedian [54] reported that using diesel fuel as a pilot injection could reduce the CR to 15.2. Furthermore, Yousefi *et al.* [55] studied the ammonia/diesel combustion in a heavy duty engine and claimed that the NO_x emissions could be reduced by up to 58% with the maximum NH_3 energy fraction of 40%. Tay *et al.* [56] reported that replacing diesel with kerosene as the high-reactivity fuel could advance the start of ignition, resulting in a more complete combustion of NH_3 near the cylinder liner and crevices. Figure 5.7 illustrates the strategies that have been implemented to promote the ammonia CI and the responses of engine performance and emission characteristics. Table 5.3 summarizes the experimental results of the recent studies on ammonia-fueled CI engines, along with their main findings and recommendations.

5.5.1 Ammonia/diesel blends

The use of diesel fuel as an ignition source is crucial due to ammonia's high auto-ignition temperature, which is significantly higher than that of conventional diesel fuel. Ammonia-diesel dual-fuel combustion offers a promising avenue for reducing greenhouse gas emissions in ICEs. Reiter and Kong [13,57] demonstrated the feasibility of this approach, highlighting the importance of careful diesel fuel injection control. By leveraging diesel's energy density, ammonia can be successfully combusted, achieving torque outputs comparable to conventional diesel operation while replacing a significant portion of diesel fuel energy without compromising performance. The research has shown that CO_2 emissions can be significantly reduced due to ammonia's carbon-free combustion. Contrary to initial expectations, NO_x emissions did not necessarily increase with ammonia combustion. Lower combustion temperatures and ammonia's potential participation in NO_x reduction reactions contributed to this. While HC emissions increased with ammonia substitution, this increase was less pronounced at low diesel fueling conditions. However, challenges remain in optimizing ammonia-diesel dual-fuel systems. Operating outside the optimal diesel fuel energy contribution range of 40%–60% can lead to incomplete combustion or lean flammability limits. Additionally, increased CO, HC, and NO_x emissions may occur, especially at higher ammonia concentrations. Exhaust ammonia concentrations can exceed regulatory limits, necessitating after-treatment solutions. In-cylinder pressure analysis has revealed increased ignition delays and reduced peak cylinder pressures with higher ammonia concentrations, due to ammonia's inherent resistance to auto-ignition and lower combustion temperature.

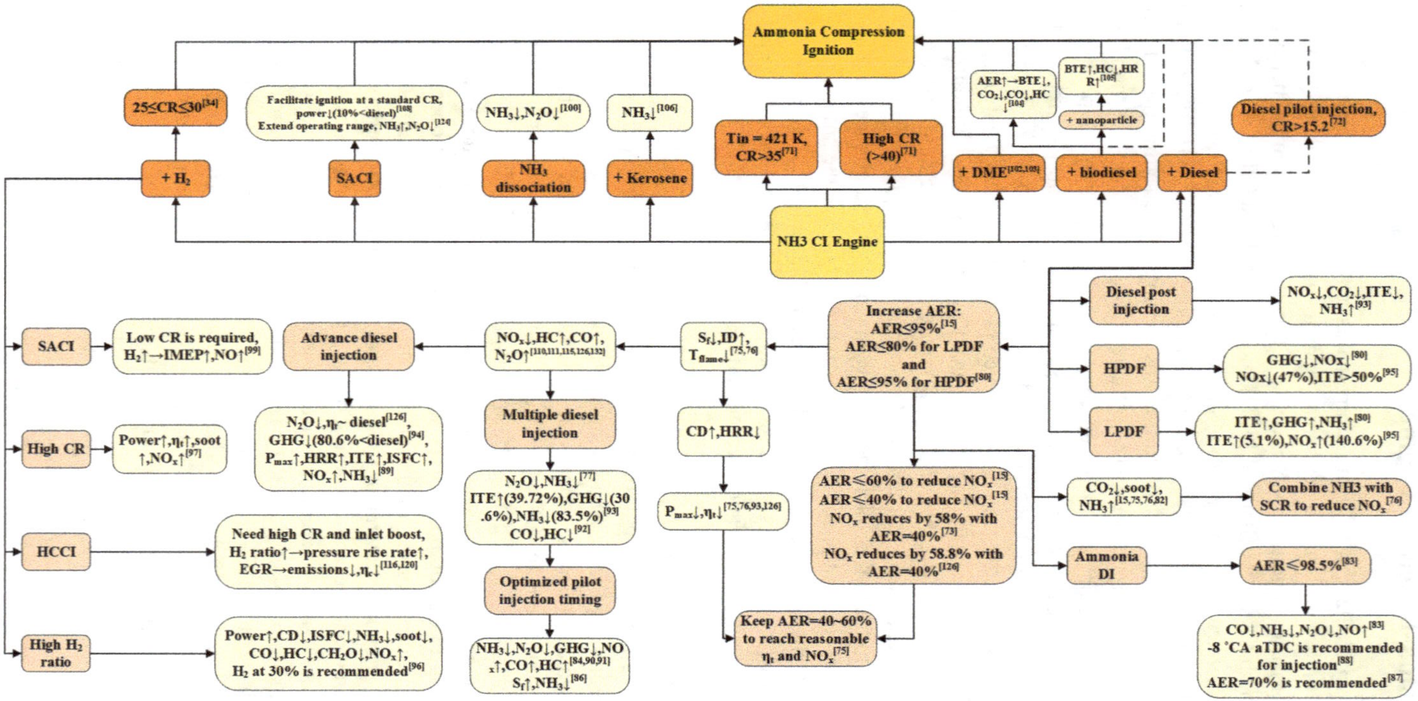

Figure 5.7 Strategies for enhancing ammonia combustion in CI engines and the responses of engine performance and emission characteristics

Table 5.3 Summary of the literature with experimental results related to compression ignition engine with ammonia in the fuel blend

Engine tested	Combustion enhancer	Main conclusions and recommendations from authors	Reference
AVDS-1790 2-cyl diesel engine CR: 18.6 Fuel: H_2, acetylene, butane and NH_3	H_2, acetylene, butane	*Main conclusions*: • Approach 1: adapt CI engine to run with NH_3+ ignition using pilot charge of diesel • Approach 2: conversion of CI to SI • Failed attempt using 30:1 CR with direct liquid injection *Recommendations:* • Use of spark ignition which is located at the center of mass • High CR (between 12:1 and 16:1) • Spherical shape combustion chamber is recommended • Supercharging to improve performance • Material should handle peak pressures up to 1200 psi	Pearsall and Garabidian [103]
CFR	H_2, acetylene, butane . . .	*Main conclusions*: • NH_3 supplied in induction air system • Existing fuel injection used for ignition via pilot charge of diesel • Preferred operation was to convert the CI to a SI engine by taking advantage of the high CR in the CI configuration • direct liquid injection of NH_3 was not considered practical • H_2 is considered best fuel promoter for remote areas. Acetylene considered excessive with 22% by weight • Radio-frequency dissociation of NH_3 not convenient due to energy intense operation • NH_3 was not detrimental to engine components apart from copper, brass, and bronze	Pearsall and Fryer [89]

(Continues)

Table 5.3 (Continued)

Engine tested	Combustion enhancer	Main conclusions and recommendations from authors	Reference
		Recommendations: • Main modifications required: increase CR, increase ignition energy, and locate spark plug near center of combustion chamber • Develop and test spark plugs with long electrode and suitable for high pressure • Design and test compact combustion chambers (close to spherical) • Use exhaust gas to vaporize ammonia • Develop NH_3 cracker to produce 1.5% H_2 by weight (for speed above 3200 rpm)	
CFR	Diesel (DI) / H_2, hydrazine	*Main conclusions*: • Test additives tendency to increase oxidation rate of ammonia and/or ignite the mixture • 45 additives were tested • Tests at different CR 20:1 to 30:1 between 900 and 1800 rpm • Combustion achieved with NH_3/H_2 mixture at CR 25:1 and 30:1 • Combustion achieved with NH_3/acetylene mixture at CR 20:1 and 25:1 • Performance improved with increasing CR and speed • Acetylene was best additive, followed by hydrogen	Gray *et al.* [90]
CFR	SACI	*Main conclusions*: • Attempt to do full CI operation: CR >35 was needed to ignite the ammonia • Add a spark plug to facilitate ignition (SACI operation) and use a standard CR • Use of liquid ammonia injection • Power was 10% lower with liquid NH_3 than with diesel • Performances were strongly depending on spark advance	Starkman *et al.* [104]

(Continues)

4-stroke single-cylinder engine Bore: 90 mm CR: 17.7 Fuel: CH_3OH, CH_4, C_2H_5OH, NH_3 and diesel	Diesel (DI)	*Main conclusions*: • Four alternative fuels tested: methanol ammonia, methane, ethanol. Ammonia less suited to CI operation • Injection method: gaseous ammonia added to intake air • Ignition method: pilot diesel fuel injection • NH_3 operation: air excess below 2 for reasonable efficiency • Problem with ammonia slip in the exhaust, high level of emissions of unburnt NH_3 • Ignition delay time was increased and lower power output due to slow combustion • NO_x emissions were increased *Emissions:* NH_3, HC, CO, NO_2	Bro and Pedersen [105]
John Deere model 4045 turbocharged diesel engine four cylinder Bore: 106 mm CR: 17 Fuel: NH_3 and diesel	Diesel (DI)	*Main conclusions*: • Study various diesel/NH_3 blend at different speeds and loads (experiment + chemical kinetic) • Setup: ammonia introduced in intake manifold, diesel direct injection in the cylinder for ignition • Use of a turbo • Modification to intake for ammonia, no other modifications • 95% of diesel could be substituted to NH_3 (energy content) • With higher NH_3 content, higher rated power can be reached depending on load • For NH_3 <60%, lower NO_x, possibly due to lower temperature • CO_2 reduction almost proportional to NH_3 ratio increase • Higher HC if high diesel ratio in the blend, possibly due to low temperature *Emissions:* CO_2, HC, NO_x	Reiter and Kong [13]
John Deere model 4045 turbocharged diesel engine four cylinder.	NH_3/DME premixed (DI)	*Main conclusions*: • Study 1: maintain constant power via various diesel/NH_3 blend • Study 2: Variable power output by changing NH_3 and use small amount of diesel	Reiter and Kong [57]

(Continues)

Table 5.3 (Continued)

Engine tested	Combustion enhancer	Main conclusions and recommendations from authors	Reference
Bore: 106 mm CR: 17 Fuel: NH_3 and diesel		• Setup: ammonia introduced in intake manifold, diesel direct injection in the cylinder for ignition *Main conclusions*: • Use of a turbo • Modification to intake for ammonia, no other modifications • Favored mixture to maintain fuel efficiency: 40%–60% energy from diesel and 60%–40% from NH_3 • CO and HC emissions higher • NO_x lower if NH_3 less than 40% (mixture energy content), otherwise, NO_x increased • Soot reduced with NH_3 increase (lower carbon content) • Ammonia slip in exhaust: 1000–3000 ppm • Poor fuel efficiency at variable power operation • NH_3 content increase: Ignition delay increase and peak cylinder pressure decrease *Recommendation:* • Optimize combustion via new strategies for ammonia/diesel injection *Emissions:* CO, HC, CO_2, NO, Soot, NH_3	
CFR Single cylinder Bore: 98.4 mm CR: 15.5 Fuel: NH_3, H_2, and diesel	Diesel	*Main conclusions*: • NH_3 injection in air-intake manifold • Diesel direct injection in cylinder for ignition • 3% (by mass) of NH_3 in air intake: 15% decrease in diesel fuel consumption and CO_2 emissions • Tested also with dissociated NH_3 and pure H_2 and best performance were for pure H_2 • At high loads, NH_3 better for stability and brake thermal efficiency than dissociated NH_3 • Lower combustion temperature with NH_3: higher HC emission • The key advantages of using dissociated NH_3 are the lower NH_3 and N_2O emissions compared to pure NH_3 addition	Gill *et al.* [80]

(Continues)

		Recommendation: ● H_2 was most promising in modes tested, dissociated NH_3 is recommended *Emissions:* CO_2, CO, HC, NO_x	
Yanmar L70V single cylinder Bore: 78 mm CR: 20 Fuel: NH_3-DME mixtures	NH_3/DME premixed (DI)	*Main conclusions*: ● Liquid ammonia and DME blended using high-pressure mixing device. DME ignite the mixture ● Modified injection system: no fuel return (blend cannot vaporize) ● Engine load limited due to NH_3 and ignition delays are longer ● Emissions: increase in CO and HC (due to lower pressure and temperature during combustion) ● Higher NO_x measured (fuel NO_x). NH_3 in the exhaust (~hundreds ppm). Soot is very low ● Higher injection pressure (by 3.0 MPa): improved performance and higher NH_3 content possible. Peak injection pressure used is 20.0 MPa ● Double injection tested with different results. Possibly not enough penetration to mix well with air *Recommendation:* ● Increase injection pressure to improve air-fuel mixing and thus combustion *Emissions:* NO_x, HC, NH_3, CO, Soot	Gross and Kong [82]
Yanmar L70V single cylinder Bore: 78 mm CR: 20 Fuel: NH_3-DME mixtures	NH_3/DME premixed (DI)	*Main conclusions*: ● Study of combustion and emissions for three blends of DME/NH_3 ● DME/NH_3 mixture directly injected in cylinder ● Engine performance decreases when NH_3 concentration increases ● For blend 40%DME and 60%NH_3 (by weight): strong variability between each cycle (but less variation with increasing load) and engine showed combustion behavior like HCCI because of the advanced timing. Higher CO and HC emissions but very low soot ● Injection timing to be advanced when increasing ammonia concentration ● NO_x increases with ammonia concentration increase	Ryu *et al.* [83]

(Continues)

Table 5.3 (Continued)

Engine tested	Combustion enhancer	Main conclusions and recommendations from authors	Reference
		Recommendation: • Exhaust after-treatment will be needed *Emissions:* CO, CO_2, NO_x, HC, O_2, NH_3	
Single-cylinder four-stroke NH_3: PFI-gaseous Bore: 112 mm CR: 18.5 Fuel: NH_3 and diesel	Diesel pilot	*Main conclusions*: • NH_3 gas introduced in the air intake via gas injector • Direct in-cylinder diesel injection • Increase NH_3 content: decrease in in-cylinder compression and maximum pressure, ignition timing was delayed • CO_2 decreased proportionally to NH_3 content increase • H_2O and N_2O increased and NO stayed at similar level • NH_3 slip in the exhaust can be combined with SCR to reduce NO. N_2O cannot be reduced in the SCR *Recommendations:* • Multi-fuel injection and SCR • Control NH_3 in the exhaust to react with NO and thus reduce pollutant emissions *Emissions*: CO_2, H_2O, CO, N_2O, NO_x	Niki *et al.* [106]
PSA DW10 single cylinder Bore: 88 mm CR: 15–25 Fuel: NH_3, H_2, CH_4, and CH_3OH	H_2 HCCI	*Main conclusions*: • HCCI combustion engine with various blends of H_2/NH_3 (H_2 needed for stable combustion) • Stable combustion attained with: 70% NH_3 (vol.) + increase of intake pressure to 0.15 MPa and intake temperature at 473 K + equivalence ratio @ 0.28. Higher CR and pressure boost needed to maximize efficiency • Pure H_2 showed very high pressure rise rate (ringing)—problem that could be solved via variable CR engine	Pochet *et al.* [107]

(Continues)

		• Pure H_2, NO_x <6 ppm., NH_3/H_2 blends: NO_x were recorded between 1000 and 3500 ppm • EGR efficient in reducing emissions however degradation of combustion efficiency *Recommendations:* • Combustion should be kept above 1400 K to reduce N_2O emissions. In-cylinder T to be kept above 1300 K to have satisfying combustion efficiency • EGR promising to reduce NO_x but SCR needed for N_2O (unaffected by EGR) • Study exhaust emissions with respect to H_2, NH_3, and NO_x *To be validated:* • Ammonia and methane are much more resistant to auto-ignition than methanol and hydrogen • Higher intake pressure increases IMEP and required intake temperature (hence further increasing IMEP) and reduces ringing intensity • Increasing the compression ratio implies to reduce the equivalence ratio of hydrogen, which is prone to ringing • CA10 control and combustion delaying for combustion intensity reduction are achievable through intake temperature reduction • Addition of hydrogen and methanol, the required intake temperature for the auto-ignition of the blend is drastically decreased and the IMEP increases by 25%	
Single-cylinder four-stroke NH_3: PFI-gaseous Bore: 112 mm CR: 18.5 Fuel: NH_3 and diesel	Diesel pilot	*Main conclusions*: • Engine with fumigated NH_3 in air intake: Develop combustion strategies to reduce NH_3 slip and N_2O emissions • Higher combustion temperature: lower N_2O • Higher torque: less N_2O produced than for low torque • Multiple diesel injection promising *Recommendations:* • Higher combustion temperature to limit N_2O and operate with higher NH_3 content to reduce PM emissions • SCR to be used to react with NH_3 slip and NO_x *Emissions:* NO_x, N_2O, CO, H_2O, CO_2, NH_3, HC	Niki *et al.* [108]

(Continues)

Table 5.3 (Continued)

Engine tested	Combustion enhancer	Main conclusions and recommendations from authors	Reference
Palmera PAHP170F single cylinder bore: 70 mm CR: 20 Fuel: NH_4OH and diesel	Diesel (DI)	*Main conclusions*: • Study ammonia fumigation effect in air intake via a carburetor for different ratios • Use of 25% NH_3 solution (with 75% H_2O) • BSFC decreased by 7.28% @5.48% ammonia (2600 rpm) • CO_2 decreases but CO, HC and NO_x increases • Best performance @2600 rpm (which is the optimal operation range of that engine) • Deterioration of parts of the carburetor and residues due to NH_3 solution *Emissions*: CO_2, CO, HC, NO_x	Şahin *et al.* [109]
Single cylinder four-stroke NH_3: PFI-gaseous Bore: 112 mm CR: 18.5 Fuel: NH_3 and diesel Pilot and post	Diesel pilot and post-injection	*Main conclusions*: • Experimental investigation of NH_3 combustion in diesel engine (NH_3 fumigated in air intake) • Diesel pilot injection and post-injection tested to see effect on reducing NH_3 and N_2O emissions • Multiple diesel injections promising to reduce N_2O and NH_3 • N_2O and NH_3 emissions decreases: (1) if diesel fuel injection timing is advanced. (2) for higher torques • Diesel pilot injection: decrease NH_3 slip but increase N_2O • Diesel post-injection: decrease NH_3 slip but slight increase of N_2O *Emissions:* NH_3, N_2O, H_2O, CO_2, CO, NO_x, GHG	Niki *et al.* [59]
Yanmar L100 V single cylinder Bore: 86 mm	H_2 HCCI	*Main conclusions*: • Study the impacts of different NH_3/H_2 ratio, different equivalence ratio, EGR and increase of intake temperature with HCCI engine with CR of 22:1 • Fuel NO_x linearly increase with ammonia concentration (max 6000 ppm)	Pochet *et al.* [110]

(Continues)

CR: 21.9 Fuel: NH_3 and H_2 Enhanced: HCCI		• EGR reduced by three NO_x emissions but negative consequence on N_2O and unburned fuel • N_2O forms below 1400 K, need combustion temperature above 1800 K for NH_3 complete combustion *Recommendations:* • Fuel NO_x primary reduction methods should be investigated and after-treatment needed • Boosted pressure combined with stroke-to-bore ratio maximization should be investigated (to optimize EGR) *Emissions:* NH_3, NO_x	
Andoria Type S320 single cylinder Bore: 120 mm CR: 17 Fuel: NH_4OH and diesel	Diesel/water ammonia solution	*Main conclusions*: • Dual fuel operation: diesel and water ammonia solution (WAS) with up to 17% WAS content. WAS made of 25% NH_3, 75% water • Study three different loads (35%, 60%, and 100%) and variation of WAS at 60% load • Operation with dual fuel: Increase of ignition delay and duration of combustion as well as HRR • Load increase: no impact on ignition delay by increase of combustion duration • At 60% load, increasing WAS causes ignition delay and combustion duration to increase • Decrease of engine stability with WAS, high pressure rise with respect to full diesel • NO_x emissions decreases, increase in CO and HC, same CO_2 • Engine efficiency increase with WAS fraction *Emissions:* NO_x, HC, CO, CO_2	Pyrc *et al.* [111]
AVL type 520 single-cylinder four-stroke NH_3: PFI-gaseous Bore: 112 mm	Diesel (RCCI)	*Main conclusions*: • Reduce NH_3 and N_2O emissions using reactivity-controlled compression ignition (RCCI) combustion and partially premixed compression ignition (PPCI) combustion (dual fuel: NH_3/diesel)	Niki [71]

(Continues)

Table 5.3 (Continued)

Engine tested	Combustion enhancer	Main conclusions and recommendations from authors	Reference
CR: 18.5 Fuel: NH_3 and diesel RCCI		• Diesel pilot injection timing advance play essential role to reduce NH_3 and N_2O emissions but lower brake thermal efficiency *Main conclusions*: • Combustion phase was delayed when early diesel pilot injection was used. Mixture believed to be more homogeneous for early injection • Ignition using diesel generates OH which decompose NH_3 • N_2O produced during NH_3 oxidation and can de decomposed at the end of NH_3 combustion when temperature is still high • RCCI: Emission of HC, NO_x and CO increased and NH_3, N_2O decreased *Recommendations:* • Investigate split diesel pilot injection to reduce CO and HC • SCR to lower emissions should be added *Emissions:* NH_3, N_2O, CO_2, CO, HC, NO_x	
AVL type 520 single-cylinder four-stroke NH_3: PFI-gaseous Bore: 112 mm CR: 18.5 Fuel: NH_3 and diesel RCCI	Diesel (RCCI)	*Main conclusions*: • Early diesel pilot injection increase CO and HC but decrease NH_3 and N_2O in the exhaust gas • RCCI combustion can be manipulated to mitigate emissions • Study effect of split injection on emission and combustion behavior • Overall GHG emissions were reduced, ammonia assist ratio of 46% reduced GHG emission by 20% • Split injection decreased CO and HC emission possibly because the diesel fuel impingement to the wall was reduced *Recommendations:* • Study N_2O formation and decomposition reaction kinetic • Optimization of the split injection required *Emissions:* NH_3, N_2O, CO_2, CO, HC, NO_x, GHG	Niki [112]

(Continues)

PSA DV6F single cylinder	SACI	*Main conclusions*: ● Study pure ammonia/air premixed in a SACI engine ● Pure ammonia adapted to be used over broad range of operation ● SACI is promising even at low load and at different engine speeds ● Early flame propagation duration is reduced thanks to increase in the CR due to SACI ● SACI allows increasing operation range at low loads with respect to SI engine ● Engine speed limit is increased in SACI mode with respect to SI operation ● Higher NH_3 emission in SACI due to higher CR and crevice mechanism ● N_2O emission <50 ppm *Recommendations:* ● Further study using CFD modeling ● Ignition method to be optimized and tested at higher speed and loads	Mounaïm-Rousselle *et al.* [113]
Ford Duratorq four-stroke single cylinder Bore: 86 mm CR: 18.3 Fuel: NH_4OH and diesel	Diesel/ammonium hydroxide (NH_4OH)	*Main conclusions*: ● Study effect of ammonium hydroxide (NH_4OH) solution direct injection in diesel engine ● Increased ignition delay, increased peak temperature but low combustion quality (incomplete combustion) ● Increase in NH_4OH content: CO emissions increased proportionally, increase of cycle to cycle variability, increase in particulates mass ● Higher peak in-cylinder pressure shows higher NO_x emissions ● Challenge when using aqueous ammonia solution *Emissions:* CO, NO_x, PM	Frost *et al.* [114]

(Continues)

Table 5.3 *(Continued)*

Engine tested	Combustion enhancer	Main conclusions and recommendations from authors	Reference
Caterpillar 3401 single-cylinder four stroke heavy-duty (HD) diesel engine Bore: 137.2 mm Fuel: NH_3 and diesel	Diesel	*Main conclusions*: ● Study effect of NH_3 content and diesel injection on performance for dual fuel operation ● Decrease of thermal efficiency due to low ammonia flame speed ● Increase of NH_3 up to 40% decreases NO_x but N_2O increases ● Advancing start of diesel injection: decrease N_2O emissions and similar thermal efficiency as for 100% diesel ● Emissions reduction achieved for NH_3/diesel: 12% lower with respect to 100% diesel ● NH_3 slip observed, which reduced NO ● Increasing ammonia energy fraction from 0% to 40% reduced the NO_x emissions by 58.8%. This is attributed to the thermal DeNO_x process where excess NH_3 inside the combustion chamber played a role for NO removal rather than for NO formation ● Increasing ammonia energy fraction increased the N_2O emissions, which offsets the benefit of lower CO_2 emissions of ADDF combustion and resulted in a higher GHG emissions compared to diesel-only mode ● Advanced SODI reduced the N_2O and GHG emissions in ADDF combustion mode ● The thermal efficiency of ADDF combustion mode at the best point of GHG emissions (i.e. ITE = 37.85%) was slightly lower than that of diesel-only mode (ITE = 38.53%). But NO_x and CO emissions of ADDF combustion were about 10% and 20%. *Emissions:* NH_3, HC, CO, NO, CO_2, N_2O, GHG	Yousefi *et al.* [115]
Caterpillar 3401 Single cylinder Bore: 137.2 mm CR: 16.25 Fuel: NH_3 and diesel Pilot		*Main conclusions*: ● Higher ITE of split diesel injection than single injection, up to 39.72% (single: 38.36%; diesel: 38.62%) ● GHG emissions reduced by 30.6% (split: 478.1 g/kwh; single: 606.2 g/kwh; diesel: 688.58 g/kwh), NH_3 reduced by 83.5% compared with single diesel injection ● Two best point: GHG reduced by 23.7% and ITE increased by 2%; GHG reduced by 30.6%, but ITE reduced by 2.2%, NO_x increased by 52.4%. both NH_3 below 900 ppm (single: 4445 ppm) *Emissions:* CO_2, N_2O, GHG, NH_3, NO_x, HC, CO	Yousefi *et al.* [60]

(Continues)

Rapid-compression-expansion-machine	Diesel	*Main conclusions*: • Study of high-pressure dual fuel direct injection of NH_3 to reduce NH_3 slip and NO_x • Focus on diesel piloted liquid ammonia sprays using a rapid-compression-expansion-machine • Study various strategies: pray interaction, charge conditions, diesel pilot quantity, injection duration • To avoid misfire: inject diesel first (otherwise, pilot mixture formation deteriorate) • Strong influence of wall quenching effect is suggested because of burnout rate correlated with ignition delay • Combustion process strongly influenced by injection duration and pilot fuel amount • Stable combustion reached for pilot with LHV fraction above 3.2% • The smaller the pilot amount, the longer the ignition delay • Short pilot injection duration decrease ignition delay	Scharl and Sattelmayer [116]
DV6F PSA single cylinder Bore: 75 mm CR: 14–17 Fuel: NH_3 and H_2		*Main conclusions*: • The improvement in early flame propagation in the SACI mode plays an important role in the combustion process and stability, extending the operating range at low loads for SACI compared to SI conditions • Unburned NH_3 emissions are higher in the SACI mode, increasing with the CR due to the increase in ammonia trapped in the crevices • Even without modifying the combustion chamber design, the SACI combustion mode is beneficial for NH_3 combustion due to the increase in the compression ratio which makes it possible to reach thermodynamic conditions that facilitate the combustion of NH_3 *Emissions:* H_2, NH_3, NO_x, N_2O	Mounaïm-Rousselle *et al.* [13]

(Continues)

Table 5.3 (Continued)

Engine tested	Combustion enhancer	Main conclusions and recommendations from authors	Reference
Lifan C186F single cylinder Bore: 86 mm CR: 16.5 Fuel: NH_3 and biodiesel		*Main conclusions*: ● Ammonia-/biodiesel-fueled engine has a lower BTE compared with pure biodiesel at the same operating point. The BTE decreased from 31.8 for pure biodiesel to 29.4 for the highest ammonia input energy ratio at full load ● Increasing ammonia input energy contribution (AEC) decreases the exhaust gas temperature. the exhaust gas temperature was reduced by 100 K when the AEC was 69.4% ● As more ammonia is introduced into the intake manifold, the combustion mode changes from diffusion to premixed combustion ● As more biodiesel was replaced by ammonia, CO_2, CO, and HC emissions decreased significantly *Emissions:* CO_2, CO, NO, HC, PM	Nadimi *et al.* [84]
MAN D2840LE V10 CR: 13.5		*Main conclusions*: ● A second ammonia injection along the expansion stroke is proposed to reduce NO_x ● Higher ammonia, ITE decreased, CO_2 emissions decreased ● CO, HC, and NO_x emissions reduced at high and low ammonia ratio *Emissions:* CO_2, NH_3, NO_x, CO, HC, N_2O	Rodríguez *et al.* [73]
Bore: 95 mm CR: 17.5 Fuel: NH_3 and diesel		*Main conclusions*: ● The maximum ammonia ratio is 80% (by energy) for LPDF, but 97% for HPDF ● The HPDF mode has similar performance with diesel mode and lower GHG and NO_x ● Higher ITE for LPDF mode due to reduced cooling loss, but higher GHG and NH_3 *Emissions:* NH_3, N_2O, NO_x, CO_2, GHG	Li *et al.* [61]

(Continues)

A two-stroke natural gas dual-fuel marine engine RTFlex50DF Bore: 500 mm CR: 12 Fuel: NH_3 (liquid and gaseous) and diesel	*Main conclusions*: • A skeletal mechanism (104 species and 573 reactions) was innovatively produced • Ammonia is a proper candidate fuel applied for two-stroke Marine engines (liquid and gaseous) • The liquid case has a higher peak pressure and faster combustion rate because of the higher swirl ratio • Higher power and ITE for liquid ammonia due to higher swirl ratio, and lower NH_3 • NO_x emissions are similar at liquid and gaseous ammonia mode • Lower temperature of lean combustion can promote the NO_x emissions level to meet Tier II limits *Emissions:* NH_3, NO_x, CO_2	Liu *et al.* [117]
CFR single cylinder Bore: 75 mm CR: 18 Fuel: NH_3 and H_2	*Main conclusions*: • Experimental investigation of spark-assisted compression-ignition with ammonia-hydrogen blends • NH_3 and H_2 port-injected at stoichiometric combustion, and blending H_2 2.5% and 5% by volume • H_2 addition benefit the IMEP of the engine • NO emission increased with the increase of H_2 addition and intake temperature *Emissions:* NH_3, NO, N_2O	Reggeti *et al.* [79]
TBD620 single-cylinder diesel engine Bore: 170 mm CR: 13.5 Fuel: NH_3, H_2, and diesel	*Main conclusions*: • With the increase in hydrogen blending ratio, the diesel engine power increases, EISFC decreases, and the combustion duration shortens. When Rh $\geq$ 30%, the ammonia escape phenomenon is eliminated • Higher H_2, NO_x increased, but soot, CO, HC, and CH_2O decreased • The best hydrogen blending ratio is 30% *Emissions:* NH_3, H_2, CO_2, CO, CH_2O, NO_x, N_2O, NO_2, Soot, HC	Wang *et al.* [76]

(Continues)

Table 5.3 (Continued)

Engine tested	Combustion enhancer	Main conclusions and recommendations from authors	Reference
Lifan C186F single cyl Bore: 86 mm CR: 16.5 Fuel: NH_3 and diesel		*Main conclusions*: • The maximum ammonia energy shares (AES) is 84.2% • With AES increasing, CO and CO_2 emissions decreased, but NO, NO_2, and N_2O emissions increased. A significant amount of unburned ammonia was measured that further investigation is need to reduced ammonia emission • The GHG emissions decreased when more than 35.9% of diesel was replaced *Emissions:* CO_2, H_2O, CO, NH_3, NO_x, N_2O, PM, GHG	Nadimi *et al.* [62]
BUKH DV 24 ME, a two-cylinder compression-ignited diesel engine Bore:85 mm CR: 16.4 Fuel: NH_3 and diesel Enhanced: GDI injection		*Main conclusions*: • The engine was run using up to 98.5% energy from ammonia with a GDI injector • For ammonia-ignition an easily ignitable ammonia–air mixture is preferable compared to a strong pilot injection • Increased ammonia energy led to reduced ammonia-slip and N_2O-emissions and increased NO-emissions, and a lower CO-formation showed better pilot fuel oxidation *Emissions:* NH_3, NO, CO, N_2O	Førby *et al.* [63]
EX340EF commercial two-stroke marine diesel engine Bore: 340 mm CR: 20 Fuel: NH_3 and diesel		*Main conclusions*: • NO_x emissions decrease if the ASR does not exceed 40% • N_2O is usually generated at the early stage of combustion due to the lower in-cylinder temperature • Pilot-fuel injection timing plays a vital role in controlling the ignition behavior and even the combustion regimes in DF combustion mode • Excessive advance in SOI show an HCCI-like combustion mode • Optimizing the pilot-fuel injection timing can reduce unburned ammonia emissions • Advanced pilot-fuel injection timing, NO_x emissions increased, N_2O decreased *Emissions:* NH_3, NO_x, CO_2, N_2O, GHG	Zhu *et al.* [64]

(Continues)

Single cylinder Bore: 137.2 mm CR:16.25 Fuel: NH_3 and diesel	*Main conclusions*: ● Ammonia energy fractions ranging from 40% to 90%, diesel injection timings from −80 to −15 CAD ● The diesel ignition region which affected the flame propagation process, unburned ammonia, and N_2O generation was changed depending on diesel injection timings ● Unburned ammonia was distributed in the periphery of the liner at the injection timing of the piston bowl targeting, and unburned ammonia was distributed in the center of the cylinder at the injection timing of the piston bowl-liner or liner targeting ● N_2O produced at ammonia flame front ● Increasing ammonia and advancing diesel injection timing reduced GHG by 80.6% compared with diesel *Emissions:* CO_2, NH_3, NO, N_2O, GHG	Shin and Park [74]
Wärtsilä medium speed, six-cylinder, four-stroke marine engine Bore: 320 mm CR:16 Fuel: NH_3 and diesel	*Main conclusions*: ● A LNG-diesel engine was modified to ammonia-diesel engine ● Diesel energy ratio increased to work (LNG: 8.5%). The GHG emissions reduced by 70%, but CO_2 increased ● N_2O emission was attributed to the slow premixed ammonia/air flame propagation and near-wall flame quenching ● NO emissions were identified as fuel-NO_x from premixed ammonia oxidation and thermal NO_x in the diesel flame region ● GHG reduced by 60%, CO by 45% and CO_2 by 89% compared with LNG (more pilot fuel injector holes, advanced diesel injection timing, higher premixed ammonia/air equivalence ratio > 0.7 and higher ammonia energy share ratio) *Emissions:* CO, NH_3, NO, N_2O, CH_4	Xu *et al.* [65]

(Continues)

Table 5.3 (Continued)

Engine tested	Combustion enhancer	Main conclusions and recommendations from authors	Reference
Kirloskar TV1 single cylinder bore: 87.5 CR: 17.5 Fuel: NH_3 and biodiesel Enhanced: TiO_2—nano-additives		*Main conclusions*: ● Ammonia blends in the diesel engine reduced the BTE irrespective of the engine loads. As the concentration of ammonia increased, there was a slight increase in the BTE. The maximum BTE has been reported for the blend with nanoparticles due to superior cetane number and latent heat of evaporation ● CO, CO_2, and HC, all the blends reduced the formation of carbon-based pollutants due to the increased oxygen and decreased carbon in the fuel ● By adding the nanoparticles, the rate of combustion increased that led to significant reduction in the formation of pollutants ● The addition of ammonia was not sustainable owing to the reduced BTE and increased NO_x emission. This drawback can be overcome by adding some of the oxygenated additives along with diesel *Emissions:* HC, CO_2, CO, NO_x	Wu *et al.* [85]
Kirloskar diesel engine 4-stroke, single-cyl Bore: 80 mm CR: 15.5 Fuel: NH_4OH and diesel		*Main conclusions*: ● Water ammonia solutions (25% NH_3) ● Pressure and heat release rates dropped when aqueous ammonia is added ● Aqueous ammonia in the ratios of 40%, 50%, and 60%, the BSFC increases by 7.15%, 10.4%, and 15.38% ● NO_x emissions reduced (25% for 40% NH_4OH, 52.32% for 50% NH_4OH, and 61.75% for 60% NH_4OH) ● Soot concentrated is reduced by 43.4%, 51%, and 49% when (40%, 50%, and 60%) of NH_4OH *Emissions:* NO_x, Soot	Al-Dawody *et al.* [118]

(Continues)

Two-stroke single-cylinder engine Bore: 150 mm CR: 13.8 Fuel: NH_3 and diesel	*Main conclusions*: ● The engine combustion phase on dual fuel mode is very sensitive to the injection timing of diesel ● Soot and CO emissions are restrained apparently because the liquid phase ammonia direct injection can enhance the interaction of spray plumes inside the current combustion chamber, which promotes the atomization and evaporation of diesel ● The total NO_x emission (thermal type NO_x and fuel type NO_x) is deteriorated compared with the pure diesel mode ● Injection timing of −8°CA ATDC is the best option for both diesel and liquid phase ammonia to balance the engine ITE and emission performance in the current study *Emissions:* HC, CO, Soot, NO_x	Zhang *et al.* [68]
TBD620 single cyl Bore: 170 mm CR: 13.5 Fuel: NH_3 and H_2	*Main conclusions*: ● After delaying the injection timing of ammonia/hydrogen mixture fuel, the power performance and economy of the engine are reduced, the ignition delay period is shortened, HC, CH_2O, CO and soot emissions are gradually increased but at a very low level, and NO_x emissions are significantly improved ● After the compression ratio is increased, the engine's power performance and economy are greatly improved, HC, CH_2O, CO, and soot emissions are slightly higher, and NO_x and N_2O emissions are gradually increased ● Under any combination of compression ratio coupling injection timing, the engine NO_x emission can meet the Tier II emission standard. When the injection timing of ammonia/hydrogen mixed fuel is 12 °CA BTDC and the compression ratio is 13.5, the NO_x emission of this operation combination of ammonia hydrogen dual fuel engine can meet the Tier III emission standard *Emissions:* NH_3, H_2, HC, CH_2O, CO, Soot, NO_x, N_2O	Wang *et al.* [77]
Inline 2 cylinders, Turbocharged, CRDI Diesel Engine Bore: 83 mm CR: 16.5	*Main conclusions*: ● An emulsification technique and ammonia solution are combined ● The ammonia solution concentration in emulsified diesel-ammonia solution (EDAS) was fixed as 10% by volume (EDAS10) with 2% v/v of surfactant to avoid excessive ammonia slip, corrosive effects, and oxides of nitrogen (NO_x) formation	Subramani *et al.* [119]

(Continues)

Table 5.3 (Continued)

Engine tested	Combustion enhancer	Main conclusions and recommendations from authors	Reference
Fuel: emulsified diesel-ammonia solution and 2% v/v of surfactant		*Main conclusions*: • EDAS10 blend derived more work output for the same total energy input at similar combustion conditions compared with diesel • HC, CO, and smoke emissions reduced by 50%, 70%, and 14% of EDAS10 blend operation compared to neat diesel operation, but NO and CO_2 increased • Corrosion effects of EDAS10 (terminal leads of the submersible fuel pump, the solenoid of the injector) *Emissions:* CO, CO_2, HC, NO, Soot	
6 cyl. Bore: 116 mm CR: 18.5 Fuel: NH_3 and diesel Pilot		*Main conclusions*: • Higher AER, lower ITE, NO emission decreased • The production and reaction of N_2O were mainly affected by temperature • N_2O will be reduced to N_2 under high temperature conditions • The pre-injected diesel can significantly improve the activity of the mixture, accelerate the combustion speed, reduce the emission of unburned ammonia and the incomplete combustion loss • By optimizing the injection timing, the ITEg of 49.18% can be achieved under the AER of 50%, and the GHG can be reduced by 14.2% *Emissions:* NH_3, NO_x, HC, CO, CO_2, GHG	Jin *et al.* [66]
2 stroke Bore: 520 mm CR: 21.8 Fuel: NH_3 and diesel		*Main conclusions*: • A new ammonia-diesel stratified injection (ADSJ) mode was explored in detail • Advanced injection timing can increase pressure while increased the diesel energy fraction can increase the peak HRR, both increase ITE and ISFC, increase NO_x, but reduce NH_3 • The diesel energy fraction of 5% shows a shorter IDT than 1% and 10% • The best injection timing and diesel energy fraction is 1°CA ATDC and 1%. ISFC and GHG reduced by 4% and 94%, combustion efficiency reached 99%, NH_3 below 150 ppm *Emissions:* NO_x, NH_3, GHG	Liu *et al.* [69]

(Continues)

4 cyl. Bore: 114 mm CR: 18 Fuel: NH_3 and diesel Pilot	*Main conclusions*: ● Pilot-injection energy ratio (PER) varied from 0% to 100% ● Pilot diesel injection reduced NH_3 emissions from 8000 ppm to 1000 ppm ● Higher PER, lower N_2O emissions ● Pilot-injection strategy could keep a high indicated thermal efficiency of about 45% ● Higher NO_x, CO, and HC emissions were disadvantages in the pilot-injection strategy *Emissions:* NH_3, NO_x, CO, HC, N_2O, CO_2, PM, GHG	Mi *et al.* [70]
TBD620 single-cylinder diesel engine Bore: 170 mm CR: 13.5 Fuel: NH_3 and diesel	*Main conclusions*: ● Higher NH_3 and N_2O at LPDF ● Engine performance and emissions are better at HPDF ● Ra = 70% can be the best ammonia blending ratio for in-cylinder direct injection *Emissions:* NH_3, CO, CH_2O, CO_2, HC, NO_x, N_2O, Soot	Wang *et al.* [67]
Marine engine Bore: 340 mm CR: 19.8 Fuel: NH_3 (Liquid and gaseous) and diesel	*Main conclusions*: ● ITEs of LPDF and HPDF are above 50% ● LPDF and HPDF both significantly reduced GHG emissions ● Effects of injector arrangement, pilot diesel energy fraction, fuel injection timing, excess air ratio, and engine load are studied ● Advanced diesel injection timing and lower excess air ratio, NH_3, and N_2O reduced ● At best LPDF point, ITE improved by 5.1%, but NO_x increased by 140.6% compared with diesel mode ● At best HPDF point, NO_x reduced by 47% and ITE is similar ● Both the best point, GHG reduced by 97% and NH_3 is small ● Higher ITE for LPDF mode due to reduced transfer loss *Emissions:* NH_3, NO_x, CO_2, N_2O, GHG	Zhou *et al.* [75]

(Continues)

Table 5.3 (Continued)

Engine tested	Combustion enhancer	Main conclusions and recommendations from authors	Reference
Marine engine Bore: 500 mm		*Main conclusions*: ● Reactivity Stratification Assisted Jet Ignition (RSAJI) ● The peak pressure increased by 27.4%, ammonia and N_2O emissions reduced by 87.1% and 99.7% ● At lower loads, the RSAJI mode increased ITE, but at higher loads, there is a tendency for knock and a decrease in thermal efficiency ● NO_x increased at low loads, NH_3 and N_2O reduced at all loads ● The injection timing should be appropriately far from the TDC (86.4 °CA and 75 °CA BTDC better) ● Higher ASR, increased ITE. The injection timing near the TDC, lower ITE *Emissions:* NO_x, NH_3, N_2O	Li *et al.* [120]

HC = Hydrocarbons, ICE = Internal Combustion Engine, CR = Compression Ratio, CFR = Cooperative Fuel Research engine, MBT = Maximum brake torque, SIT = Spark ignition timing, SCR = Selective Catalytic Reduction, ER = Equivalence Ratio, TDC = Top Dead Center, CA = Crank Angle, EGR = Exhaust Gas Recirculation, BSFC = brake specific fuel consumption, LHV = Lower Heating Value, HRR = Heat Release Rate.

Experiments by Niki *et al.* [58,59] found that increasing ammonia injection led to decreased cylinder compression, maximum pressures, and ignition timing, due to ammonia's lower combustion temperature and slower burning velocity. While ammonia can reduce CO_2 emissions, it may increase N_2O emissions. Multiple diesel fuel injection strategies, including pilot and post-injections, were shown to be effective in reducing NH_3 and N_2O emissions. Yousefi *et al.* [55,60] explored the effects of ammonia energy fraction and injection strategies on ADDF engine performance and emissions. They found that increasing ammonia energy fraction can lead to lower GHG emissions but may also increase NO_x and N_2O emissions. Split diesel injection strategies were shown to be effective in managing emissions and improving combustion efficiency. Li *et al.* [61] compared LPDF and HPDF modes for ammonia/diesel dual-fuel engines, and the results showed that the former case could achieve a higher ITE (see Figure 5.8(a)), while the latter case significantly reduced the emissions of NH_3, NO_x and GHG (see Figure 5.8(b)).

Several studies have investigated the effects of ammonia energy share (AES) on ammonia-diesel dual-fuel engine performance and emissions. Nadimi *et al.* [62] found that increasing AES can lead to improved thermal efficiency, but may also increase NO_x and unburned ammonia emissions. Førby *et al.* [63] explored the potential of high ammonia energy contributions, demonstrating increased combustion efficiency and reduced ammonia slip and N_2O emissions. Zhu *et al.* [64] and Xu *et al.* [65] investigated the effects of ammonia substitution ratio (ASR) on dual-fuel combustion. They found that increasing ASR can lead to intensified premixed combustion and reduced CO_2 emissions, but also increased NO_x and N_2O

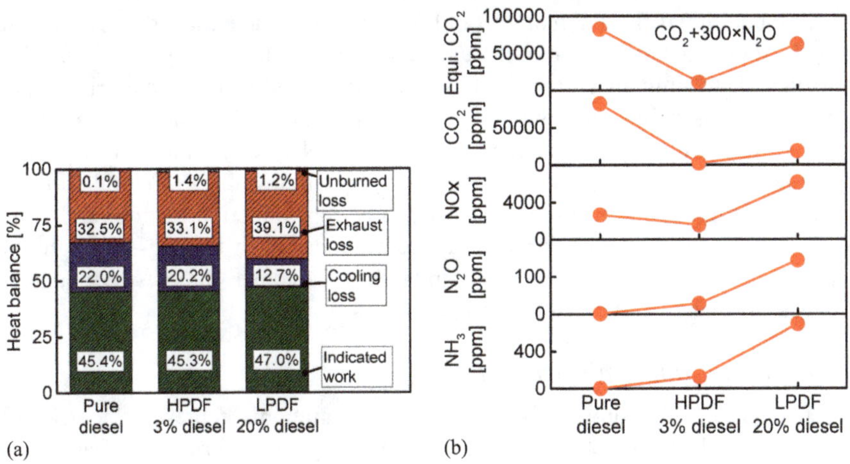

Figure 5.8 Comparisons of the energy balances (a) and emission characteristics (b) among the pure diesel, HPDF and LPDF modes. (SODI = −8°CA aTDC, N = 1000 rpm, Pin = 1.3 bar and Tin = 322 K). Reproduced, with permission, from [61].

emissions. Jin *et al.* [66] explored the effects of AER and fuel injection strategy on ADDF engine combustion and emissions. They found that higher AER decreased ITEg and increased unburned ammonia emissions. NO_x emissions decreased due to the denitrification effect of amino groups, while N_2O emissions were influenced by temperature. The dual injection strategy improved mixture activity, accelerated combustion, and reduced unburned ammonia emissions. Wang *et al.* [67] investigated the effects of ammonia mixing ratio and injection location. They found that in-cylinder direct injection of ammonia provided a more controlled combustion process with minimal ammonia slip and lower emissions.

Some advancements on the injection strategies also showed potentials in enhancing combustion performance and reducing emissions of ADDF combustion. Zhang *et al.* [68] explored the Jet Controlled Compression Ignition (JCCI) concept, demonstrating that diesel injection timing can effectively regulate the combustion phase, leading to increased ITE. However, NO_x emissions may increase with higher ammonia energy fractions and advanced injection timings. Liu *et al.* [69] proposed the ammonia-diesel stratified injection (ADSJ) mode for marine low-speed engines. ADSJ demonstrated superior ignition characteristics, higher ITE, and reduced ignition delay time. NO_x emissions were lower across all conditions, and unburned ammonia emissions were effectively controlled. Mi *et al.* [70] investigated the impact of diesel pilot-injection strategies. Pilot-injection significantly reduced ammonia and N_2O emissions, maintained high ITE, and achieved stable combustion. However, it increased NO_x, CO, and hydrocarbon emissions. Niki [71,72] explored the effects of early diesel pilot injection on ammonia-assisted diesel engine emissions and combustion performance. Early diesel pilot injection significantly reduced unburned NH_3 and N_2O emissions, but also decreased BTE due to over-advanced combustion phasing. The split injection approach mitigated the rise in HC and CO emissions.

Computational fluid dynamics (CFD) simulations have played a crucial role in understanding ammonia-diesel dual-fuel combustion. Rodríguez *et al.* [73] utilized CFD to simulate ammonia-diesel dual-fuel combustion in a marine diesel engine, demonstrating the potential for reducing CO_2, CO, and HC emissions while highlighting challenges such as increased NO_x emissions and ammonia slip. Shin and Park [74] and Zhou *et al.* [75] employed CFD simulations to investigate the effects of ammonia energy fraction, injection timing, and combustion modes on engine performance and emissions. Their findings highlight the importance of optimizing these parameters to achieve the desired balance between efficiency and emissions reduction.

5.5.2 Blend ammonia with H_2

The integration of hydrogen as a co-fuel with ammonia represents a strategic approach to enhance combustion efficiency and engine performance of ammonia fueled CI engines. Hydrogen, known for its high reactivity and rapid flame speed, acts as an effective combustion promoter. When combined with ammonia, which can be partially dissociated into hydrogen, the resulting mixture exhibits enhanced reactivity due to hydrogen's lower ignition energy and higher flame velocity compared to pure ammonia. This synergistic effect facilitates easier ignition and faster combustion, addressing the challenges posed by ammonia's high auto-ignition temperature and slow flame speed. Moreover, hydrogen-enriched mixtures promote a more complete and stable burn, thereby increasing the engine's thermal efficiency. Additionally, hydrogen-enriched blends can mitigate issues associated with the slow oxidation kinetics of ammonia, potentially reducing emissions of unburned ammonia and NO_x. This approach leverages hydrogen not only as a combustion enhancer but also as a carrier for decarbonizing transportation fuels, aligning with sustainable and carbon-neutral powertrain development goals.

An early project report by Pearsall *et al.* [21] demonstrated the efficacy of hydrogen additives in improving ammonia combustion characteristics in I engines. Their study highlighted hydrogen's role in lowering ignition temperatures and enhancing oxidation rates, crucial for stable combustion in CI engines. Additionally, Wang *et al.* [76] conducted comprehensive research on blending ammonia with hydrogen, focusing on combustion dynamics, emissions profiles, and the pathway to achieving net-zero carbon emissions. Their findings underscored that a 30% hydrogen blending ratio significantly boosted engine performance metrics such as in-cylinder pressure and power output, while also reducing ammonia slip by 99.1%. This blending strategy concurrently moderated NO_x emissions and improved soot and hydrocarbon (HC) emissions, underscoring its potential to optimize combustion efficiency. Further exploring the dual-fuel concept, Wang *et al.* [77] investigated ammonia/hydrogen blends as a viable alternative to conventional fossil fuel-based engines. Their computational simulations evaluated various injection strategies and CRs to optimize combustion efficiency and emissions performance. The study identified that adjusting injection timing and CRs influenced combustion stability, engine power, and emission profiles. Specifically, optimal conditions at a 13.5 CR and a 12°CA BTDC injection timing effectively met stringent Tier III emission standards, highlighting the promise of ammonia/hydrogen blends in achieving sustainable and compliant engine operations.

Innovative combustion modes have also been explored for ammonia/hydrogen blends. Pochet *et al.* [78] conducted comprehensive experimental analyses on an ammonia-hydrogen blend in a homogeneous charge compression ignition (HCCI) engine with a 22:1 CR. Their study focused on varying parameters such as ammonia blending ratio, equivalence ratio, intake temperature, and exhaust gas recirculation (EGR) to assess engine performance and emissions. The research highlighted that while ammonia's resistance to auto-ignition necessitated higher intake temperatures, blending with hydrogen enhanced combustion stability and increased indicated mean effective pressure (IMEP) by 50% compared to pure hydrogen (see Figure 5.9). However, elevated ammonia flow rates correlated with increased NO_x emissions, mitigated effectively by EGR. Notably, temperatures above 1,800 K were essential to prevent N_2O formation during combustion, underscoring the critical role of temperature control in ammonia combustion. Reggeti *et al.* [79] explored ammonia's application in spark-assisted compression-ignition (SACI) engines, leveraging its high CR benefits and enhancing combustion with hydrogen addition. Their experimental investigations demonstrated that SACI combustion achieved optimal

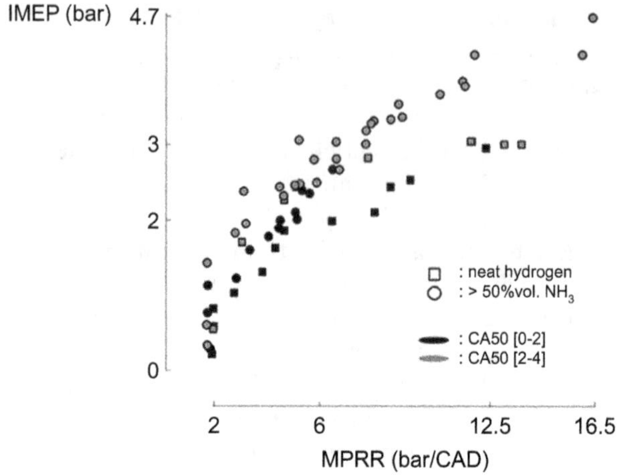

Figure 5.9 The achievable IMEP is tapering off as the MPRR increases due to an increased heat and friction losses (all experimental points). Ammonia allows higher IMEP due to a reduced MPRR for a given fuel loading and the major share of IMEP rise occurs between 15% and 50%vol. of ammonia. Reprinted, with permission, from [78]. License: CC BY 4.0.

combustion phasing and comparable burn durations to conventional SI engines, despite ammonia's slower flame speed. Hydrogen blending improved gross indicated mean effective pressure (gIMEP) at MBT timing, highlighting its role in enhancing combustion stability across diverse operating conditions. However, increased intake temperatures and hydrogen content correlated with higher NO emissions, with accompanying increases in N_2O emissions at higher engine speeds.

Additionally, dissociation of ammonia into hydrogen has been explored to further optimize combustion efficiency. Gill *et al.* [80] investigated the effects of pure and dissociated ammonia on diesel engine performance and emissions. They demonstrated that substituting 3% of the air intake with ammonia reduced diesel consumption and CO_2 emissions by 15%. Pure hydrogen addition exhibited superior combustion efficiency and emission characteristics, while co-fueling with dissociated ammonia balanced thermal efficiency with emission quality, showcasing potential for waste-heat recovery strategies.

5.5.3 SACI

The SACI approach represents an innovative combustion strategy that combines elements of both SI and CI engines to optimize performance when utilizing alternative fuels such as ammonia.

Starkman *et al.* [22] pioneered the use of ammonia as a fuel in CI engines with SI initiation. Their early work highlighted the sensitivity of engine performance to spark advance and identified optimal injection timings for ammonia around

150–180 CAD BTDC. They observed that while ammonia's specific fuel consumption aligned closely with theoretical predictions, its power output was approximately 10% lower than diesel due to ignition challenges and slower combustion rates.

Mounaïm-Rousselle *et al.* [81] investigated ammonia's performance in SACI engines, which utilize SI to initiate combustion followed by auto-ignition. Their study elucidated the operational limits, combustion characteristics, and emission profiles of ammonia in SACI mode. They found that pure ammonia combustion was viable across a wide range of operating conditions, demonstrating stable operation even at low intake pressures of 0.8 bar. Increasing the CR improved combustion efficiency, although hydrogen addition slightly enhanced gIMEP and increased NO_x emissions. SACI engines exhibited lower NO_x emissions compared to traditional SI engines, with minimal N_2O emissions and manageable levels of unburned ammonia.

Reggeti *et al.* [79] further explored SACI combustion using ammonia, highlighting its compatibility with high CRs to achieve optimal combustion phasing despite ammonia's inherently slower flame speed. Their findings underscored SACI's potential to exploit ammonia's high knock-resistance without requiring extensive hydrogen blending. Unburned ammonia emissions were slightly mitigated by blending hydrogen in the fuel or increasing the intake temperature. However, these adjustments led to increased NO emissions. N_2O emissions did not show a clear trend with fuel hydrogen content or intake temperature but increased at higher engine speeds, as shown in Figure 5.10.

Overall, the high knock-resistance of ammonia enables SACI, which could be a solution to its low flame speed without blending large amounts of hydrogen. Operating with a high CR could also improve thermal efficiency.

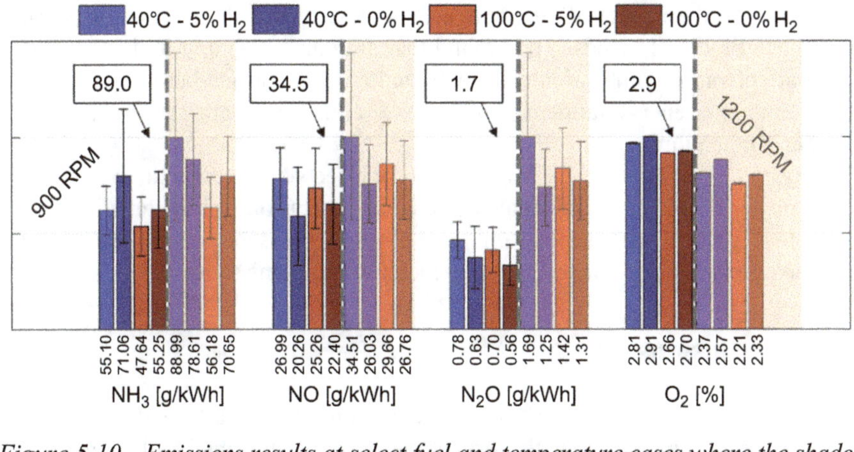

Figure 5.10 Emissions results at select fuel and temperature cases where the shaded and unshaded regions correspond to 1200 rpm and 900 rpm, respectively. Reprinted, with permission, from [79]. License: CC BY 4.0.

5.5.4 Blend with other high-reactivity fuels

To address the challenges of directly using ammonia in CI engines, such as its high auto-ignition temperature and low combustion rate, researchers also employed other high-reactivity fuels apart from the above-mentioned hydrogen and diesel, including DME, biodiesel, and kerosene, to facilitate ammonia combustion initiation and improve overall engine performance.

Gross and Kong [82] conducted experimental investigations into CI engine performance using mixtures of ammonia and DME. Their study revealed that while ammonia's inclusion in the fuel blend prolonged ignition delays and restricted engine load conditions due to its combustion characteristics, stable operation was achieved with ammonia-DME mixtures. Despite challenges, the engine demonstrated comparable combustion efficiency to diesel fuel, albeit with increased NO_x emissions attributed to fuel NO_x formation and higher CO and HC emissions due to lower combustion temperatures. Soot emissions remained minimal under all tested conditions, underscoring the potential of ammonia-DME blends in decarbonizing transportation sectors. Ryu *et al.* [83] expanded on this by evaluating CI engine performance using varying concentrations of ammonia and DME mixtures. Their findings indicated diminishing engine performance with increasing ammonia concentration, with significant cycle-to-cycle variations observed at higher ammonia levels. Advanced injection timing was necessary to optimize torque output, highlighting ammonia's resistance to auto-ignition. Despite these challenges, the 40% DME–60% NH_3 mixture demonstrated HCCI characteristics, albeit with increased CO, HC, and NO_x emissions. The study underscored the role of DME as a co-fuel and advanced injection systems in managing ammonia combustion challenges.

Nadimi *et al.* [84] explored the use of ammonia as a primary fuel with biodiesel. They found that up to 69.4% of biodiesel input energy could be replaced by ammonia, albeit with a slight reduction in BTE. The dual-fuel operation significantly reduced carbon-based emissions but increased NO_x emissions. Wu *et al.* [85] evaluated the impact of ammonia as a secondary fuel in diesel engines using microalgae biodiesel blends. They found that ammonia blending initially reduced BTE but enhanced combustion efficiency and reduced carbon-based emissions at higher ammonia energy ratios. However, NO_x emissions increased due to higher cylinder temperatures.

Tay *et al.* [56] numerically investigated the effects of injection timing and pilot fuel type on ammonia-diesel dual-fuel engine combustion characteristics. They found that advancing injection timing and using kerosene as a pilot fuel can enhance combustion efficiency but may increase NO_x emissions.

5.6 Conclusions

The imperative to curtail greenhouse gas emissions, particularly within the transportation sector, underscores the urgent need for sustainable fuel alternatives. Renewable energy sources offer a promising pathway to mitigate climate change. While hydrogen presents potential, its energy density and logistical challenges limit its widespread adoption. Ammonia emerges as a compelling candidate due to its carbon-free nature, favorable energy density, and ease of storage and transportation. This chapter comprehensively examined recent advancements in ammonia combustion research, drawing upon a substantial dataset of ammonia utilization in internal combustion engines. Both SI and CI engine configurations were explored, highlighting key challenges and opportunities associated with ammonia as a fuel.

The findings underscore the potential of ammonia as a viable fuel for internal combustion engines, albeit with significant technical hurdles. While dual-fuel strategies and engine modifications have shown promise in overcoming ammonia's combustion limitations, further research and development are essential to optimize engine performance, efficiency, and emissions. Addressing challenges such as ammonia's low flame speed, high auto-ignition resistance, and NO_x emissions will be crucial for the successful commercialization of ammonia-fueled internal combustion engines.

Continued research and development efforts focused on improving ammonia combustion characteristics, optimizing engine design, and developing advanced aftertreatment systems are necessary to realize the full potential of ammonia as a sustainable transportation fuel.

Acknowledgment

The authors gratefully acknowledge the support from the ESPRC through the program SAFE AGT (EP/T009314/1), OceanREFuel (EP/W005018/1), MariNH3 (EP/W016656/1), FLEXnCONFU (EU no. 884157), and optimal fuel blends for ammonia-fueled thermal propulsion systems (EP/T033800/1). The first author would like to express gratitude for the support provided by the Alexander von Humboldt Foundation. The authors extend their gratitude to Prof. Christine Mounaïm-Rousselle for her invaluable suggestions for the chapter preparation.

References

[1] Lamb WF, Wiedmann T, Pongratz J, *et al.* A review of trends and drivers of greenhouse gas emissions by sector from 1990 to 2018. *Environ Res Lett* 2021;16:073005.

[2] Valera-Medina A, Xiao H, Owen-Jones M, David WIF, and Bowen PJ. Ammonia for power. *Prog Energy Combust Sci* 2018;69:63–102.

[3] Cardoso JS, Silva V, Rocha RC, Hall MJ, Costa M, and Eusébio D. Ammonia as an energy vector: current and future prospects for low-carbon fuel applications in internal combustion engines. *J Clean Prod* 2021; 296:126562.

[4] Kandemir T, Schuster ME, Senyshyn A, Behrens M, and Schlögl R. The Haber–Bosch process revisited: on the real structure and stability of "ammonia iron" under working conditions. *Angew Chem Int Ed* 2013;52:12723–6.

[5] Zamfirescu C, and Dincer I. Ammonia as a green fuel and hydrogen source for vehicular applications. *Fuel Process Technol* 2009;90:729–37.

[6] Rault TM, Clees S, Figueroa-Labastida M, *et al.* Multi-speciation and ignition delay time measurements of ammonia oxidation behind reflected shock waves. *Combust Flame* 2024;260:113260.

[7] Cai T, and Zhao D. Overview of autoignition and flame propagation properties for ammonia combustion. *AIAA J* 2023;61:2754–78.

[8] Zitouni S, Brequigny P, and Mounam-Rousselle C. Turbulent flame speed and morphology of pure ammonia flames and blends with methane or hydrogen. *Proc Combust Inst* 2023;39:2269–78.

[9] Li J, Zhang R, Pan J, Wei H, Shu G, and Chen L. Ammonia and hydrogen blending effects on combustion stabilities in optical SI engines. *Energy Convers Manag* 2023;280:116827.

[10] Tornatore C, Marchitto L, Sabia P, and De Joannon M. Ammonia as green fuel in internal combustion engines: state-of-the-art and future perspectives. *Front Mech Eng* 2022;8:944201.

[11] Ma F, Guo L, Li Z, *et al.* A review of current advances in ammonia combustion from the fundamentals to applications in internal combustion engines. *Energies (Basel)* 2023;16:6304.

[12] Zamfirescu C, and Dincer I. Using ammonia as a sustainable fuel. *J Power Sources* 2008;185:459–65.

[13] Reiter AJ, and Kong SC. Demonstration of compression-ignition engine combustion using ammonia in reducing greenhouse gas emissions. *Energy Fuels* 2008;22:2963–71.

[14] Kang L, Pan W, Zhang J, Wang W, and Tang C. A review on ammonia blends combustion for industrial applications. *Fuel* 2023;332:126150.

[15] Chiong MC, Chong CT, Ng JH, *et al.* Advancements of combustion technologies in the ammonia-fuelled engines. *Energy Convers Manag* 2021; 244:114460.

[16] Frigo S, and Gentili R. Analysis of the behaviour of a 4-stroke Si engine fuelled with ammonia and hydrogen. *Int J Hydrogen Energy* 2013; 38:1607–15.

[17] Koch E. Ammonia–a fuel for motor buses. *J Inst Pet* 1945;31:498.

[18] Cornelius W, Huellmantel LW, and Mitchell HR. Ammonia as an engine fuel. *SAE Trans* 1966;74:300–26.

[19] Starkman ES, Newhall HK, Sutton R, Maguire T, and Farbar L. Ammonia as a spark ignition engine fuel: theory and application. *SAE Trans* 1967;75: 765–84.

[20] Smith DS, Sawyer RF, and Starkman ES. Oxides of nitrogen from gas turbines. *J Air Pollut Control Assoc* 1968;18:30–5.

[21] Pearsall TJ. *Ammonia Application to Reciprocating Engines*. CFSTI; 1967.

[22] Starkman ES, James GE, and Newhall HK. Ammonia as a diesel engine fuel: theory and application. *SAE Trans* 1968;76:3193–212.

[23] Leeladhar R, Grannell S, Bohac S, and Assanis D. Comparison of ammonia/gasoline and ammonia/ethanol fuel mixtures for use in IC engines. *NH3 Fuel Association*, 2007.

[24] Petela K, and Proniewicz M. Ammonia as carbon free fuel for internal CombusTion engIne driVen AgriculTural vEhicle (ACTIVATE). 2022. https://ammoniaengine.org/wp-content/uploads/2022/11/ACTIVATE_REPORTS.D4.2.pdf.

[25] Wärtsilä to coordinate EU-funded program for ammonia engine development – Ammonia Energy Association [Internet]. [cited 2024 Aug 18]; Available from: https://ammoniaenergy.org/articles/wartsila-to-coordinate-eu-funded-program-for-ammonia-engine-development/ (accessed August 10, 2024).

[26] Solutions MANE. *Engineering the Future Two-Stroke Green-Ammonia Engine.* Copenhagen: MAN Energy Solutions; 2019. https://fathom.world/wp-content/uploads/2020/05/engineeringthefuturetwostrokegreenammoniaengine1589339239488.pdf.

[27] Dimitriou P, and Javaid R. A review of ammonia as a compression ignition engine fuel. *Int J Hydrogen Energy* 2020;45:7098–118.

[28] Lhuillier C, Brequigny P, Contino F, and Rousselle C. Combustion characteristics of ammonia in a modern spark-ignition engine. In: *SAE Technical Paper*; 2019.

[29] Mashruk S, Vigueras-Zuniga MO, Tejeda-del-Cueto ME, *et al.* Combustion features of $CH_4/NH_3/H_2$ ternary blends. *Int J Hydrogen Energy* 2022; 47:30315–27.

[30] Chai WS, Bao Y, Jin P, Tang G, and Zhou L. A review on ammonia, ammonia-hydrogen and ammonia-methane fuels. *Renew Sustain Energy Rev* 2021;147:111254.

[31] Dinesh MH, Pandey JK, and Kumar GN. Study of performance, combustion, and NO_x emission behavior of an SI engine fuelled with ammonia/hydrogen blends at various compression ratio. *Int J Hydrogen Energy* 2022; 47:25391–403.

[32] Van Blarigan P. Advanced internal combustion electrical generator. *In: Proceedings of the Citeseer*; 2002. pp. 1–16.

[33] Lhuillier C, Brequigny P, Contino F, and Mounaïm-Rousselle C. Experimental study on ammonia/hydrogen/air combustion in spark ignition engine conditions. *Fuel* 2020;269:117448.

[34] Lhuillier C, Brequigny P, Contino F, and Mounaïm-Rousselle C. Experimental investigation on ammonia combustion behavior in a spark-ignition engine by means of laminar and turbulent expanding flames. *Proc Combust Inst* 2021;38:5859–68.

[35] Mounaïm-Rousselle C. Operating limits for ammonia fuel spark-ignition engine. *Energies (Basel)* 2021;14:4141.

[36] Grannell SM, Assanis DN, Bohac SV, and Gillespie DE. The fuel mix limits and efficiency of a stoichiometric, ammonia, and gasoline dual fueled spark ignition engine. *J Eng Gas Turbine Power* 2008;130(4):042802.

[37] Koike M, Miyagawa H, Suzuoki T, and Ogasawara K. Ammonia as a hydrogen energy carrier and its application to internal combustion engines. *In: Sustainable Vehicle Technologies*. Gaydon: Elsevier; 2012. pp. 61–70.

[38] Frigo S, and Gentili R. Analysis of the behaviour of a 4-stroke Si engine fuelled with ammonia and hydrogen. *Int J Hydrogen Energy* 2012;38: 1607–15.

[39] Grannell SM, Assanis DN, Gillespie DE, and Bohac SV. Exhaust emissions from a stoichiometric, ammonia and gasoline dual fueled spark ignition engine. *In: ASME 2009 Internal Combustion Engine Division Spring Technical Conference*. ASMEDC; 2009. pp. 135–41.

[40] Westlye FR, Ivarsson A, and Schramm J. Experimental investigation of nitrogen based emissions from an ammonia fueled SI-engine. *Fuel* 2013; 111:239–47.

[41] Ryu K, Zacharakis-Jutz GE, and Kong SC. Effects of gaseous ammonia direct injection on performance characteristics of a spark-ignition engine. *Appl Energy* 2014;116:206–15.

[42] Haputhanthri SO, Maxwell TT, Fleming J, Austin C Ammonia and gasoline fuel blends for spark ignited internal combustion engines. *J Energy Resour Technol* 2015;137:062201.

[43] Oh S, Park C, Kim S, Kim Y, Choi Y, and Kim C. Natural gas–ammonia dual-fuel combustion in spark-ignited engine with various air–fuel ratios and split ratios of ammonia under part load condition. *Fuel* 2021;290:120095.

[44] Comotti M, and Frigo S. Hydrogen generation system for ammonia–hydrogen fuelled internal combustion engines. *Int J Hydrogen Energy* 2015; 40:10673–86.

[45] Salek F, Babaie M, Shakeri A, Hosseini SV, Bodisco T, and Zare A. Numerical study of engine performance and emissions for port injection of ammonia into a gasoline\ethanol dual-fuel spark ignition engine. *Appl Sci* 2021;11:1441.

[46] Koike M, Suzuoki T, Takeuchi T, Homma T, Hariu S, and Takeuchi Y. Cold-start performance of an ammonia-fueled spark ignition engine with an on-board fuel reformer. *Int J Hydrogen Energy* 2021;46:25689–98.

[47] Lhuillier C, Brequigny P, Contino F, and Rousselle C. Combustion characteristics of ammonia in a modern spark-ignition engine. In: *SAE Technical Papers;* 2019.

[48] Xin G, Ji C, Wang S, Hong C, Meng H, and Yang J. Experimental study on the effect of hydrogen substitution rate on combustion and emission characteristics of ammonia internal combustion engine under different excess air ratio. *Fuel* 2023;343:127992.

[49] Xin G, Ji C, Wang S, *et al.* Experimental study of the effect of variable valve timing on hydrogen-enriched ammonia engine. *Fuel* 2023; 344:128131.

[50] Xin G, Ji C, Wang S, Meng H, Chang K, and Yang J. Effect of ammonia addition on combustion and emission characteristics of hydrogen-fueled engine under lean-burn condition. *Int J Hydrogen Energy* 2022;47:9762–74.

[51] Zhang R, Chen L, Wei H, Li J, Chen R, and Pan J. Understanding the difference in combustion and flame propagation characteristics between ammonia and methane using an optical SI engine. *Fuel* 2022;324:124794.

[52] Kurien C, Varma PS, and Mittal M. Effect of ammonia energy fractions on combustion stability and engine characteristics of gaseous (ammonia/methane) fuelled spark ignition engine. *Int J Hydrogen Energy* 2023; 48:1391–400.
[53] Gray Jr JT, Dimitroff E, Meckel NT, and Quillian Jr RD. Ammonia fuel–engine compatibility and combustion. *SAE Trans* 1967;75:785–807.
[54] Pearsall TJ, and Garabedian CG. Combustion of anhydrous ammonia in diesel engines. *SAE Trans* 1968;76:3213–21.
[55] Yousefi A, Guo H, Dev S, Liko B, and Lafrance S. Effects of ammonia energy fraction and diesel injection timing on combustion and emissions of an ammonia/diesel dual-fuel engine. *Fuel* 2022;314:122723.
[56] Tay KL, Yang W, Chou SK, *et al.* Effects of injection timing and pilot fuel on the combustion of a kerosene-diesel/ammonia dual fuel engine: a numerical study. *Energy Procedia* 2017;105:4621–6.
[57] Reiter AJ, and Kong SC. Combustion and emissions characteristics of compression-ignition engine using dual ammonia-diesel fuel. *Fuel* 2011; 90:87–97.
[58] Niki Y, Yoo DH, Hirata K, and Sekiguchi H. Effects of ammonia gas mixed into intake air on combustion and emissions characteristics in diesel engine. *In: Internal Combustion Engine Division Fall Technical Conference*. New York: American Society of Mechanical Engineers; 2016. p. V001T03A004.
[59] Niki Y, Nitta Y, Sekiguchi H, and Hirata K. Diesel fuel multiple injection effects on emission characteristics of diesel engine mixed ammonia gas into intake air. *J Eng Gas Turbine Power* 2019;141:061020.
[60] Yousefi A, Guo H, Dev S, Lafrance S, and Liko B. A study on split diesel injection on thermal efficiency and emissions of an ammonia/diesel dual-fuel engine. *Fuel* 2022;316:123412.
[61] Li T, Zhou X, Wang N, *et al.* A comparison between low-and high-pressure injection dual-fuel modes of diesel-pilot-ignition ammonia combustion engines. *J Energy Inst* 2022;102:362–73.
[62] Nadimi E, Przybyła G, Lewandowski MT, and Adamczyk W. Effects of ammonia on combustion, emissions, and performance of the ammonia/diesel dual-fuel compression ignition engine. *J Energy Inst* 2023;107:101158.
[63] Førby N, Thomsen TB, Cordtz RF, Bræstrup F, and Schramm J. Ignition and combustion study of premixed ammonia using GDI pilot injection in CI engine. *Fuel* 2023;331:125768.
[64] Zhu J, Zhou D, Yang W, Qian Y, Mao Y, and Lu X. Investigation on the potential of using carbon-free ammonia in large two-stroke marine engines by dual-fuel combustion strategy. *Energy* 2023;263:125748.
[65] Xu L, Xu S, Bai XS, Repo JA, Hautala S, and Hyvönen J. Performance and emission characteristics of an ammonia/diesel dual-fuel marine engine. *Renew Sustain Energy Rev* 2023;185:113631.
[66] Jin S, Wu B, Zi Z, Yang P, Shi T, and Zhang J. Effects of fuel injection strategy and ammonia energy ratio on combustion and emissions of ammonia-diesel dual-fuel engine. *Fuel* 2023;341:127668.

[67] Wang B, Wang H, Hu D, Yang C, Duan B, and Wang Y. Effect of different ammonia mixing methods for diesel ignition on combustion and emission performance of high pressure common rail engine. *J Energy Inst* 2023; 111:101402.
[68] Zhang Z, Long W, Dong P, *et al.* Performance characteristics of a two-stroke low speed engine applying ammonia/diesel dual direct injection strategy. *Fuel* 2023;332:126086.
[69] Liu L, Wu J, Liu H, Wu Y, and Wang Y. Study on marine engine combustion and emissions characteristics under multi-parameter coupling of ammonia-diesel stratified injection mode. *Int J Hydrogen Energy* 2023; 48:9881–94.
[70] Mi S, Wu H, Pei X, *et al.* Potential of ammonia energy fraction and diesel pilot-injection strategy on improving combustion and emission performance in an ammonia-diesel dual fuel engine. *Fuel* 2023;343:127889.
[71] Niki Y. Reductions in unburned ammonia and nitrous oxide emissions from an ammonia-assisted diesel engine with early timing diesel pilot injection. *J Eng Gas Turbine Power* 2021;143:091014.
[72] Niki Y. Experimental investigation of effects of split diesel-pilot injection on emissions from ammonia-diesel dual fuel engine. *In: Internal Combustion Engine Division Fall Technical Conference*. New York: American Society of Mechanical Engineers; 2021. p. V001T01A002.
[73] Rodríguez CG, Lamas MI, Rodríguez J de D, and Abbas A. Possibilities of ammonia as both fuel and NO_x reductant in marine engines: a numerical study. *J Mar Sci Eng* 2022;10:43.
[74] Shin J, and Park S. Numerical analysis for optimizing combustion strategy in an ammonia-diesel dual-fuel engine. *Energy Convers Manag* 2023; 284:116980.
[75] Zhou X, Li T, Wang N, Wang X, Chen R, and Li S. Pilot diesel-ignited ammonia dual fuel low-speed marine engines: a comparative analysis of ammonia premixed and high-pressure spray combustion modes with CFD simulation. *Renew Sustain Energy Rev* 2023;173:113108.
[76] Wang B, Yang C, Wang H, Hu D, and Wang Y. Effect of diesel-ignited ammonia/hydrogen mixture fuel combustion on engine combustion and emission performance. *Fuel* 2023;331:125865.
[77] Wang B, Yang C, Wang H, Hu D, Duan B, and Wang Y. Study on injection strategy of ammonia/hydrogen dual fuel engine under different compression ratios. *Fuel* 2023;334:126666.
[78] Pochet M, Jeanmart H, and Contino F. A 22:1 compression ratio ammonia-hydrogen HCCI engine: combustion, load, and emission performances. *Front Mech Eng* 2020;6:43.
[79] Reggeti S, Kane S, and Northrop W. Experimental investigation of spark-assisted compression-ignition with ammonia-hydrogen blends. *J Ammonia Energy* 2023;1:91–105.

[80] Gill SS, Chatha GS, Tsolakis A, Golunski SE, and York APE. Assessing the effects of partially decarbonising a diesel engine by co-fuelling with dissociated ammonia. *Int J Hydrogen Energy* 2012;37:6074–83.

[81] Mounaïm-Rousselle C, Mercier A, Brequigny P, Dumand C, Bouriot J, and Houillé S. Performance of ammonia fuel in a spark assisted compression ignition engine. *Int J Engine Res* 2022;23:781–92.

[82] Gross CW, and Kong SC. Performance characteristics of a compression-ignition engine using direct-injection ammonia–DME mixtures. *Fuel* 2013;103:1069–79.

[83] Ryu K, Zacharakis-Jutz GE, and Kong SC. Performance characteristics of compression-ignition engine using high concentration of ammonia mixed with dimethyl ether. *Appl Energy* 2014;113:488–99.

[84] Nadimi E, Przybyła G, Emberson D, Løvås T, Ziółkowski Ł, and Adamczyk W. Effects of using ammonia as a primary fuel on engine performance and emissions in an ammonia/biodiesel dual-fuel CI engine. *Int J Energy Res* 2022;46:15347–61.

[85] Wu Y, Zhang Y, Xia C, *et al.* Assessing the effects of ammonia (NH_3) as the secondary fuel on the combustion and emission characteristics with nano-additives. *Fuel* 2023;336:126831.

[86] Xin G, Ji C, Wang S, Meng H, Chang K, and Yang J. Effect of different volume fractions of ammonia on the combustion and emission characteristics of the hydrogen-fueled engine. *Int J Hydrogen Energy* 2022; 47:16297–308.

[87] Cornelius W, Huellmantel LW, and Mitchell HR. Ammonia as an engine fuel. SAE Technical Paper 650052, 1965, https://doi.org/10.4271/650052

[88] Starkman ES, Newhall HK, Sutton R, Maguire T, and Farbar L. Ammonia as a spark ignition engine fuel: theory and application. In: *1966 Automotive Engineering Congress and Exposition*. SAE Technical Paper; 1967. p. 660155.

[89] Pearsall T, and Fryer L. *Ammonia Application to Reciprocating Engines* vol. 1.Continental Aviation and Engineering Corp., Detroit, MI; 1967.

[90] Gray DS, Domke CJ, Meguerian GH, and Mieville RL. *Ammonia Application to Reciprocating Engines* vol. 2. Continental Aviation and Engineering Corp., Detroit, MI; 1967.

[91] Mørch CS, Bjerre A, Gøttrup MP, Sorenson SC, and Schramm J. Ammonia/hydrogen mixtures in an SI-engine: engine performance and analysis of a proposed fuel system. *Fuel* 2011;90:854–64.

[92] Ryu K, Zacharakis-Jutz GE, and Kong SC. Performance enhancement of ammonia-fueled engine by using dissociation catalyst for hydrogen generation. *Int J Hydrogen Energy* 2014;39:2390–8.

[93] Woo Y, Jang JY, Lee YJ, and Kim JN. Recent progress on the Ammonia-Gasoline and the Ammonia-Diesel dual fueled internal combustion engines in Korea. *In: 11th NH3 Fuel Conference*. 2014.

[94] Koike M, and Suzuoki T. In-line adsorption system for reducing cold-start ammonia emissions from engines fueled with ammonia and hydrogen. *Int J Hydrogen Energy* 2019;44:32271–9.

[95] Lhuillier C, Brequigny P, Contino F, and Rousselle C. Performance and emissions of an ammonia-fueled SI engine with hydrogen enrichment. In: *SAE Technical Papers;* 2019.

[96] Mercier A, Mounaïm-Rousselle C, Brequigny P, Bouriot J, and Dumand C. Improvement of SI engine combustion with ammonia as fuel: effect of ammonia dissociation prior to combustion. *Fuel Commun* 2022;11:100058.

[97] Oh S, Park C, Oh J, *et al.* Combustion, emissions, and performance of natural gas–ammonia dual-fuel spark-ignited engine at full-load condition. *Energy* 2022;258:124837.

[98] Brequigny P, Pacaud E, and Mounaïm-Rousselle C. Combustion of a syngas from sewage sludge gasification enriched with ammonia in a spark-ignition engine. In: *30th European Biomass Conference & Exhibition*. 2022.

[99] Liu Z, Zhou L, Zhong L, and Wei H. Reactivity controlled turbulent jet ignition (RCTJI) for ammonia engine. *Int J Hydrogen Energy* 2023; 48:12519–22.

[100] Liu Z, Zhou L, Zhong L, and Wei H. Enhanced combustion of ammonia engine based on novel air-assisted pre-chamber turbulent jet ignition. *Energy Convers Manag* 2023;276:116526.

[101] Liu Z, Zhou L, and Wei H. Experimental investigation on the performance of pure ammonia engine based on reactivity controlled turbulent jet ignition. *Fuel* 2023;335:127116.

[102] Liu Z, and Liu J. Machine learning assisted analysis of an ammonia engine performance. *J Energy Resour Technol* 2022;144:112307.

[103] Pearsall TJ, and Garabedian CG. Combustion of anhydrous ammonia in diesel engines. In: SAE Technical paper; 1967. p. 670947.

[104] Starkman ES, James GE, and Newhall HK. Ammonia as a diesel engine fuel: theory and application. *In: National Fuels and Lubricants, Powerplants, Transportation Meetings*; 1968. p. 670946.

[105] Bro K, and Pedersen PS. Alternative diesel engine fuels: an experimental investigation of methanol, ethanol, methane and ammonia in a D.I. diesel engine with pilot injection. SAE Technical Paper 770794, 1977, https://doi.org/10.4271/770794.

[106] Niki Y, Yoo DH, Hirata K, and Sekiguchi H. Effects of ammonia gas mixed into intake air on combustion and emissions characteristics in diesel engine. *In: ASME 2016 Internal Combustion Engine Division Fall Technical Conference*. New York: American Society of Mechanical Engineers; 2016.

[107] Pochet M, Dias V, Jeanmart H, Verhelst S, and Contino F. Multifuel CHP HCCI engine towards flexible power-to-fuel: numerical study of operating range. *Energy Procedia* 2017;105:1532–1538.

[108] Niki Y, Nitta Y, Sekiguchi H, and Hirata K. Emission and combustion characteristics of diesel engine fumigated with ammonia. *Large Bore Engines; Fuels; Advanced Combustion* vol. 1. New York: American Society of Mechanical Engineers; 2018. p. V001T03A016.

[109] Şahin Z, Ziya Akcanca İ, and Durgun O. Experimental investigation of the effects of ammonia solution (NH_3OH) on engine performance and exhaust emissions of a small diesel engine. *Fuel* 2018;214:330–41.

[110] Pochet M, Jeanmart H, and Contino F. A 22:1 compression ratio ammonia-hydrogen HCCI engine: combustion, load, and emission performances. *Front Mech Eng* 2020;6:43.

[111] Pyrc M, Gruca M, Jamrozik A, Tutak W, and Juknelevičius R. An experimental investigation of the performance, emission and combustion stability of compression ignition engine powered by diesel and ammonia solution (NH_4OH). *Int J Engine Res* 2020;22:2639–53.

[112] Niki Y. Experimental investigation of effects of split diesel-pilot injection on emissions from ammonia-diesel dual fuel engine. *In: ASME 2021 Internal Combustion Engine Division Fall Technical Conference*. New York: American Society of Mechanical Engineers; 2021. p. V001T01A002.

[113] Mounaïm-Rousselle C, Mercier A, Brequigny P, Dumand C, Bouriot J, and Houillé S. Performance of ammonia fuel in a spark assisted compression ignition engine. *Int J Engine Res* 2021;23:781–92.

[114] Frost J, Tall A, Sheriff AM, Schönborn A, and Hellier P. An experimental and modelling study of dual fuel aqueous ammonia and diesel combustion in a single cylinder compression ignition engine. *Int J Hydrogen Energy* 2021;46:35495–510.

[115] Yousefi A, Guo H, Dev S, Lafrance S, and Liko B. Combustion performance and emissions of an ammonia/diesel dual-fuel engine: effects of ammonia substitution ratio and diesel injection timing. *Proceedings of Combustion Institute – Canadian Section Spring Technical Meeting,* 2022, pp. 1–6.

[116] Scharl V, and Sattelmayer T. Ignition and combustion characteristics of diesel piloted ammonia injections. *Fuel Commun* 2022;11:100068.

[117] Liu L, Wu Y, and Wang Y. Numerical investigation on the combustion and emission characteristics of ammonia in a low-speed two-stroke marine engine. *Fuel* 2022;314:122727.

[118] Al-Dawody MF, Al-Obaidi W, Aboud ED, *et al.* Mechanical engineering advantages of a dual fuel diesel engine powered by diesel and aqueous ammonia blends. *Fuel* 2023;346:128398.

[119] Subramani AK, Duraisamy G, Govindan N, and Hossain AK. An innovative method of ammonia use in a light-duty automotive diesel engine to enhance diesel combustion, performance, and emissions. *Int J Hydrogen Energy* 2024;49:38–58.

[120] Li Z, Zhang Z, Fan Y, *et al.* Fuel reactivity stratification assisted jet ignition for low-speed two-stroke ammonia marine engine. *Int J Hydrogen Energy* 2024;49:570–85.

[121] Zhang Y, Li G, Zhang Z, *et al.* Mechanistic investigation of ammonia adsorption and reforming over Ni, Ru loaded catalysts toward green H_2 production of marine engines: DFT calculation and experimental evaluation. *J Alloys Compd* 2024;1001:175124.

[122] Dong D, Wei F, Long W, *et al.* Optical investigation of ammonia rich combustion based on methanol jet ignition by means of an ignition chamber. *Fuel* 2023;345:128202.
[123] Wei F, Lu M, Long W, *et al.* Optical experiment study on Ammonia/Methanol mixture combustion performance induced by methanol jet ignition in a constant volume combustion bomb. *Fuel* 2023;352:129090.
[124] Dong D, Wei M, Zhang Z, *et al.* Enhanced ignition possibilities of ammonia by the prechamber fueled methanol: rich, stoichiometric and lean combustion evaluations. *Sustain Energy Technol Assess* 2024;64:103723.

Chapter 6

Combustion systems for ammonia-fuelled gas turbines and other propulsion devices

H. Shi[1], *S. Mashruk*[1], *M. Alnajideen*[1] *and A. Valera-Medina*[1]

6.1 Introduction

Gas turbines are systems designed to provide large power outputs employing advanced thermodynamics, fluid mechanics, material science and combustion principles, amongst many other complex fields that are involved in the design and development of these units. Therefore, talking about gas turbines could cover an entire book on its own rights. However, this chapter solely attempts to elucidate the advantages of using gas turbines for power and propulsion generation whilst ensuring that ample emphasis on the combustion system of these units is properly addressed. Further, the chapter approaches the challenges of using ammonia as a fuelling vector, raising the potential of the molecule as a gas turbine fuel that can potentially support the transition to the decarbonisation of both the power and aerospace sectors.

The chapter is divided into an introduction where the concept of gas turbines for power and propulsion is briefly explored. Following sections present some details of current combustion systems used in these units, whilst also bringing up the complexities of using ammonia in gas turbine combustors. Various technologies capable of supporting this potential application are discussed. The units employ state-of-the-art techniques that are under assessment worldwide. The chapter ends with details of several programmes that are currently ongoing to stabilise and mitigate unwanted emissions from gas turbines powered with ammonia. Conclusions around future trends are also presented as guidance for readers interested in pursuing this exciting field of ammonia to power.

6.2 Why gas turbines

Amongst developers, engineers, scientists and users, there has always been a constant debate about the use of various powergen technologies – 'gas turbines are

[1]College of Physical Sciences and Engineering, Cardiff University, UK

more efficient at greater load…', 'internal combustion engines are more robust and can also cope with heavy loads…', etc. It is not uncommon to hear prolonged debates around the subject. However, there is a reason for that. Mainly, each system will possess its strengths depending on the conditions where they are intended to be commissioned. Operating at high altitude locations will have an impact on these units, whilst humidity, fuel availability and turnover time are all major issues that need to be considered when selecting a gas turbine or an internal combustion engine.

Along those lines, companies have dedicated considerable efforts to depicting the best scenarios for each of their technologies. Whilst this brings a great variety of players to the discussion, it is clear that there is a certain number of manufacturers that would be promoting both technologies. In Europe, Siemens Energy is one of these companies that has profoundly documented the issue, thus giving guidance to their customers when selecting a power unit. In a paper entitled 'Natural Gas Cleanliness' [1], and further details presented by Siemens-Energy [2], the company uses natural gas as the fuelling vector to evaluate these two technologies. In the assessment, they find several strengths in the use of one or the other technology (Table 6.1 and Figure 6.1).

Table 6.1 Advantages in the use of gas turbines or internal combustion engines [2]

High combined cycle efficiency	**Open cycle efficiency**
High-temperature cogeneration or combined heat and power (CHP): steam and hot air	Low-temperature CHP: hot water
Low combustion emissions	Frequent starts and stops
Gas fuel flexibility	Liquid fuel flexibility
Lowest maintenance costs	

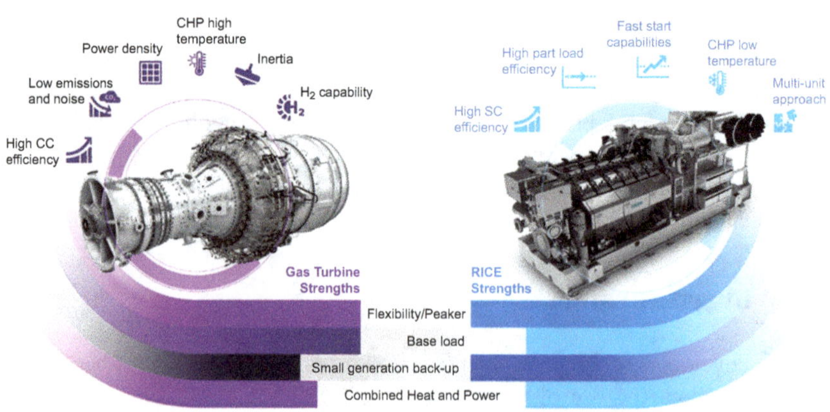

Figure 6.1 Comparison between gas turbines and internal combustion engines strengths. Courtesy of Siemens Energy [2].

As the analysis continues, several points are considered for the comparison in large power generation plants,

- Emissions: Emissions of NOx, CO, particulates and many other molecules (such as methane, which has a global warming potential (20 years) 84 times greater than CO_2) are emitted in a significantly lower quantity by gas turbines compared to gas engines. The reason for this is the different combustion principle: whilst in internal combustion engines, like in cars, thousands of single explosions with very high temperatures in the cylinders generate the power, gas turbines have a continuous combustion process at a lower and more evenly distributed temperature profile.
- Fuel flexibility: This parameter is essential for the decarbonisation of the energy sector, as new low or zero-carbon fuels will be employed for the generation of power. Gas turbines can operate using a wide range of fuels including non-conventional fuels (such as hydrogen and biogas) or conventional fossil-based fuels such as natural gas, LPG and diesel. Similarly, gas engines can run on very low heating value (LHV) fuels like syngas (4.5 MJ/Nm^3). They can also burn biogas, landfill and higher LHV gases such as propane and LPG (110 MJ/Nm^3). Recently, it has been demonstrated that both units can operate using ammonia, a point that has been raised previously for internal combustion engine technologies, and that will be further expanded for gas turbines in the next sections.
- Efficiency: Operating at part-load or full load are considerations that will not only impact the economics of the plant but also the generation of CO_2 and other pollutant emissions. When it comes to small plants with low power outputs, engines deliver the best electrical efficiency (i.e. 40–45% efficiency in the range of 300–2000 kW [2]). Meanwhile, gas turbines in combined cycles can reach efficiencies above 60%. Similarly, gas engines can provide an availability average of over 96%, whilst industrial and aero-derivative gas turbines can average over 97% availability [2]. Furthermore, gas turbine maintenance tends to be lower on a €/MWh basis.
- Flexibility: This parameter depends on the particular application and location that is under consideration. Since future systems will require fast reactions to the change in the energy grid (i.e. as a consequence of the intermittency of renewable energies), in general, the starting and loading phases from warm standby are similar for gas turbines and gas engines, usually 5–10 minutes. However, a combined cycle plant's start-up time is much longer than that of simple cycle plants [2].
- Grid balancing: For future grids, synchronous inertia becomes a paid good. Thus, in general, the larger the synchronous electrical generator is, the less equipment in operation is necessary to have a significant impact on the grid stability. Therefore, gas turbines have an advantage over engines since the former have higher inertia due to their higher speeds and larger power trains. This is especially important when operating under combined cycle modes,

which as previously raise, also helps in the reduction of emissions. Since engines have very low inertia due to lighter crankshafts and slower rotation, grid balancing is better accomplished using gas turbines.

Therefore, it becomes evident that there are several niche applications in which gas turbines outstand internal combustion engines in the energy sector.

Similarly, when a comparison is done for aerospace applications, the use of gas turbines is also critical, as there are important differences between systems as a consequence of their efficiencies and propulsion outputs, which are dependent on the particular application. For example, engines are still used in small aircraft, not only because these units are cheaper and easier to maintain, but also because they tend to suffer lower inefficiencies when flying at lower altitudes. However, turbines (for turboprop systems) have added reliability, whilst also being compact and lightweight.

In terms of larger units and fast-delivering devices, gas turbines surpass engine-based propulsion, with larger civil aircraft and military fighters being propelled using gas turbine technologies. Their superior performance at high altitudes, with their power-to-weight advantage, make them the conventional unit for these applications. Further, gas turbines can also be employed in helicopters, military ships, locomotives and tanks, whilst large cargo and passenger cruises might also benefit from the use of gas turbines due to their fast reaction and higher rotational speeds.

Therefore, gas turbines are technologies that will stay for the future, ensuring that novel fuels such as ammonia are properly consumed whilst reducing emissions and increasing fuel flexibility.

6.3 Power and propulsion cycles using gas turbines

6.3.1 Brayton cycle

The Brayton cycle models the operation processes of a simple gas turbine engine. The cycle was named after George Bailey Brayton, an engineer who patented a constant-pressure reciprocating oil-burning engine in 1872. Chen *et al.* [3] illustrated that the Brayton cycle is advantageous over other powering cycles regarding efficiency.

1. Idealised Brayton cycle

The idealised Brayton cycle consists of four thermodynamic processes, as shown in Figure 6.2.

1–2: Isentropic compression. The fresh air at initial state (1) enters the compressor, where the steady flow is pressurised to achieve higher pressure and temperature (2).

2–3 Isobaric heat addition (normally supplied by fuel combustion). The compressed air (2) proceeds into the combustion system, where the fuel is injected and mixed with the air flow. The mixture is burned at very high temperatures and the pressure is kept constant (3).

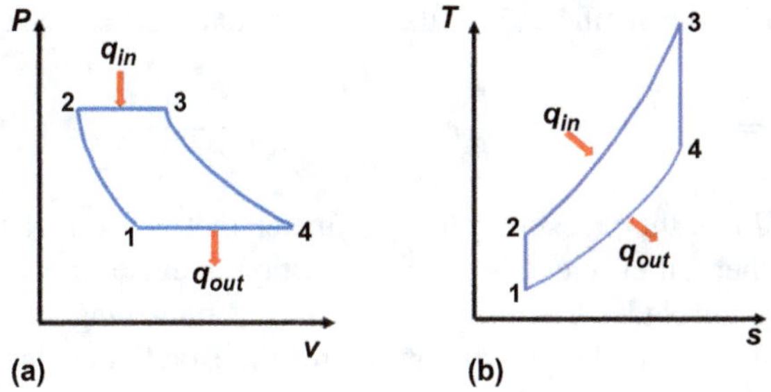

Figure 6.2 Idealised Brayton cycle: (a) P–v diagram and (b) T–s diagram

3–4 Isentropic expansion in the turbine and exhaust nozzle. The combustion products (3) at high pressure and temperature expand through the turbine, spin the rotating blades and turn the generator to produce electricity. Part of the work extracted from the turbine is used to drive the compressor. After that, the expanded exhaust gas (4) leaves the turbine.

4–1 Isobaric heat rejection. After interacting with the surroundings, the exhaust gas (4) is restored to the compressor inlet state (1) after going through a heat exchanger where heat is rejected to the surroundings.

According to the *P*–*v* and *T*–*s* diagrams in Figure 6.2, the values of heat addition (process 2–3) and heat rejection (process 4–1) can be calculated as

$$q_{in} = h_3 - h_2 = C_p(T_3 - T_2) \tag{6.1}$$

$$q_{out} = h_4 - h_1 = C_p(T_4 - T_1) \tag{6.2}$$

where q is energy (out or in from the cycle), h enthalpy, C_p heat capacity and T temperature. The numbers refer to Figure 6.2. The thermal efficiency of the ideal Brayton cycle is given as

$$\eta_{th,Brayton} = 1 - \frac{q_{out}}{q_{in}} = 1 - \frac{C_p(T_4 - T_1)}{C_p(T_3 - T_2)} = 1 - \frac{T_1}{T_2}\frac{T_4/T_1 - 1}{T_3/T_2 - 1} \tag{6.3}$$

For the isentropic processes 1–2 and 3–4,

$$\frac{T_1}{T_2} = \left(\frac{P_1}{P_2}\right)^{(k-1)/k} \tag{6.4}$$

$$\frac{T_4}{T_3} = \left(\frac{P_4}{P_3}\right)^{(k-1)/k} \tag{6.5}$$

where k is the specific heat ratio of the gases in the process. As $P_2 = P_3$ and $P_1 = P_4$,

$$\frac{T_1}{T_2} = \frac{T_4}{T_3} \tag{6.6}$$

Therefore, the thermal efficiency of the ideal Brayton cycle can be expressed as [4]:

$$\eta_{th,\ Brayton} = 1 - \frac{T_1}{T_2} = 1 - \left(\frac{P_1}{P_2}\right)^{(k-1)/k} = 1 - \frac{1}{r_p{}^{(k-1)/k}} \tag{6.7}$$

where $r_p = P_2/P_1$ is the pressure ratio, parameter that is a direct function of the increase on the thermal efficiency of the ideal Brayton cycle. It must be remember that the gas turbine blades can only withstand a certain range of the maximum temperature, which limits the pressure ratios for the Brayton cycle.

As presented in Figure 6.3, for the fixed minimum (inlet) and maximum temperatures, the network output of the Brayton cycle first increases then decreases with the pressure ratio. Therefore, the compromise between the work and the pressure ratio normally needs to be made. For most gas turbines, the pressure ratio ranges from 11 to 16 [5].

Compared to the steam turbine (Rankine cycle) and diesel propulsion systems (Diesel cycle), the gas turbine operating with the Brayton cycle provides higher power density, better reliability, longer service life and more convenient operation. For example, the boot time for a typical steam turbine is normally around 4 hours, whilst it can be reduced to less than 2 minutes for a gas turbine. To be specific, the benefits offered by the Brayton cycle are listed as follows [6]:

1) High efficiency. For a steam-based Rankine cycle, the system efficiency can reach 40% when operating at 25 MPa and 620 °C [7], which are limited by material characteristics. In comparison, a current gas-based Brayton cycle can operate at a higher temperature between 1300 °C and 1500 °C with a maximum efficiency of around 60% [8].

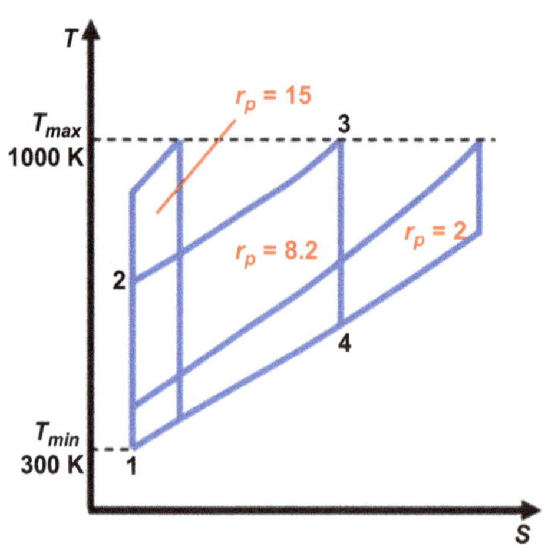

Figure 6.3 T–s diagram for Brayton cycles with different pressure ratios

2) Attractive electrical output characteristics. The Brayton cycle can operate at a constant power without degradation, allowing the use of AC power transmission. Additionally, the alternator rotor of the Brayton unit can operate at high speeds to provide high-frequency output (800–1600 Hz) and high voltage from the windings.
3) Scalability to high power. For the system that runs with the Brayton cycle, heat exchanger scales and turbo machinery sizes are not proportional to the power level. Therefore, machinery equipment can be made compact and lightweight according to specific requirements. This property makes the Brayton cycle very attractive for scaling power applications.
4) Ease of manufacturing. In contrast to the cumbersome assembly process of solar PV arrays, the modules of Brayton cycle have great potential for automated production. This contributes to cost and schedule benefits. Recently, many components can be manufactured using 3D printing, reducing single pieces, bolts and nuts, whilst increasing complexity.
5) High reliability, long life and radiation tolerance. The high reliability potential of the Brayton cycle is evidenced by thousands of long-run aircraft APUs and power generation units. The constant power output of the Brayton cycle ensures a stable and reliable power supply used for aerospace missions or small/medium/large power generation.

2. Deviation of actual gas-turbine cycles from ideal Brayton cycle

Due to gas turbine irreversibilities such as frictions within the compressor, gas turbine and heat exchanger, there are pressure drops for the working fluid with the inherited specific entropy increasing when passing through different components. As presented in Figure 6.4, these gas turbine irreversibilities and losses reduce the output work produced by the turbine and raise the work input of the compressor, hence they decrease the net work output of the power plant.

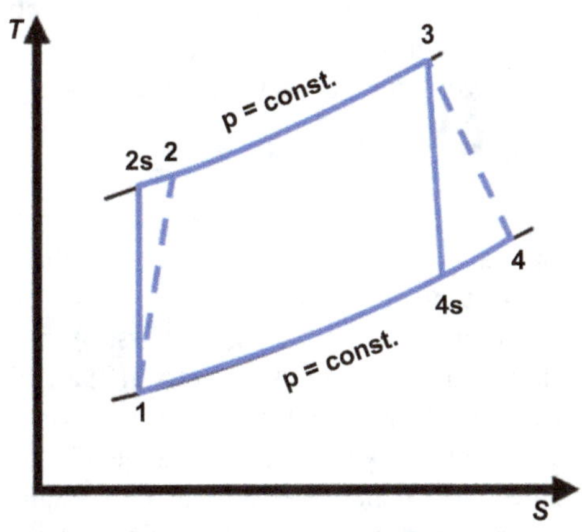

Figure 6.4 T–s diagram for real Brayton cycles

Although the combustion irreversibility is significant, the irreversibilities from the turbine and compressor also impose critical effects on the gas turbine efficiency. After the developments for decades, the efficiencies of the turbines and compressors can reach 80% to 90%. According to Figure 6.4, the isentropic turbine and compressor efficiencies in real Brayton cycle are given by,

$$\eta_t = \frac{h_3 - h_4}{h_3 - h_{4s}} \tag{6.8}$$

$$\eta_c = \frac{h_{2s} - h_1}{h_2 - h_1} \tag{6.9}$$

where states 2s and 4s are the corresponding states for the ideal Brayton cycle, whilst 2 and 4 are the actual exit states of the compressor and the turbine, respectively.

Many studies have been carried out to analyse losses and optimise the power and efficiencies of Brayton cycles. For instance, Radcenco *et al.* [9], Wang *et al.* [10] and Chen *et al.* [11] incorporated the irreversibilities into the power plant model, which resulted from the pressure drops along the flow path. Irreversibilities generated by heat transfer and fluid flows consequence of thermal resistances and pressure drops were investigated by Bejan [12] and Chen *et al.* [13]. Moreover, Angelino and Invernicci [14] carried out an exhaustive thermodynamic analysis and designed the most suitable fluid to operate the real gas Brayton cycles at the critical point, hence maximising efficiency. Overall, there are three main ways to improve the efficiencies of Brayton-based systems [15]: (1) better energy harvesting from heat sources; (2) increasing the inherent cycle efficiency; (3) reducing the power requirements of equipment (e.g. generator, heat exchanger, control system).

Brayton cycles can run with other fluids such as carbon dioxide and hydrocarbons, concepts that have been proposed and extensively studied [16–18]. Results on these fuels and fluids reported promising findings to use low to medium-grade heat for power generation, with systems being more compact and efficient.

6.3.2 Combined cycles

Combined cycles refer to a type of thermodynamic cycle that merge two or more cycles to generate more heat, power or both, thus improving overall efficiency and reducing fuel costs and exhaust gas emissions. The first thermodynamic combined power cycle was developed in the early 20th century [19] and the most widespread forms of this cycle are cogeneration [20], the Brayton–Rankine (gas-steam) and trigeneration [21] cycles. Combined cycles present a range of potential applications in industrial settings where a particular energy source is required, along with the requisite amount of process steam, to meet specific energy demands. Such demands may include the operation of specialised processes within the plant or the provision

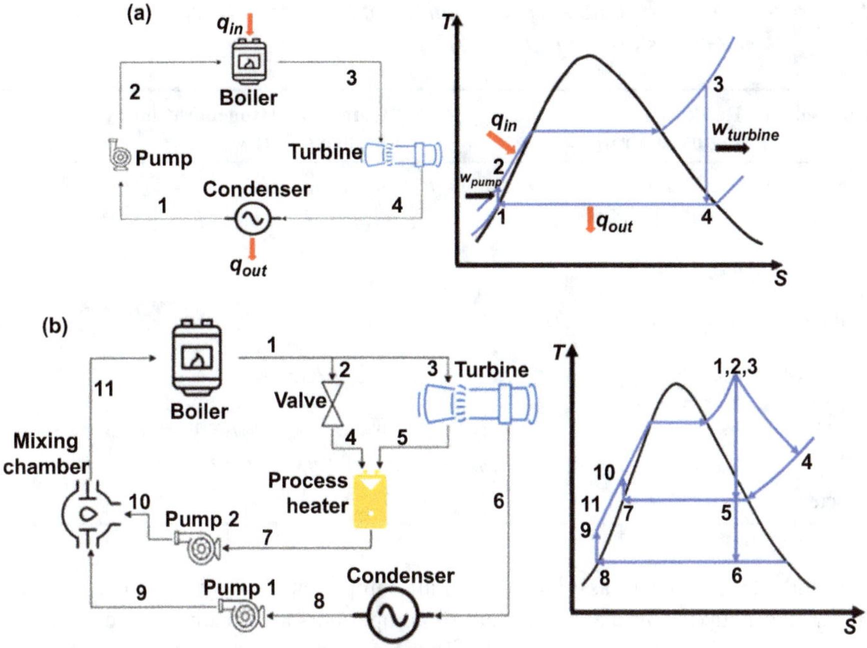

Figure 6.5 Schematic and T–s diagram of (a) the ideal Rankine cycle and (b) the combined cycle (cogeneration)

of space heating to the facilities. The schematic and *T–s* diagrams for the Rankine (steam) and combined (cogeneration) cycles are illustrated in Figure 6.5.

It can be seen from Figure 6.5 that with some predetermined adjustments, the combined (cogeneration in this example) cycle can contribute to overall efficiency, which is given as

$$\eta_{th} = \frac{w_{net}}{q_{in}} = 1 - \frac{q_{out}}{q_{in}} \tag{6.10}$$

$$\epsilon_{\mu} = \frac{w_{net} + q_p}{q_{in}} = 1 - \frac{q_{out}}{q_{in}} \tag{6.11}$$

where η_{th} and ϵ_{μ} are the thermal efficiency of the Rankine cycle and the utilisation factor for the cogeneration cycle, respectively. w, q_{in}, q_{out} and q_p are the net work (for the cycle), heat transferred to the system, heat rejected from the system and process heat supplied by the system, respectively.

Basically, the thermodynamic processes in either Rankine or cogeneration cycle are (a) an isentropic process in pumps and turbines and (b) constant pressure in boilers, condensers and process heaters. In order to distinguish between the cycles shown in Figure 6.5 mathematically, the energy equations of the

Table 6.2 The steady-state energy equations of the thermodynamic processes of the cycles in Figure 6.5

Component	Rankine (Steam) cycle, Figure 6.5(a)	Combined (Cogeneration cycle), Figure 6.5(b)
Pump ($q = 0$)	$w_{pump,\ in} = h_2 - h_1 = v_1(P_2 = P_1)$	$w_{pumpI,\ in} = h_9 - h_8 = v_8(P_9 = P_8)$ $w_{pumpII,\ in} = h_{10} - h_7 = v_7(P_{10} = P_7)$
Boiler ($w = 0$)	$q_{in} = h_3 - h_2$	$q_{in} = h_1 - h_{11}$
Turbine ($q = 0$)	$w_{turbine,\ out} = h_3 - h_4$	$w_{turbine,\ out} = (h_2 - h_5) + (h_5 - h_6)$
Condenser ($w = 0$)	$q_{out} = h_4 - h_1$	$q_{out} = h_6 - h_8$
Network	$w_{net} = q_{in} - q_{out}$ $= w_{turbine,\ out} - w_{pump,\ in}$	$w_{net} = q_{in} - q_{out} = w_{turbine,\ out} - w_{pump,\ in}$ $q_p = h_4 + h_5 + h_7$
Efficiency	$\eta_{th} = \frac{w_{net}}{q_{in}} = 1 - \frac{q_{out}}{q_{in}}$	$\epsilon_\mu = \frac{w_{net} + q_p}{q_{in}} = 1 - \frac{q_{out}}{q_{in}}$

thermodynamic process associated with all components are systematically listed in Table 6.2, taking into account that all components are steady-flow devices.

Cogeneration can be implemented using a variety of topping and bottoming cycle configurations [22,23]. Examples of such configurations include, but are not limited to, combining nuclear power plants with process heaters or steam generators, cogeneration systems that use organic waste materials (i.e. biomass resources), paper mills, refineries and chemicals (i.e. syngas) to generate heat and power, or incorporating renewable sources such as solar or geothermal energy with gas or steam turbines [22–27].

The cogeneration cycle is also called combined heat and power (CHP), which refers to the generation of a by-product whilst making the main product [28]. This concept of CHP offers a more efficient way to utilise fuel by recovering the thermal energy that is usually lost during electricity generation for heating purposes. The concept enables CHP plants to achieve overall efficiency levels of up to 80–90%, compared to the approximately 30–50% efficiency of conventional power plants that do not recover waste heat [29]. In the UK, as an example, CHP can save around 20% on energy bills and reduce carbon emissions by up to 30% compared to the separate means of conventional generations via boiler and power stations [30].

The use of CHP has become an increasingly important goal in the fight against climate change. Governments around the world have recognised the benefits of CHP and have implemented policies and incentives to encourage their development and use. For instance, the European Union has set a target for CHP to account for 18% of total electricity production by 2050, whilst in the United States, the Department of Energy has established programmes to promote the adoption of CHP in commercial, industrial and institutional sectors [30–33].

The feasibility of CHP can be significantly enhanced, where opportunities for trigeneration cycles exist. In such a scenario, the heat produced by the CHP plant is utilised as a primary energy source to provide cooling through the use of an absorption chiller. Trigeneration, or combined cooling, heating and power (CCHP), can be also associated with water distillation/desalination or hydrogen production [34].

Trigeneration cycles, also known as CCHP, is a type of combined cycle that produces three forms of energy: electricity, heating and cooling [35]. In this cycle, a primary energy source such as natural gas or biogas is used to drive a generator that produces electricity. The heat generated during this process is recovered and used for heating purposes, and the remaining heat is then used to generate chilled water for cooling applications.

The trigeneration cycle offers a more efficient and sustainable approach to energy production and management as it can reduce overall energy consumption and greenhouse gas emissions. For instance, the heat recovered from the power generation process can be used to provide space heating in buildings or to drive absorption chillers that generate chilled water for air conditioning. This approach can result in energy efficiencies of up to 90%, compared to the approximately 35–50% efficiency of conventional power plants that do not recover waste heat [36,37].

Trigeneration systems can be tailored to meet the energy demands of various applications, such as hospitals, data centres and manufacturing facilities. They are particularly effective in locations with high cooling demands, such as in hot and humid climates [38]. Trigeneration systems can also provide energy resilience, where a facility generates its own electricity, heating and cooling, reducing reliance on external energy sources.

One of many examples of the trigeneration cycles was carried out by Guteša Božo *et al.* [21,39]. A two-shaft gas turbine plant facility was developed by incorporating a basic humidified Brayton cycle and a reversed Brayton cycle with steam generation. This system works as a trigeneration cycle, utilising the benefits of ammonia through fuel preheating, ammonia-cooled inlet air at the compressor and steam generation for addition to the combustion chamber. The resulting system is shown in Figure 6.6.

The study found that the use of a two-shaft gas turbine system with recirculated steam, fuel preheating and cooled inlet air was capable of delivering efficiencies of approximately 60%, which is competitive with fossil fuel-based cycles. The study also found that the production of pollutants such as NO_x and N_2O, as well as unburned traces of ammonia, was minimal. However, further development is needed to address issues related to ammonia/hydrogen reactivity, as different models predict different results. Overall, the use of this system offers high performance, efficient employment of stranded renewable energy as backup power and parallel streams employed to support further heating applications, increasing the overall efficiency of the cycle. As energy prices continue to rise and concerns about climate change grow, the use of trigeneration systems is expected to increase in the coming years [40,41].

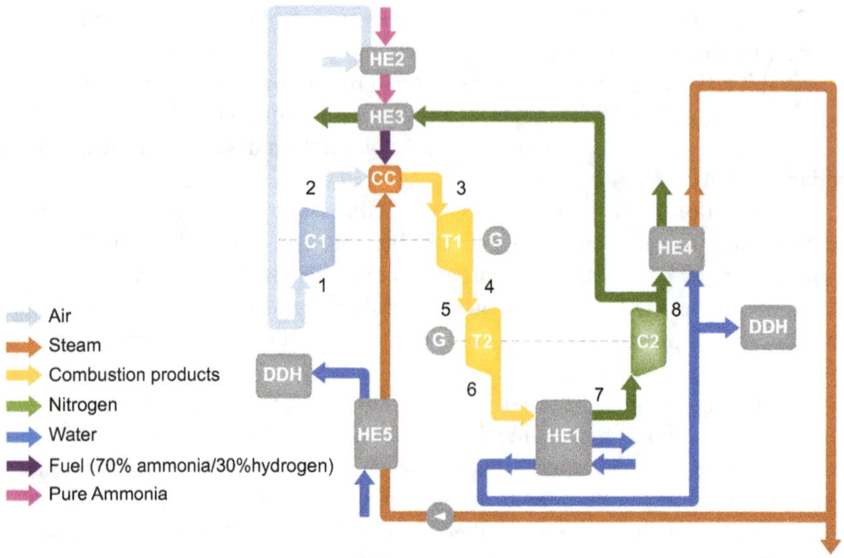

Figure 6.6 Ammonia trigeneration cycle. Reprinted, with permission, from [39].

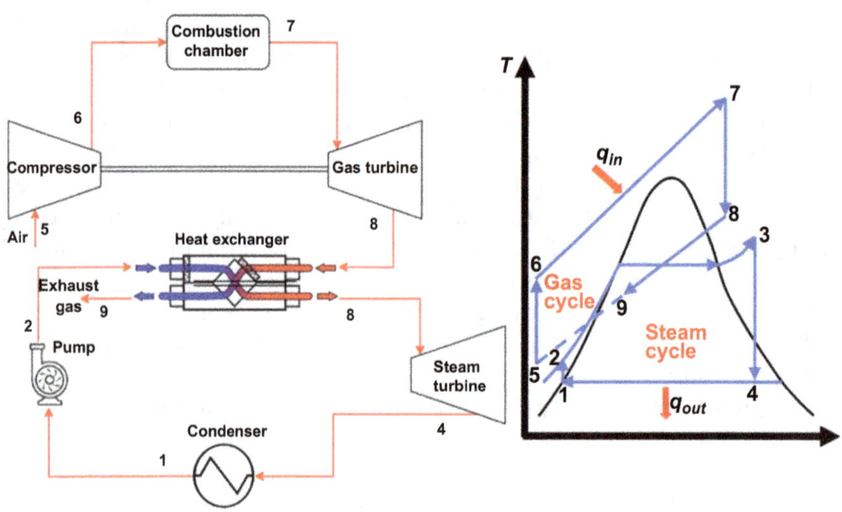

Figure 6.7 Schematic and T–s diagram of combined Brayton–Rankine (gas-steam) cycle

The most common type of combined cycle is the combined Brayton–Rankine cycle. This cycle combines two thermodynamic cycles: the Brayton (gas-turbine) cycle is combined with the Rankine (steam-turbine) cycle as shown in Figure 6.7. In the combined cycle, the gas turbine from the Brayton cycle is used to generate electricity through the combustion of fuel, usually natural gas. The resulting high-temperature exhaust gases from the gas turbine (~620 °C) are directed to a heat exchanger in the Rankine cycle, where they heat high-pressurised water, producing superheated steam. This high-pressure high-temperature steam is then utilised to generate electricity by driving a steam turbine. The heat exchanger is also known as the 'Heat Recovery Steam Generator, HRSG'.

The advantages of the combined Brayton–Rankine cycle are that it allows for higher overall efficiency and greater flexibility in power generation without increasing the capital costs greatly [42,43]. Technically, the gas turbine can be turned on and off quickly (compared to the steam turbine in a pure Rankine cycle that is slower to respond to changes in demand), allowing for a more rapid response to changes in electricity demand. The combined power cycle also provides steady power output over a long period.

Nowadays, many power plants operate on combined cycles, and many more existing gas or steam power plants are being converted to combined-cycle power plants due to the higher efficiency that is well over 40% as a result of conversion and improvements [5]. Combined-cycle power plants have seen significant advancements over the years, which have led to increased efficiency and reduced emissions. Some of the key advancements in combined-cycle power plants include the use of high-efficiency gas turbines, high-efficiency HRSGs, advanced control systems to optimise performance and reduce emissions, and the flexibility for using the most cost-effective fuel source [44,45].

6.3.3 Humidified cycles

Humidified cycles are thermodynamic cycles that involve adding water vapour to the working fluid in a power plant. This can be done in several ways, including by injecting steam into the combustion chamber or by using evaporative cooling to cool the compressor inlet air as previously mentioned in [39]. Humidified cycles have several benefits, including increased efficiency and reduced emissions. The added water vapour in the working fluid allows for higher compression ratios and increased turbine inlet temperatures, which results in increased efficiency. Additionally, the added water vapour can help reduce NOx emissions by lowering the combustion temperature and providing additional oxygen for combustion [46].

Humidified cycles are commonly used in gas turbine power plants, particularly in combined cycle power plants. In these processes, the exhaust heat from the gas turbine is used to generate steam, which is then fed to the turbine to generate additional power. By humidifying the working fluid in the gas turbine cycle, the efficiency and power output of the combined cycle can be increased. A detailed examination of diverse humidified gas turbines was carried out by Jonsson and Yan [47] through an extensive review of the literature. The following thermodynamic conclusions have been drawn:

- Humidified gas turbines have the potential to achieve electrical efficiencies equivalent to or greater than combined cycle efficiencies.

- Amongst humidified gas turbine cycles, the evaporative gas turbine with a humidification tower has the highest efficiency.
- Humidified gas turbines have higher specific power outputs than dry gas turbine cycles.

Regarding the cost of humidified gas turbines, the following conclusions have been made:

- The specific investment cost of a mature humidified gas turbine is expected to be lower than that of a combined cycle since the cost of the steam turbine in the bottoming cycle can be avoided. The difference in specific investment cost is greater for small power outputs.
- Evaporative cycles with part-flow humidification offer lower specific investment costs than evaporative cycles with full-flow humidification whilst maintaining the same or higher electrical efficiency.

Further development work and potential applications for humidified gas turbines have also been discussed:

- Currently, only steam-injected gas turbines are commercially available. The evaporative gas turbine has been demonstrated in a pilot plant, whilst the water-injected cycle has had a less experimental investigation.
- Humidification of the gas turbine working fluid results in a volumetric flow rate mismatch between the compressor and expander. New gas turbines that can accommodate the increased flow rate are still under development.
- Intercooling and recuperation are required for humidified gas turbines to achieve the highest efficiencies. Further development is necessary before these heat exchangers become standard components in gas turbine power cycles.
- Aeroderivative gas turbines are better suited for operation in humidified cycles than industrial gas turbines.
- Humidified gas turbines can recover water from the exhaust gas and reuse it in the cycle, making it a promising option for the cogeneration of power and district heating. The specific investment costs and costs of electricity and heat should be significantly lower than for combined cycles.
- Power production integrated with carbon dioxide removal is feasible for humidified gas turbines since the CO_2 removal system can be smaller than for a combined cycle, resulting in lower costs. Additionally, humidification can reduce the generation of NO in the combustion process.

Overall, humidified cycles are a promising technology for improving the efficiency and reducing the emissions of power plants, particularly in applications where gas turbines are used. How ammonia fuelling can fit into humidified power cycles has also been addressed by the work of Guteša Božo *et al.* [21,39], who recognised the potential of recovering water from flue gases, using this water for humidifying the combustion process, Figure 6.7. The outcomes of that work show that humidification with ammonia/hydrogen blends provide not only stable combustion at water/fuel rates up to 40%, but also an acute reduction in NOx emissions potentially a consequence of O/H/OH pools and lower reaction temperatures within the ammonia DeNOxing zone (between 1350 °C and 1450 °C).

6.3.4 Pulse detonation engines (PDE)

Another method that is employed using ammonia, and that is relatively recent, is the use of Pulse Detonation systems for gas turbines and rocket propulsion. Although more complex in nature than conventional gas turbines, as a consequence of the much higher operating pressures, the concept promises greater efficiencies.

Self-propagating combustion waves are classified depending on flame velocity and pressure variation. Thus, they are classified into two types, Deflagration and Detonation.

6.3.4.1 Detonation and deflagration

When reactants are ignited, a combustion wave is generated, propagating away from the ignition source. As this combustion wave passes through the reactants, these are transformed into combustion products by breaking chemical bonds and releasing stored chemical energy. The energy is then converted into thermal and kinetic energy. Further, the high energy release enables a self-sustained process unless the balance is lost (i.e. energy losses and instabilities have a greater impact on the combustion process).

Flame speed, which is the speed at which the flame front travels in an unburned mixture, is related to many factors such as air-fuel-ratios, geometrical aspects and obstacles in the path of the waves. The speed will determine if the propagation wave scenario is deflagration (low speed, conventional flames), deflagration-to-detonation transition (DDT) or detonation (fast speed, supersonic propagation). Although deflagration and detonation are similar processes, they represent opposite limits of the spectrum of reactive flow phenomena [48].

The detonation phenomenon was discovered by the end of the Nineteenth century when the diagnostic tools developed by the time enabled observation of rapid combustion phenomena and measurement of combustion waves propagation velocity. The first theory predicted the detonation velocity, which is based on Rankin and Hugoniot analyses of conservation equations across a shock wave [49]. The theory was formulated independently by Chapman and Jouguet [50] shortly after the discovery of the phenomenon. Zeldovich [51], von Neuman [52] and Doring [53] progressed this theory by taking into account the time that radicals take to form and initiate the reaction using Rankin-Hugoniot curves, thus developing an improved theory called ZND [54].

The detonation wave in the Chapman–Jouguet (C–J) theory has been considered as a discontinuity with infinite reaction rates, with all energy added instantaneously in the chemical reaction zone that follows the shock wave. Both reactants and products are modelled as perfect gases.

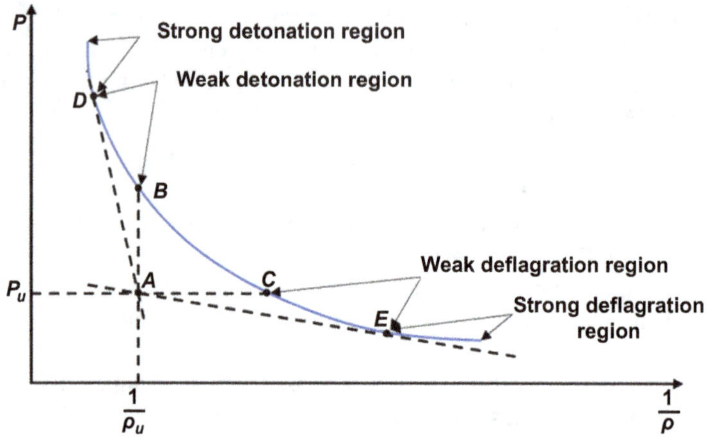

Figure 6.8 A schematic of the Rankine–Hugoniot curve and Rayleigh lines showing the regions distinguished along the curve. The subscript 'u' stands for 'unburned' reactants.

Although the detonation wave speed has two possible solutions, i.e. strong and weak detonation, these two solutions converge when the speed of detonation goes to a minimum, and there is no solution provided for detonations below this minimum velocity. For the strong detonation, the flow downstream of the detonation wave is subsonic and pressure and density are higher than those for weak detonation with supersonic downstream flows [55]. Thus, these regions can be depicted in the Rankine–Hugoniot curve (Figure 6.8), where the pressure (P) and densities (ρ) are compared. The lower part of the curve represents the deflagration solutions, whilst the upper part represents the detonation solutions [54].

In the ZND approach, a shock wave followed and strengthened by the reaction region, will increase the pressure and temperature of the reactants upstream of the shock to ensure a high-rate reaction so that the energy released is enough to keep the shock moving at a high speed whilst assuming no reaction takes place in the shock wave region. This can be depicted in Figure 6.9. Four regions are defined for this phenomenon, namely (1) unburned gases, (2) shock wave, (3) induction zone and (4) heat addition zone. As can be observed, the sudden change occurs at the shock wave, whereas the physical properties of the induction zone are almost flat due to its low reaction rate. As the reaction rate soars in the heating addition zone, physical properties will change sharply, and the C–J state will be reached by the end of this regime.

The key advantage of detonation combustion is that it generates pressure gain in the system, compared to the pressure loss produced by deflagration combustion.

Pulse detonation engines operate by injecting propellants into long cylinders that are open on one end and closed on the other. When gas fills a cylinder, an

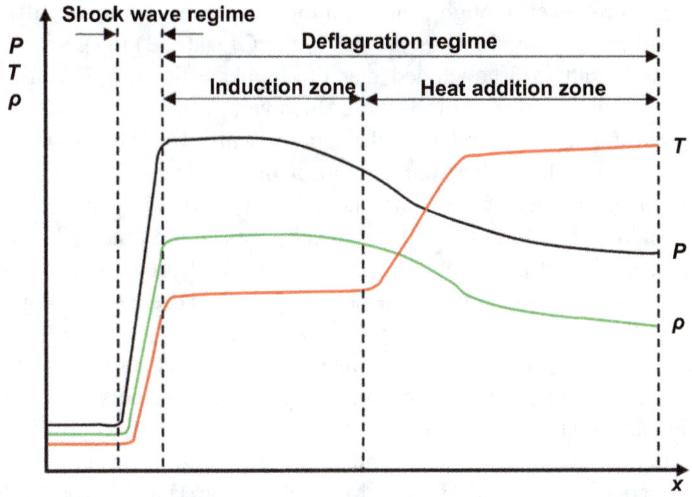

Figure 6.9 Physical properties behaviour through a ZND detonation wave

igniter–such as a spark plug–is activated. Fuel begins to burn and rapidly transitions to a detonation. The shock wave travels through the cylinder at ten times the speed of sound, so combustion is completed before the gas has time to expand [56].

The cycle is based on the following steps: Purging the gases enables a reduced pressure (usually atmospheric or a slight vacuum) (1), allowing the intake of both fuel and air under pressure (this initial pressure will have great implications in the detonation parameters and efficiency of the process, as discussed in the following sections) (2). Ignition follows (3), hence creating a deflagration wave that will transition to detonation. The detonation wave propagates along the tube (4) until it reaches the exit, exhausting the products of the reaction and producing thrust (5). This thrust can be used for direct propulsion purposes or to enable rotation of turbomachinery, which can be fully used for power generation or turbo-propeller systems.

As previously mentioned, PDE systems generate greater pressures, hence making the combustion process via detonation more efficient whilst avoiding other heavy, costly components. The reduction in the need for pressurisation provides a reduction in the weight of the pressurisation tank or turbopumps to be used in the spacecraft. Thanks to the work gained from the thermodynamic cycle, the thrust value obtained per unit of fuel weight increases.

Furthermore, the ZND cycle is the most efficient amongst other options, Figure 6.10. The gas is shock compressed, followed by exothermic chemical reactions. The shock compression is assumed to proceed along the inert Rankine-Hugoniot (2–2') which is equivalent to assuming local thermodynamic equilibrium (i.e. leading shock wave, no reaction). Subsequent heat release from exothermic chemical reactions, even though a nonequilibrium process, is identical to that of Rayleigh heating (2'–3) and brings the gas to the CJ state (3) [57].

Heiser and Pratt [58] concluded that the higher efficiency is caused by the lower entropy increase during combustion. However, further work [59] showed that the processes are periodic and time-dependent, with also the problem of work extraction starting at the highest temperature point (3). Thus, the work proposed a simplified model that delivers an equivalent steady thermodynamic state that can provide good comparisons to those developed by Heiser and Pratt [58]. This model does not account for entropy generation due to shock in detonations, but it is a good approximation of the physical phenomena taking place in these technologies, which are best modelled by the Humphrey cycle [60].

Different to the ZND cycle, where the process (1–2) is assumed to follow the inert Ranking–Hugoniot curve, the Humphrey cycle considers this part of the process to be an isentropic compression. Further, the combustion process between cycles is also different, with the Humphrey cycle assuming a constant volume process instead of a detonation [61]. Meanwhile, the Brayton (or Joule) cycle assumes a constant pressure process (Figure 6.10), which in turn leads to more entropy generation and less turbine work.

Therefore, the specific impulse that these systems can deliver is superior to other technologies, Figure 6.11. Also, it must be noticed that the selected fuel is of great importance to reach a higher specific impulse. Comparisons in the literature have been made between hydrogen and other hydrocarbon fuels. However, ammonia is still under basic research and has not reached this maturity for these types of systems. However, as explained in the following sections, the chemical has

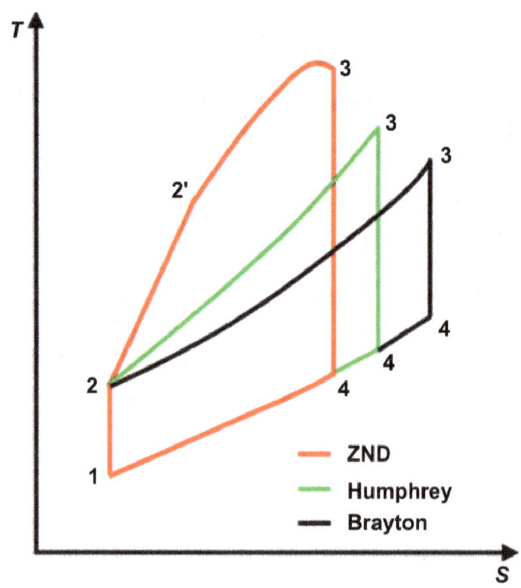

Figure 6.10 T–s diagrams of various power cycles, including the ZND cycle

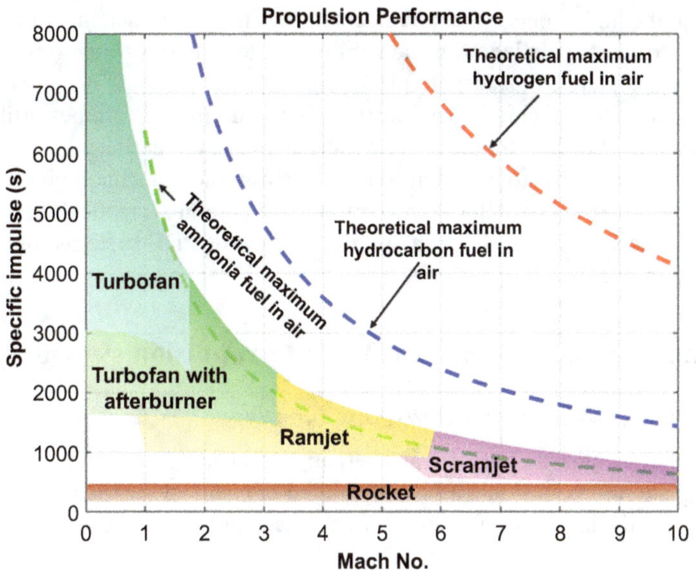

Figure 6.11 Specific impulse and Mach number based on fuel and technology. Courtesy of the Aerodynamic Research C [63].

that potential and can provide unique opportunities for energy storage usable in PDE systems for propulsion and power generation.

A variant of the PDE is the rotating detonation engine (RDE). The basic concept is that the detonation wave moves along a circular channel (ring) instead of a straight tube. Although similar in design to a pulsed detonation engine, the rotating one is more efficient because the waves continuously rotate around the chamber, whilst the impulse units require the chambers to be purged after each pulse [62].

There are many barriers to the use of PDE systems though. One of the main challenges in the practical implementation of pressure gain combustors (PGC) into systems such as gas turbines is the lack of turbomachinery that can efficiently harvest work from the PGC exhaust gas, an area of active research [60]. To address this challenge, one can follow two methods, (a) a combustor outlet geometry designed to adapt the exhaust stream of a PGC, with a turbine operating at its design efficiency; (b) a dedicated turbine design that could directly expand the outlet flow of a PGC. A much more rational approach would be to optimise the combination of a PGC outlet geometry and an adapted turbine design to achieve the maximum possible work extraction [60]. Another challenge is the impact that turbine cooling would have on the efficiency of the process (which seems to reduce efficiencies by 1%–3% points [60]).

Another factor affecting the practical implementation of PDEs is the difficulty in achieving consistent detonations within the combustion chamber using a short tube due to the higher energy requirements to start the phenomenon. A more useful method is to start a deflagration and transition to a detonation using obstacles (spirals, orifice plates, nozzles, etc.) [63].

Therefore, PDE technologies and PGC systems in gas turbines still require further research and development. However, as presented earlier, the use of detonation of ammonia/hydrogen blends is viable for these technologies, reducing inefficiencies in the overall cycle. Emissions and other practical barriers still require further assessment that can support the use of PDE systems in aerospace and even power generation systems.

6.4 Ammonia as a fuel for GTs and propulsion systems

6.4.1 Swirling flows with gaseous blends

The use of ammonia as a potential alternative fuel in gas turbines has received increasing attention, as it has a high hydrogen content, low carbon footprint (i.e. when produced via renewable sources) and the potential for low-emission combustion. However, the application of ammonia as a fuel suffers from some challenges such as low flame speed, low reactivity and high NOx emissions.

Several studies in recent years showed that ammonia-air flames could be successfully stabilised using swirling flows, main technology used in gas turbine combustors, which creates internal recirculation zones to move hot products and active radicals from downstream to the upstream zones of the main combustion field, generating full mixing of the gaseous blends [64]. The recirculation facilitates the unburned gas preheating and promotes the reaction rates of the mixture. Moreover, the recirculation and enhanced fluid dynamics anchor the flame and prolong the residence time of radicals, allowing sufficient time for the reactions [65].

Hayakawa *et al.* [66] reported the NH_3/air flame could be stabilised with swirling flows at different inlet velocities and equivalence ratios without any additives, Figure 6.12. Syred [67] also stressed how strong swirl flows recirculate the heat and radicals to the base of the flame and enhance its stability. Similarly, Okafor *et al.* [64] used a swirler in an ammonia MGT combustor to promote the air/fuel mixing and fluid residence time, leading to efficient combustion and low NOx emissions. For industrial applications, Kurata *et al.* [68] successfully achieved power generation at 50 kW using turbulent non-premixed NH_3/air flame in a micro gas turbine, which applied a high swirl number (0.88) and a jet flow of ammonia.

Somarathne *et al.* [69,70] numerically studied the premixed NH_3/air swirl flames in a gas turbine-like combustor at elevated pressure, and the results showed the pure ammonia flame could be stabilised over a wide range of pressures, equivalence ratios and turbulence intensities up to 39.1 m/s, Figure 6.13. Further numerical work by Mashruk *et al.* [71] investigated an industry-scale (~10 MW) humidified Rich-Quench-Lean (RQL) system using five different reaction mechanisms and demonstrated that $70/30_{VOL.\%}$ NH_3/H_2 blend can potentially have

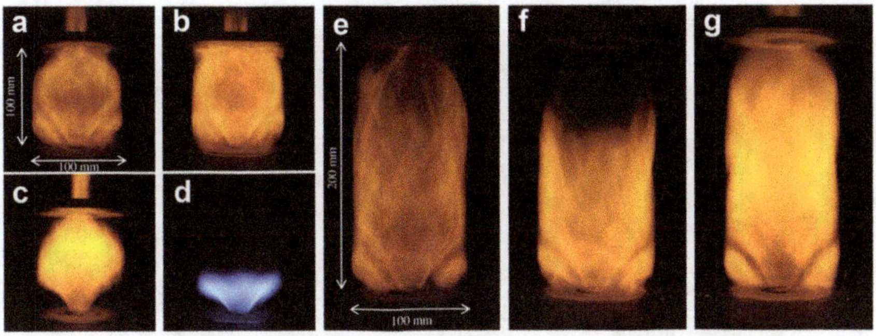

Figure 6.12 Stabilised flames in the swirl burner with different fuel blends, equivalence ratios (∅), cylindrical liner lengths (h) and mixture inlet velocities (Uin): (a) NH_3/air flame, ∅ = 1.0, h = 100 mm, Uin = 3.14 m/s; (b) NH_3/air flame, ∅ = 1.2, h = 100 mm, Uin = 3.14 m/s; (c) NH_3/air flame, ∅ = 1.2, h = 100 mm, Uin = 3.35 m/s; (d) CH_4/air flame, ∅ = 1.0, h = 100 mm, Uin = 3.14 m/s; (e) NH_3/air flame, ∅ = 0.8, h = 200 mm, Uin = 3.14 m/s; (f) NH_3/air flame, ∅ = 1.0, h = 200 mm, Uin = 3.14 m/s; (g) NH_3/air flame, ∅ = 1.2, h = 100 mm, Uin = 3.14 m/s. Reprinted, with permission, from [66].

flue gases with 99.97% water, nitrogen and oxygen with only 100 ppmv (15% O_2) NOx emissions, hence paving the way to the use of ammonia for clean, large power applications.

For the use of ammonia swirling flames in gas turbines, challenges remain related to the NOx emissions at different conditions. Valera-Medina *et al.* [72], suggested that stratified injection with a weak swirl could be employed to optimise the power generation of ammonia fuel blends. The study demonstrated that CO and NO_x emissions were significantly decreased when the equivalence ratio was above 1.1 for the NH_3/CH_4/air flame. Yang *et al.* [73] investigated the emissions of NH_3/CH_4/air flames in a tangential swirl burner, and they proposed an ammonia co-combustion technology with a central ammonia jet surrounded by a group of separately tangential swirl methane jets, which could effectively reduce NO_x emissions with ammonia blending ratios at 20% and equivalence ratios between 0.5 and 0.9. Kunkuma *et al.* [69] introduced a secondary air injection at elevated pressure for turbulent NH_3/air premixed swirl flames and the results indicated that NO and unburnt NH_3 were reduced with increasing pressure, especially at fuel-rich conditions, Figure 6.14. Similar conclusions were drawn by Xiao *et al.* [74]. This phenomenon was explained by Chai *et al.* [75] that showed how the increasing pressure promoted some three-body reactions $H + OH + M = H_2O + M$ and $H + O_2 + M = HO_2 + M$, thus suppressing NOx emissions.

Recently, Ilbas *et al.* [76] investigated the use of ammonia-assisted kerosene swirling flames with pure oxygen in a gas turbine combustor, the results showed that both medium swirling flow and ammonia addition changed the flame position. Zhang *et al.* [77] demonstrated that increasing the swirl number of the NH_3/CH_4

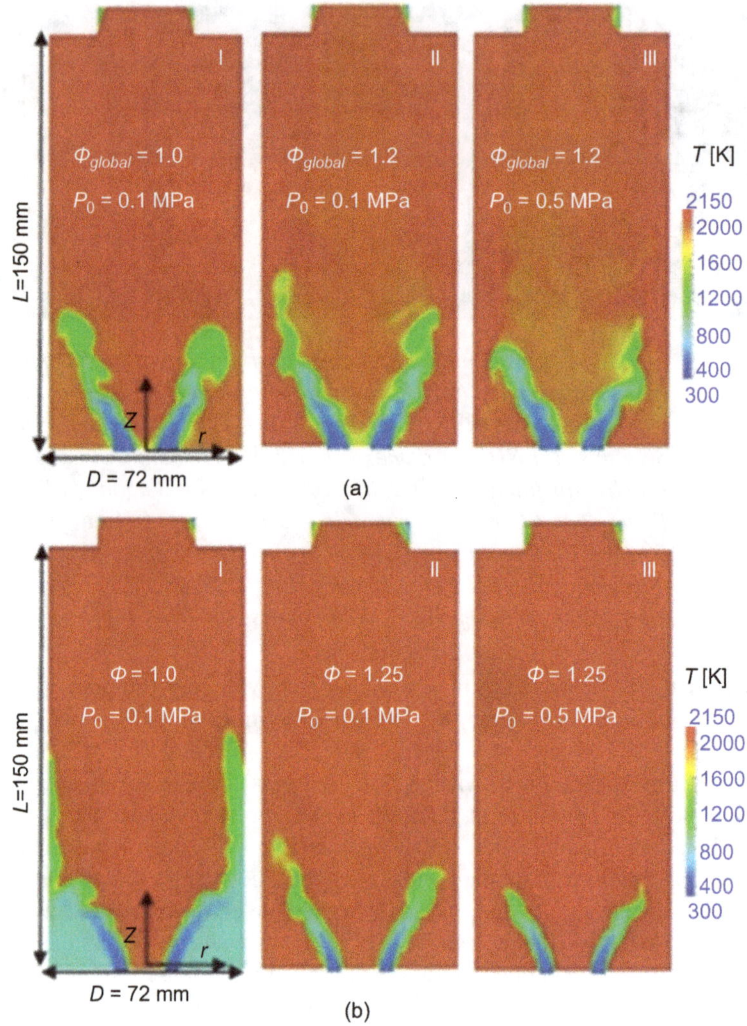

Figure 6.13 Temperature distributions in terms of global (or) and: (a) non-premixed; (b) premixed flames. Reprinted, with permission, from [70].

(50/50 wt%) flame could extend its lean blow-off limit. Valera-Medina *et al.* [78] studied the effects of equivalence ratio, ambient temperature and bypass air on the NO_x emissions in an optical generic swirl-burner. The results implied that high inlet air temperature raised NO_x and OH reactivity, whilst reducing the flame size. Khateeb *et al.* [79] studied the combustion performance of NO emission of NH_3/CH_4/air swirl flames (Figure 6.15), using different blends and equivalence ratios. They demonstrated that mixing ammonia with methane could expand stable combustion ranges and suppress the flashback propensity. Some other

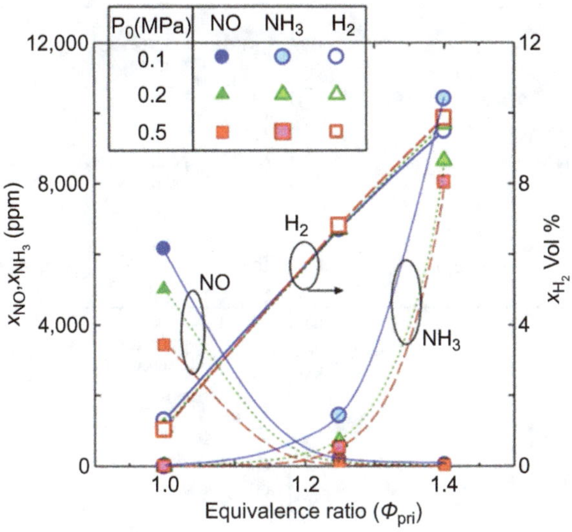

Figure 6.14 NO, NH_3 and H_2 of NH_3/air premixed flames in terms of pressure and equivalence ratio. Reprinted, with permission, from [69].

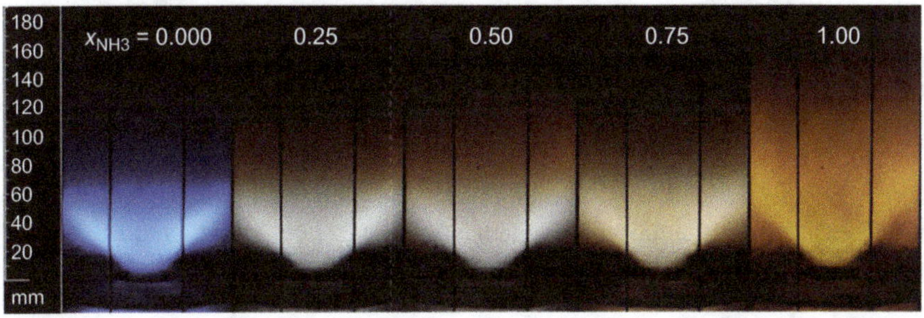

Figure 6.15 Time-averaged broadband images of ammonia-methane-air swirl flames for five different ammonia additions ranging from x_{NH3} = 0.00–1.00 (Re = 5000). Reprinted, with permission, from [79].

studies [78,80] also indicated that hydrogen could be introduced to stabilise the turbulent ammonia swirl flames. Additionally, Mong *et al.* [81] numerically examined swirl combustion characteristics of premixed ammonia/biogas, finding that the peak Damkhöler number is reduced with higher ammonia proportion whilst the produced CO_2 could decrease the NH_3/CH_4 mixture reactivity and cause more reacting flow fluctuations.

Studies by Khateeb *et al.* [82,83] and Zhu *et al.* [84] looked at the possibilities of ammonia/hydrogen swirling flames at various lean conditions. Results showed significant NO reduction at lean conditions for low turbulent swirling flames but the researchers did not report any NH_3, N_2O and NO_2 emissions. These findings led

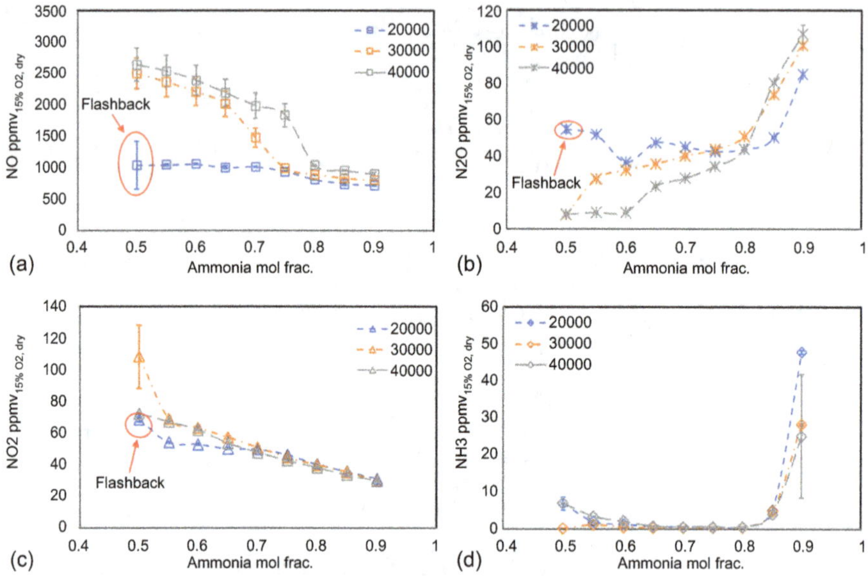

Figure 6.16 Sampled NO (a), N_2O (b), NO_2 (c) and NH_3 (d) emissions at different Re and Φ *= 0.65. Reprinted, with permission, from [85].*

Mashruk *et al.* [85] to investigate various ammonia/hydrogen blends (~50/50–90/10) at a fixed equivalence ratio of 0.65 in an industrial scale swirl burner at different thermal powers and Re conditions (Figure 6.16).

The results have shown a substantial amount of N_2O emissions across all fuel blends which has ~300 times GWP_{20} compared to CO_2 [86], with an increasing trend as ammonia content increased. Few other studies [87–89] investigated various ammonia/hydrogen swirling flames at a wide range of equivalence ratios (0.55–1.4). In general, these studies reported NO and NO_2 emissions peaking at around $\Phi \approx 0.85$, dropping more severely at rich conditions, whilst a significant amount of N_2O emissions were reported for equivalence ratios $(\Phi) < 0.7$, whilst leading to unburnt ammonia emissions at rich conditions due to the unavailability of oxidisers. The authors have identified decreased availabilities of H radicals and the reduced reactivity of the third body reaction $N_2O(+M) \leftrightarrow N_2 + O(+M)$ at the post-flame zone due to decreased flame temperature as main reasons for high N_2O emissions at these lean conditions. Mashruk *et al.* [90] also investigated premixed ammonia-methane-air swirling flames using chemiluminescent-based technologies and reported certain radical ratios can be used to infer ammonia mole fractions, equivalence ratios as well as NO emissions at highly turbulent flames.

Investigations are ongoing to explore the potential of using ammonia as a fuel for gas turbines. Efforts are focused on improving reaction processes, combustion systems and cycles, to make ammonia usage more efficient. Currently, most trials and demonstrations are being carried out using methane, and hydrogen as a blending agent, leading to a reduction in carbon emissions. Despite this progress, challenges such as NOx emissions, stability issues, material properties and cycle persist; however, the academic and industrial communities remain optimistic that ammonia will play a significant role in the energy sector in the future. New programmes across Europe, Asia and North America are also targeting the use of 'cracked' ammonia (H_2 and N_2) for gas turbine systems. The obtained blends will have traces of the three gases, $NH_3/H_2/N_2$, concept conceived to tackle costly retrofitting whilst using demonstrated technologies such as dry low NOx (DLN) emissions.

6.4.2 Liquid injection and jets

The interest in using ammonia in liquid stage has been growing with the development of various technologies. Since the chemical is normally stored in its liquid form, the use of multi-phase injection becomes feasible and potentially more efficient and economic with the reduction of evaporators and shorter start-up times.

Early attempts to use liquid ammonia were performed by the US military in collaboration with Solar [91]. Part of the development plan tried to use liquid ammonia to replace kerosene. Unfortunately, it was found that the system performance was poor and the use of liquid injection was discarded for the rest of the programme. Further work was also conducted by Verkamp *et al.* [92] and Bull [93] who also found that using liquid injection leads to unstable regimes as a consequence of the high latent heat of vaporisation of ammonia, which combined with its low flame speed and LHVs led to blow-off instabilities. Hence, it was suggested that ammonia could be co-fired using other more reactive fuels such as hydrogen or methane, as previously described.

Moving forward, it would take decades before further works on ammonia liquid jets would be employed in gas turbines. Tests and experiments have been conducted for internal combustion engines (as raised in other chapters). However, it was until recently that the National Institute in Advanced Industrial Science and Technology (AIST) took over the challenge and demonstrated the applicability of the concept, which was implemented in IHI engines later. Okafor *et al.* [94] developed, in collaboration with industrial partners, a system capable of delivering liquid ammonia into AIST's micro gas turbine (Figure 6.17).

Initial results denoted the reduced stability as the ammonia content was increased, whilst the additional injection of ammonia also led to a greater amount of NO_x emissions. The works are testimony that liquid injection is possible, although there are still many challenges related to its more efficient injection to mitigate both emissions and unstable regimes.

Further studies followed, and the group employed during that same campaign Mie Scattering to visualise the characteristic patterns of ammonia sprays directly used within a gas turbine combustion chamber. The ammonia pipeline had to be cooled down to avoid the fuel's vaporisation, a problem that can lead to cavitation at the nozzle. Results also showed that the airflow could enhance the spray dispersion. Furthermore, co-firing the sprayed ammonia with other more reactive

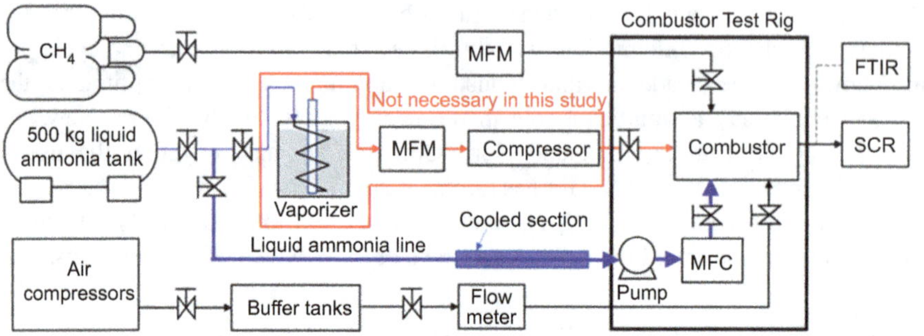

Figure 6.17 Liquid ammonia injection system. Emissions monitoring is performed on site. Reprinted, with permission, from [94].

fuels, i.e. methane, could also enhance the flame stability. This multi-phase injection could also prove to be highly beneficial in other combustion systems such as large boilers or furnaces. Another recent work [95] investigated a two-stage liquid ammonia spray combustion in a 230 kW gas turbine combustor, and it was reported that sufficient compensation for heat extracted from droplet vaporisation was necessary to improve efficiency in the primary combustion zone, thereby reducing the quantity of unburned NH_3 that was oxidised in the secondary zone.

For direct ammonia injection, the jet flow typically exists in a two-phase state after fuel injection at engine-like conditions due to its low boiling point [96]. On one hand, atomised droplets travel downwards in the liquid phase, and the droplet size is critical for fuel vaporization, combustion and emissions. On the other hand, ammonia flash boiling occurs when the subcooled liquid is depressurised promptly below its saturation pressure, leading to a phase change and forming a more uniform air/fuel mixture. This phenomenon has three stages: bubble nucleation, bubble growth and the two-phase flow [97]. Pelé *et al.* [98] applied Schlieren imaging to capture both liquid and liquid/vapour ammonia sprays (Figure 6.18), and compared the ammonia spray penetration length with those of biofuel, ethanol and gasoline.

The works can be coupled with other fundamental studies conducted on ammonia sprays. Li *et al.* [99] investigated the characteristics of superheated ammonia sprays at various temperatures and pressure using a constant-volume chamber. Results showed that spray bubble explosion intensity in the near-field is strongly dominated by superheating degrees and fuel viscosity in the entire flashing region, whilst the evolution of NH_3 spray penetration is similar to diesel under critical and initial flashing conditions. Similarly, Cheng *et al.* [100] showed that ammonia penetration is greater than the one observed in methanol and ethanol, with a larger cone angle and a faster evaporation rate. Pressure, as expected, led to shorter penetration and smaller spray areas.

When it comes to jets, due to the relatively poor reactivity of NH_3, there are significant challenges associated with flame stabilisation. Thus, there is also limited research studying the application of NH_3 in co-fired, non-premixed jets. The

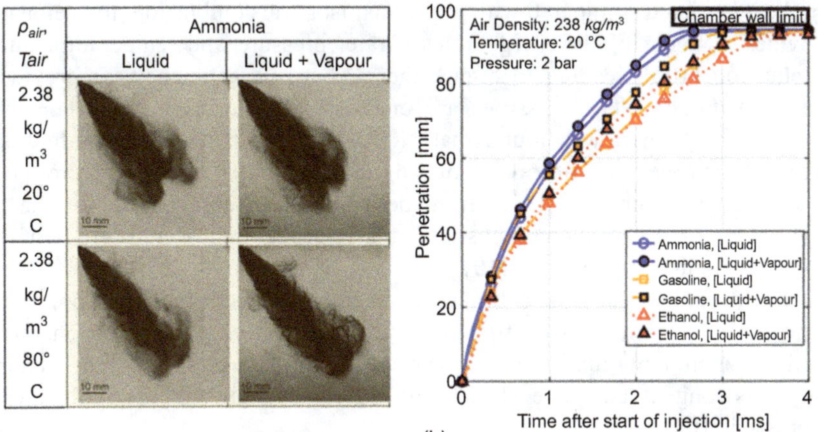

(a) (b)

Figure 6.18 Comparisons of liquid and liquid/vapour NH_3 sprays: (a) Spray shape and (b) Spray penetration (adjusted from [98])

work of Sullivan *et al.* [101] reported a combined experimental and numerical investigation into NOx reduction in a CH_4/NH_3 flame. However, the quantity of NH_3 introduced to the flame was low (up to 1000 ppm), with the study focussing on emission reduction pathways. NH_3 conversion to NO was shown to depend on fuel fraction, and reasonable agreement was demonstrated between the experimental results and modelled data. Woo *et al.* [102] also investigated NO_x emissions in co-flow CH_4/NH_3 jet flames, however, this time increasing the NH_3 fraction in the fuel to 15% and changing O_2/N_2 ratios within the oxidiser. In the case of pure oxygen, NO_x concentrations increased steadily with an NH3 in the fuel. However, a non-monotonic trend was observed as N_2 was introduced to the oxidant, which was replicated in corresponding simulations. Colson *et al.* [103,104] investigated the impact of NH_3 on flame stability, characterising operating regimes with flame lift-off and reattachment using a vertical non-premixed jet surrounded by an air co-flow. It was observed that once the NH_3 fraction was increased above 50% (molar), the flame was no longer stable and could not be sustained. Heat transfer to the burner was also studied [104], finding that when ammonia is compared to non-premixed methane, the flame position moves towards the jet and downstream as ammonia is increased. Temperatures at the burner also decreased with ammonia addition. However, the lack of further understanding of the fundamental characteristics of ammonia jets requires more research in this field.

6.4.3 Detonation

Ammonia was considered as a stable fuel because of its zero carbon emissions, excellent energy density and low liquefaction pressure. However, its wide application was limited by poor combustion characteristics and NO_x emissions.

On the other hand, compared with traditional isobaric combustion, the detonation provides a much higher energy release rate, pressure and temperature during combustion [105], leading to higher thermodynamic efficiency, as previously raised. Although NO_x emissions are one of the major concerns when using ammonia as fuel, it was found that NO_x emissions from ammonia-hydrogen mixtures can be significantly reduced at medium to high pressure [106]. Therefore, these characteristics from detonation can compensate for the disadvantages of ammonia combustion and improve the overall efficiency of ammonia-fuelled gas turbines [107].

Additionally, ammonia detonation can cause damages that are more serious than those by explosion or fire [108]. Consequently, the detonation engine fuelled by ammonia may provide a more efficient and cleaner way to generate power as long as adequate Health and Safety concerns of the molecule are addressed [109].

Since the 1960s, some attempts had been made to the liquid ammonia to power rocket engines (Figure 6.19). At the time, ammonia demonstrated better fuel performance and stability compared to other fuel mixtures [110]. Recently, more studies on ammonia detonation have been carried out to get deeper understanding of detonation parameters and DDT processes of ammonia, exploring the possibility of using ammonia-hydrogen mixtures as fuel in detonation engines. Huang *et al.* [111] first introduced the ammonia/oxygen-enriched air to achieve continuous rotating detonation in a hollow chamber where the oxygen mass fraction (OMF) needs to be kept above 43%. The captured detonation processes are presented in Figure 6.20.

Some numerical studies were implemented to analyse the end-gas autoignition and DDT phenomena in ammonia-hydrogen-oxygen/air mixtures. For example,

Figure 6.19 The ammonia-powered XLR99 rocket engine. Reprinted, with permission, from [151].

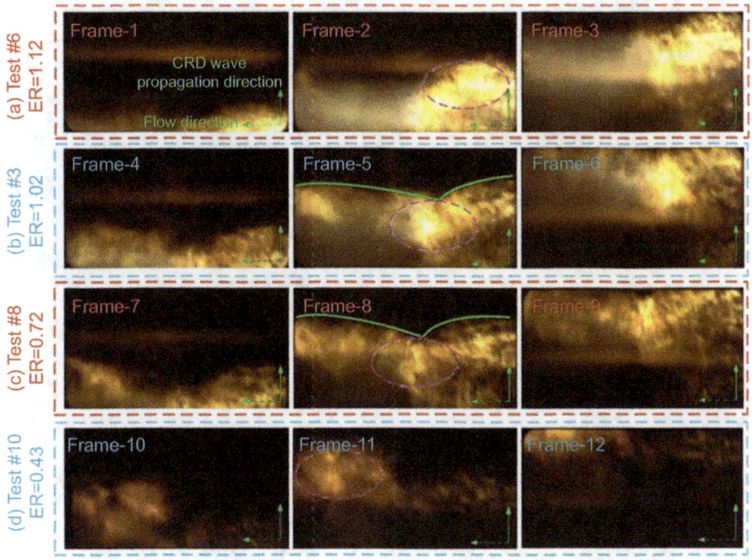

Figure 6.20 The captured ammonia detonation in Tests #3, #6, #8 and #10. Reprinted, with permission, from [111].

Yu and Zhang [112] observed different end-gas combustion modes in a confined reactor and analysed the evolutions of thermochemical states regarding different mixing ratios of ammonia/hydrogen fuel. According to the chemical explosive mode analysis, NH_2 and H_2NO are the most critical nitrogen-containing species when the hydrogen mole ratio is low, whilst they become less important at high hydrogen mole ratios. Figure 6.21 shows the developing detonation processes in a NH_3/H_2/air end-gas at the same hydrogen mole ratio whilst different initial temperatures. As shown by the detonation front (DF) in Figure 6.21(a), the pressure waves couple with the autoignition front (AF) in the residual end-gas near the wall, hence the detonative combustion proceeds and leads to high local pressure and heat release rate.

Zhu *et al.* [113,114] numerically modelled the flame propagation and DDT processes of NH_3/H_2/air flame. Despite the stable detonation could be obtained under favourable conditions, the detonation characteristics were influenced by the fuel composition. For example, the hot spot triggers the DDT before the flame front at high ammonia content, whilst the local explosion occurs along the flame front for DDT at high hydrogen content. Additionally, the explosions may be triggered again when the residual NH_3/H_2 mixes with the fresh air. Similarly, Sun *et al.* [107] simulated the detonation developments for NH_3/H_2 mixed with oxygen-enriched air in a RDE. The effects of hydrogen concentration and equivalence ratio on the rotating detonation performance were studied. According to the results, the detonation can stay stable only when the hydrogen concentration is above 0.2. With increasing the hydrogen concentration, the height of the detonation wave decreases,

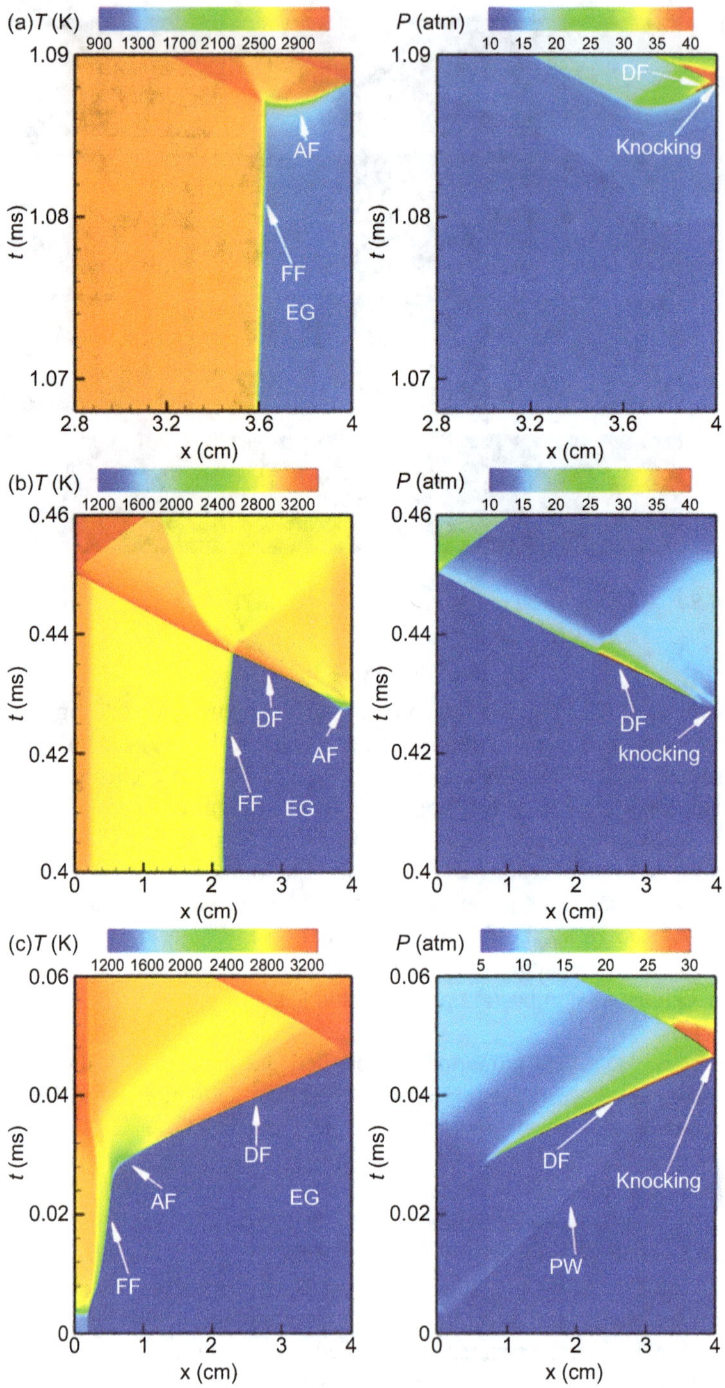

Figure 6.21 x–t diagrams of temperature and pressure at the same hydrogen mole ratio ($a_{H_2} = 0.9$) and different initial temperatures: (a) 900 K, (b) 1000 K and (c) 1100 K. Reprinted, with permission, from [112].

whilst the wave velocity goes up. The NO emission is more dependent on the equivalent ratio rather than hydrogen concentration under these conditions.

Moreover, Jing *et al.* [115] explored the flame propagation and detonation processes of ammonia/oxygen mixtures at different equivalence ratios in a large-scale tube. The results showed that the DDT process of NH_3/O_2 mixtures contained four stages: slow propagation stage, flame and pressure wave acceleration stage, fast propagation and detonation wave formation stage and detonation wave self-sustained propagation stage. The maximum detonation velocity and over-pressure for stoichiometric NH3/O_2 mixtures were 2480 m/s and 28.7 bar, respectively. These values were very similar to those of stoichiometric CH_4/O_2 mixtures.

6.5 Applications of ammonia in GT over the years

Over the years, there have been several programmes that have attempted to use ammonia for gas turbines. Here are some of those programmes that have served to the advancement of this technology and potential future units operating with such as fuel.

6.5.1 Solar GT/US Navy (post-WWII)

Works during the post-war period enabled the analysis of various systems for the use of ammonia in gas turbine engines to address issues with fuel scarcity, especially at the war front. The U.S. Army in collaboration with Solar worked on the study of ammonia-fuelled engines to determine the feasibility of new fuelling methods [91]. The programme, developed in various phases, initially tried to use liquid ammonia similarly as kerosene is employed in jet engines. Results denoted extremely poor performance, thus liquid injection was immediately discarded. Vaporised and catalytic combustion were assessed further. High altitude environments (i.e. −30 °C and 2440 m) and various combustor sizes were employed. Results denoted how the combustor size had an impact on the flame profile as a consequence of surface-to-volume ratio quenching, although it was recognised that higher wall temperatures would reduce this effect as occurs with hydrocarbon fuels. Therefore, new designs were attempted. Vaporised ammonia was evaluated using upstream and downstream injection, the former being superior. Particularly, the catalytic combustor, which burned some ammonia and decomposed the rest into hydrogen and nitrogen, needed to be kept at high temperatures for the reaction to be maintained, thus providing combustion efficiencies between 92% and 96%, whilst the vapour combustor delivered almost constant efficiencies ranging from 93.5% to 94.5% [91]. However, after 30 hours of tests, the catalyst suffered considerable deterioration. It was observed that complex control systems would be needed for this design. A schematic of the system can be seen in Figure 6.22.

Further analyses were conducted using a 250 hp Solar T-350 engine in Phase II of the programme (Figure 6.23). Two burner designs were employed, and as

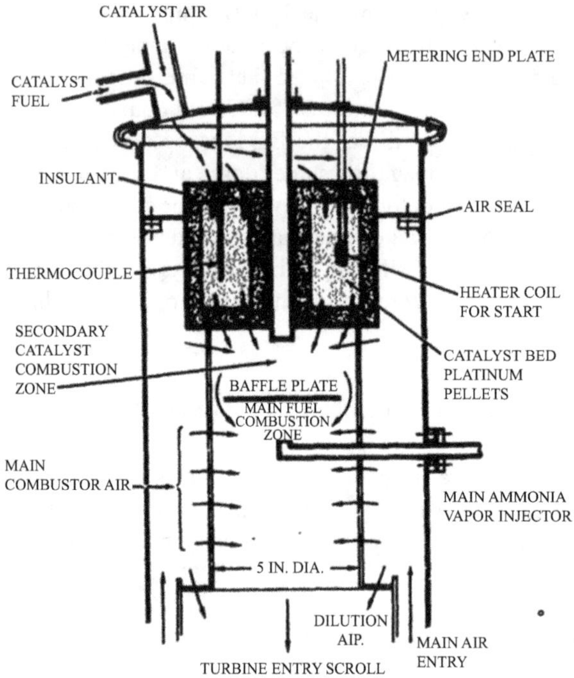

Figure 6.22 Oxidation catalyst combustor. Reprinted, with permission, from [91].

expected, a pre-heating period was needed especially for the catalyst to reach optimal operating conditions. During tests, the major issues were related to the ignition of the blend, which was extremely difficult for the case with vaporised ammonia. Although final tests were obtained, limited operability was observed due to the large size of the system and low flame velocities of the blend. Similarly, deterioration of the catalytic bed combined with unstable control was noted in the catalytic combustor. An interesting finding was that for other components, Inconel 600 showed no deterioration whilst being used in nozzles, scrolls or liners. Inconel X and 713C showed also satisfactory resistances. Transient tests were inconclusive for comparison purposes [91].

A complex fuel delivery system was designed for the unit (Figure 6.24). Final performance analyses were done comparing both systems with JP-4 hydrocarbon fuel. The highest discrepancy between ammonia and JP-4 tests was the lower turbine inlet temperature at rated loads, thus confirming that the use of ammonia at 2.35 times the rate of diesel would result in cooler turbine temperatures for a given power output. Therefore, the gain in power output using ammonia at the same inlet temperature would be between 9% and 13% for the vapour and catalytic systems, respectively. Moreover, there seems to be an improvement in diesel equivalent fuel flow, reflecting an improvement in thermal efficiency of ~2.5%. Another

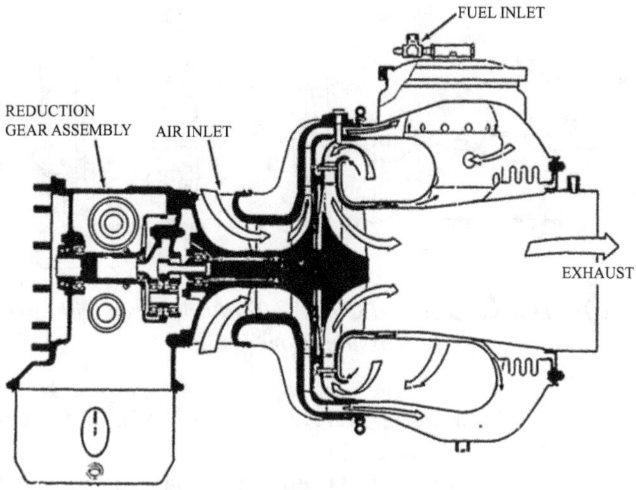

Figure 6.23 T-350 gas turbine cross-section. Reprinted, with permission, from [91].

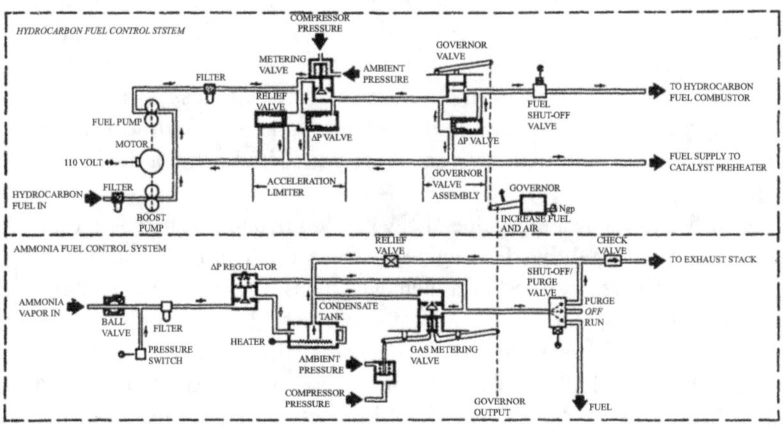

Figure 6.24 Fuel and control system schematic. Reprinted, with permission, from [91].

interesting point was that the efficiency of the catalytic system during performance analyses was above 100% with higher temperatures than those expected. It was theorised that the formation of NOx species could be culprit of the phenomenon by increasing the calorific value of the blend by ~8% [91]. Thus, it was concluded that ammonia is a satisfactory substitute for hydrocarbon fuels in simple-cycle gas turbines with no major increase in cost or complexity for these units. However, further research was recommended to develop the technology [91].

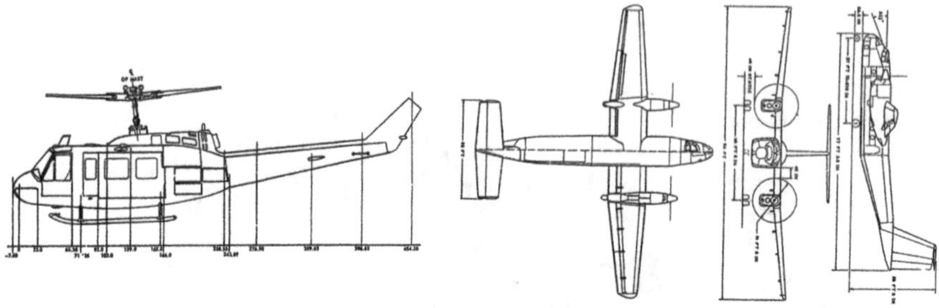

Figure 6.25 UH-1D helicopter and CV-7A aircraft powered with ammonia. Reprinted, with permission, from [91].

The results from these works were employed by the U.S. Army for the application of the Nuclear Powered Energy Depot Concept. Ammonia-fuelled gas turbines were evaluated in a UH-1D helicopter, and in the CV-7A fixed-wing aircraft (Figure 6.25) [116]. Different operating weights were used in combination with multi-fuel capabilities (ammonia and hydrocarbons). The missions required warm-up, takeoff, cruise, landing in a remote base and getting back to the initial point with a reserve allowance. The maximum permitted filling density of 87.5% (vol) was also used for the ammonia tanks. Fibreglass (small tanks) and steel-based materials (large tanks) could be employed for the manufacture of storage tanks. However, due to the long-distance missions, Inconel-600 vessels (with higher strength properties at high temperatures and less corrosion response to nitrogen) were used. Low and high-pressure tanks with cylindrical and spherical geometries were studied. Designs were conducted using a design-operating ratio of 2.5.

Due to the lower heating value (LHV) of ammonia, regeneration was recommended. Results showed that for engines such as the T53-L-11 and T64-GE-10 a maximum power increase of 15% was achieved by using ammonia (a consequence of the higher flow rates and increased heat capacities of the combustion products). Similarly, the use of regeneration only decreased the Specific Fuel Consumption (SFC) by 7.7% (from 1.10 to 1.02 at maximum power) compared to a 28.1% reduction using hydrocarbons. As per the UH-1D helicopter, investigations were performed at the constant gross weight (3630 kg) for endurance and maximum speed (up to 189 km/h). Regenerative ammonia conditions showed 34% less Specific Range values (nautical miles per pound of fuel) compared to hydrocarbon fuels without regenerative systems, whilst the ammonia regenerative system provided only one-third range capability of the hydrocarbon non-regenerative system, which improved to one-half when using the low-pressure storage tanks (i.e. 29% more than the high-pressure system) which showed higher performance than the high-pressure system. Further results depict that a regenerative ammonia system would require twice as much for a 40 miles helicopter mission, whilst missions longer than 60 miles would not be practical when transferring cargo [91]. The study also highlighted that maintenance would be similar to conventional units, although more complexity would be expected in control and fuelling systems, with an emphasis to avoid spillage to aluminium surfaces. Finally, it was estimated that costs related to the development, procurement and installation, and production of ammonia would be higher than for hydrocarbon engines. Conclusions suggested the withdrawal of ammonia fuelling for the Nuclear Powered Energy Depot Concept, raising that further implementations could employ ammonia gelling for better handling. More research was recommended in several areas but the use of ammonia in gas turbines was stopped for years.

6.5.2 mGT Tohoku

To address the threats posed by global warming, the Japanese government initiated research projects on energy carriers in 2013 and announced at the COP21 that the country would cut greenhouse gas emissions by 26% by 2030. Due to the high costs of transportation and storage of hydrogen, ammonia has attracted more attention as the potential energy carrier for this transition, a consequence of its high energy density and low liquefication pressure. The Japan Science and Technology Agency (JST) has supported R&D projects on ammonia production and utilization, including ammonia-fuelled power generation in gas turbines.

Since the direct combustion of ammonia in thermal power stations normally produces a large amount of NOx, Tohoku University, Toyota and the National Institute of Advanced Industrial Science and Technology (AIST) worked on a retrofitted micro gas turbine and redesigned the combustor to show the possibility of ammonia combustion in a such a unit [120]. The development programme was divided into three phases [121],

Phase 1 (2013–2014): Kerosene-ammonia co-firing to generate electricity with a simple ammonia supply facility.

Phase 2 (2014–2016): Using the improved ammonia supply facilities to implement 100% ammonia combustion and methane-ammonia co-firing for power generation.

Phase 3 (2016–2019): Redesign of the combustor to reduce NO_x emissions.

As the first step, AIST demonstrated a 50 kW class gas turbine powered by kerosene mixed with 30% ammonia. This facility was equipped with an SCR unit to reduce the generated NOx. After that, AIST and Tohoku University developed new ammonia combustion technologies to achieve flame stability (through the use of diffusion combustion) with lower NO_x emissions [122]. After the achievement of stable power outputs, the micro gas turbine was remodelled to burn pure ammonia for power generation [120].

As shown in Figure 6.26, due to the allowable regenerator temperature, unburned ammonia concentration and the rotating speed limits for remodelling the generator, the micro gas turbine system was tested to achieve 47 kW of power [121].

Figure 6.27 shows the unburned ammonia and NO emissions when the mGT system operated at different conditions, and it demonstrates how these emissions were influenced by the combustion air temperature. The unburned ammonia was negligible when the combustion air temperature was above 550 °C, whilst the NO went up linearly at high combustion air temperatures [121]. To address NOx emissions, the combustor was re-designed to apply two-stage (rich-lean) combustion. The effect halved NOx

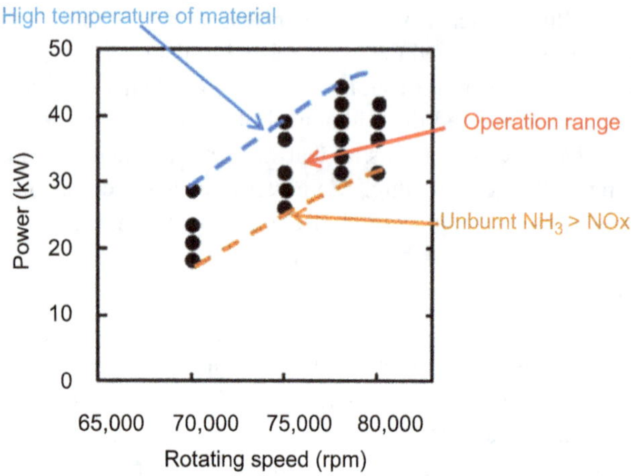

Figure 6.26 Operation range for NH_3–air combustion. Reprinted, with permission, from [121].

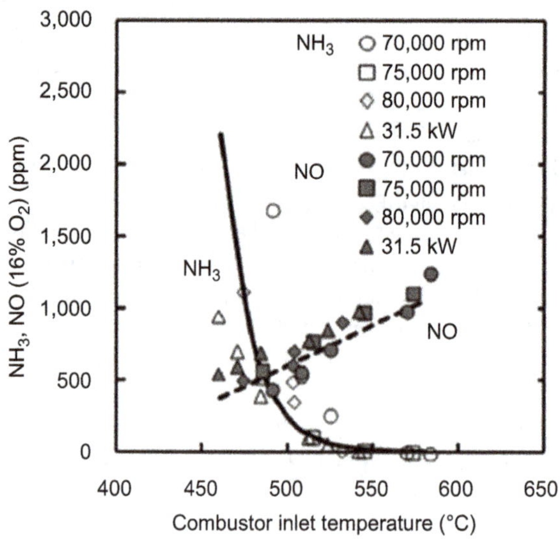

Figure 6.27 Variations of emitted NO and NH_3 concentrations with combustor inlet temperature for the NH_3–air combustion. Reprinted, with permission, from [121].

emissions. Later the combustor design was further optimised to promote the mixture uniformity and the NO concentrations were decreased below 200 ppm [121].

Further, Toyota Energy Solutions Inc. (TE) worked jointly with Tohoku University and AIST on the ammonia direct combustion of the mGT. TE manufactured a 300 kW

Table 6.3 Specification of 300 kW mGT. Permission from Springer [123].

Model		300 kW mGT
Generator output		295 kW (25 °C)
Power generation efficiency		17.4%
Motor	Type	Simple open cycle Single-shaft gas turbine
	Rotating speed	40,000 rpm
	Air flow rate	2.59 kg/s
	Pressure ratio	6.6
	Combustor inlet temperature	300 °C
	Fuel consumption (LNG)	0.054 kg/s
	Fuel consumption (converted to NH_3)	328.9 kg/h
	Exhaust gas temperature	496 °C
Generator type		Synchronous generator

mGT with 100% ammonia-only combustion for industrial use [123]. The 300 kW mGT was included in a power generation package, which also included a synchronous generator, a reduction gear, a fuel gas compressor, a lubricant and cooling system, a control panel, a high-voltage grid interconnection panel and a selective catalytic reductor for any unburned ammonia. The mGT specifications are listed in Table 6.3 [123].

To provide sufficient ammonia for the 300 kW mGT system, an ammonia supply facility was developed, which consisted of liquid ammonia tank (tank capacity: 50 t) and a vaporiser (vaporising capacity: 600 kg/h). Moreover, a gas compressor was installed between the ammonia supply unit and the mGT to raise the ammonia supply pressure and deliver the fuel directly to the unit [123].

To decrease NO_x levels, the SCR NOx removal unit was installed downstream of the mGT, which was capable of reducing the NO_x concentrations from 1000 to 30 ppm. This unit consists of 576 catalysts within a honeycomb structure (8 pieces × 8 pieces × 9 stages). The materials were activated at around 500 °C which corresponds to the exhaust temperature of the mGT [123].

Figure 6.28 shows the test results of ammonia-only combustion in the 300 kW mGT. With increasing the engine speed and load, the ammonia flow rate was gradually increased with the engine output increasing to the rated output (295 kW), whilst the liquid natural gas (LNG) flow rates were decreased automatically, reaching 100% ammonia-only combustion with support from the feedback control. The engine performance and exhaust gas properties are listed in Table 6.4 [123].

Future works will attempt to have the mGT optimised to reduce further NOx emissions. This includes concepts such as combustor cooling, fuel supply and waste heat recovery. The durability and reliability of the facilities under long-hour operation will be evaluated [123].

6.5.3 IHI 2MW gas turbine

Japanese IHI Corporation Ltd. demonstrated tests of ammonia/natural gas co-firing power generation with a 2 MW class gas turbine unit [124]. The testing apparatus

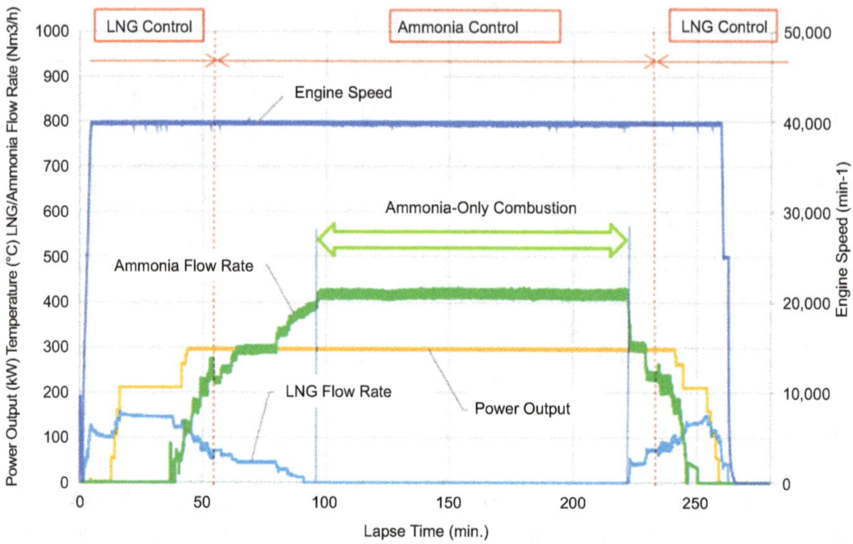

Figure 6.28 300 kW MGT operation with 100% ammonia combustion. Reprinted, with permission, from [123].

Table 6.4 Test result at 100% ammonia combustion. Permission from Springer [123].

Engine performance		**Measurement value**
Power output	kW	294.7
Intake air temperature	°C	14.6
Turbine outlet temperature	°C	427.2
Compressor discharge pressure	kPa	651
Compressor discharge temperature	°C	281.7
LNG flow rate	Nm^3/h	0.0
Ammonia flow rate	Nm^3/h	419.8
Exhaust gas properties (catalyst outlet)		
O_2	%	17.4
CO_2	%	0.04
NO_x	ppm	10.6
NO_x (@ 16% O_2)	ppm	14.7
CO	ppm	0

comprises various components, namely an ammonia supply unit, a natural gas compressor, a gas turbine and a NO_x removal device (de-nitration) as shown by the schematic diagram of the system in Figure 6.29.

In the ammonia supply unit, liquid ammonia is pressurised by a pump at 2 MPa and gasified with a hot water evaporator at a temperature of 70 °C before it is

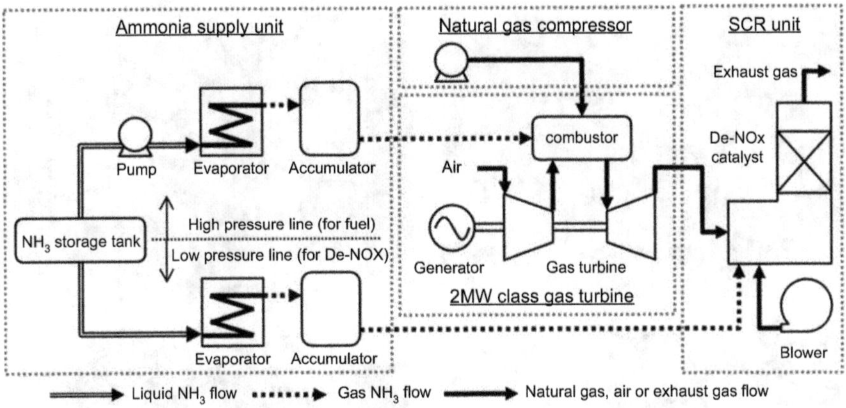

Figure 6.29 System flow diagram. Reprinted, with permission, from [124].

supplied to the gas turbine at 400 kg/h. Low-pressure ammonia gas is also supplied to the de-nitration unit at 60 kg/h at 0.3 MPa and ambient temperature. The gas turbine used is the IM270 model [125] manufactured by IHI Corporation, with modifications made only to the combustor to enable co-firing of ammonia and natural gas [126,127]. The turbine outlet is directly connected to the de-nitration device. The de-nitration device employs a selective catalyst reduction (i.e. commonly used in natural gas-firing power plants) to lower the concentration of NOx in the gas turbine's exhaust. A low-pressure ammonia gas, functioning as a NOx-reducing agent, is provided by the ammonia supply unit. Once the gas has undergone de-nitration, it is discharged into the atmosphere via the exhaust tower, which is linked to the device's outlet. Figure 6.30 presents the photographs of the ammonia supply unit and the gas turbine used in the test.

The experimental method was conducted in two stages. Firstly, the gas turbine was operated using natural gas, and the power output was verified to be stable at 2 MW. Subsequently, fuel gas ammonia was supplied whilst keeping the power output constant. The natural gas flow rate was automatically controlled according to the ammonia flow rate. During the experiment, the ammonia mixing ratio was defined as the ratio of the heat quantity of the supplied fuel ammonia to the total heat quantity of the supplied fuel which ranged from 0% LHV to 20% LHV. The heat quantity was based on the LHV. The tests conducted have demonstrated that successful power generation of 2 MW can be achieved with an ammonia mixing ratios of 20% LHV. Additionally, CO2 emissions can be reduced by approximately 23%. Whilst the NO_x concentration at the turbine outlet is greater than that of a natural gas-fired gas turbine, it has also been shown that the NO_x removal apparatus can bring it down to 7 ppm or less.

It is to be noted that to safely handle ammonia, this testing facility has implemented safety measures following the Electricity Business Act [128]. These measures include a gas leak detection alarm and sprinkler in the ammonia supply unit which are designed to quickly adsorb leaked ammonia with water.

Figure 6.30 IHI 2MW-class gas turbine test facility; (a, b) ammonia supply unit: tank and evaporator side; and (c, d) gas turbine apparatus and engine body © Combustion Society of Japan. Reprinted, with permission, from [124].

Additionally, the secondary and vent piping of the safety valve is connected to a dilution tank that is water-sealed to eliminate any ammonia discharge. The ammonia water produced from the aforementioned measures is then stored in a drainage tank and considered industrial waste. By implementing these safety measures, the facility can effectively handle ammonia whilst minimising any potential harm or risk to its workers or the environment.

In June 2021, an agreement has been made between IHI and General Electric to collaborate on creating an ammonia gas turbine business roadmap. The primary goal of this collaboration is to advance technology and markets in Asia for ammonia gas turbines [129]. According to GE Power's director of Emergent Technologies [130], Japan is an ideal case study because it is actively pursuing ammonia as a decarbonised fuel for power generation. The current targets set by Japan require that by 2050, 10% of the country's electricity should come from hydrogen or ammonia, and by 2030, at least 1% (which is approximately 1 GW) should be generated using these fuels.

6.5.4 NTNU oil/gas retrofitting

A team of researchers from Norway and Sweden [131] conducted an experimental study to determine the most effective NH_3-H_2-N_2 fuel blends for use in gas turbines and to provide manufacturers with recommendations for their use in retrofit and new combustion applications. The study utilised the industrial burner found in the SGT-750 Siemens gas turbines, which was tested in the SINTEF high-pressure combustion facility. The study's results indicate that NO_x emissions can be reduced when the primary flame zone is fuel-rich, despite the presence of ammonia in the fuel. Furthermore, increasing pressure has a positive impact on NO_x formation. However, when ammonia is fully decomposed to 75% H_2 and 25% N_2, the opposite effect is observed. In conclusion, the study suggests that either a low rate of decomposition or full decomposition are both recommended options.

6.5.5 Trigeneration cycle Cardiff

Previous analysis demonstrated that the basic Brayton cycle with humidified ammonia could achieve overall plant efficiencies of around 34%, which needed to be further improved to make these units competitive with current fossil fuel power plants. Amongst all the new technologies, Božo *et al.* [132] applied the trigeneration cycle to reduce emissions and improve combustion performance. This technology is based on multi-staged combustion with the Rich-Quench-Lean (RQL) system.

The trigeneration cycle is achieved by introducing the basic gas turbine cycle into a two-shaft gas turbine plant facility, and it consists of a basic humidified Brayton cycle and a reversed Brayton cycle. Apart from the fuel preheating, air compression and combustion, it is critical to apply steam generation for this trigeneration cycle.

As shown in Figure 6.31, after the combustion in the chamber (CC), the exhaust gas was led into two gas turbines (T1, T2) as a part of the reversed Brayton cycle. The high-temperature combustion products were used to produce additional power from the expansion to negative pressures in the turbine. After going through the second gas turbine (T2), the high-temperature flue gas needed to be cooled by a special regenerative heat exchanger (HE1), and the steam condensate was extracted

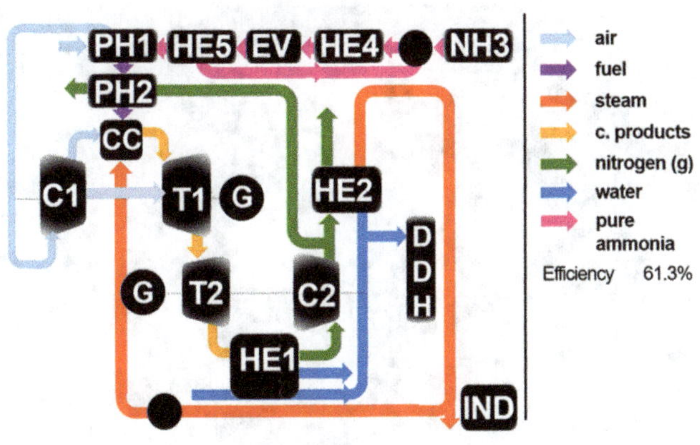

Figure 6.31 Conceived cycle to employ ammonia for the production of power, steam, heating and cooling. PH – Preheater; HE – Heat Exchanger; G – Generator; T – Turbine; DDH – District Distributed Heat; IND – Industry; C – Compressor; EV – Evaporator.

from the mixture. As the exhaust gas pressure is lower than the atmospheric pressure, the gas was compressed in another compressor (C2) to atmospheric pressure. Then the compressed flue gas was used to preheat the fuel in an exchanger (PH2) to further recover waste heat. The waste heat from the compressed gas is used for district heating, preparation of warm water or other applications. Before the inlet air was compressed by the first compressor (C1), it was cooled by the ammonia fuel (PH1) to improve the air volumetric efficiency. Specific off-design work of compression was calculated as a function of compressor pressure ratio, inlet temperature and specific heat [133]. Potential efficiencies of this cycle range between 55% and 61%, with the potential of producing cooling, heating, power and even water from the condensates obtained from the flue gas streams. Further work is underway at Cardiff University to demonstrate the concept experimentally by 2024.

6.5.6 Mitsubishi 40 MW

In March 2021, Mitsubishi Power began developing a gas turbine with a capacity of 40 MW. The unit will be capable of directly burning pure ammonia as fuel by 2025 [134,135]. This novel gas turbine will be derived from the company's H-25 series and, upon completion, will be the first in the world to make exclusive use of ammonia as fuel in a system of this scale. According to Mitsubishi Power, this gas turbine will help promote the decarbonization of small- to medium-scale power stations for industrial applications and on remote islands [134,136].

In addition, Mitsubishi Heavy Industries (MHI) has signed an agreement with Institut Teknologi Bandung (ITB) [137], a research entity in Indonesia, to collaborate on the research and development of an ammonia-fired gas turbine. As part of the agreement, the joint research will eventually lead to a demonstration of the technology on an H-25 gas turbine, Figure 6.32. MHI has initiated a collaboration with Singapore to conduct

Figure 6.32 Mitsubishi Power's H-25 Series gas turbine will run with 100% ammonia, producing up to 40 MW of power. Courtesy of Mitsubishi Heavy Industries [134].

a feasibility study on the development of power plants solely fuelled by ammonia. They are also working with JERA to investigate the establishment of a combined cycle power plant that exclusively uses ammonia as fuel in Jurong Island, Singapore.

Mitsubishi Power recently announced that due to certain characteristics of ammonia combustion, the fuel may be more suitable for smaller gas turbines [134,138]. This is because ammonia has a slower combustion speed, which requires a larger combustor. Additionally, as ammonia contains nitrogen, any system using it as a fuel needs to address the 'fuel NOx' it generates. Although Mitsubishi Power has explored ways to lower NO_x through two-stage combustion, larger gas turbines present technical challenges, including the complication of the combustor and its size. However, the development of a 40 MW ammonia-capable gas turbine suggests that Mitsubishi Power is now looking to directly combust ammonia, rather than relying on a multistage ammonia-cracking approach. To mitigate the production of NOx resulting from the oxidation of nitrogen in ammonia during combustion, Mitsubishi Power plans to incorporate a newly developed combustor that reduces NOx emissions along with selective catalytic reduction technology in its commercialised gas turbine system. The H-25 series gas turbines, Figure 6.32, from Mitsubishi Power, has been sold commercially to utilities and industrial users since 1988.

6.5.7 Baker Hughes' high-pressurised tests

As previously raised, FLEXnCONFU [139] is a Horizon 2020 sponsored programme made up of a consortium of 21 partners including Cummins, Baker Hughes, Proton Ventures, Eindhoven University of Technology, etc., with Cardiff University's combustion laboratories leading the combustion work package. The main objectives of the programme include: (i) design, develop and test a gas turbine combustion system able to burn hydrogen and/or ammonia, (ii) demonstrate the concept 'power to ammonia' at lab scale and (iii) demonstrate economic and safety feasibility, social viability and environmental sustainability of the novel solutions.

As part of the works on combustion, Cardiff University in collaboration with Baker Hughes has conducted a great variety of tests, many of them under high pressurised and high inlet temperature conditions representative of industrial gas turbine systems. To account for non-adiabatic conditions in the system (i.e. which are known to produce higher NOx and lower NH_3 consequence of higher reactivity), tests were conducted in a High-Pressure Optical Combustor (HPOC) using a generic tangential swirl burner (Figure 6.33). The experimental system, detailed elsewhere [139], enabled experimentation that served to establish the boundary conditions of medium pressure studies. At elevated conditions, temperature (up to 550 K) and pressure (up to 6 bar) were increased in such a way that a constant T/P ratio was maintained, retaining quasi-steady nozzle outlet velocities and combustor residence time for the same Φ were also used for validation purposes.

Combustor exhaust gas emissions were sampled downstream of the quartz confinement using a nine-hole equal-area probe, water-conditioned with a heat exchanger to regulate sample temperature (433 K) following specifications in ISO-11042 [140]. Nitric oxide concentrations were quantified using heated vacuum chemiluminescence

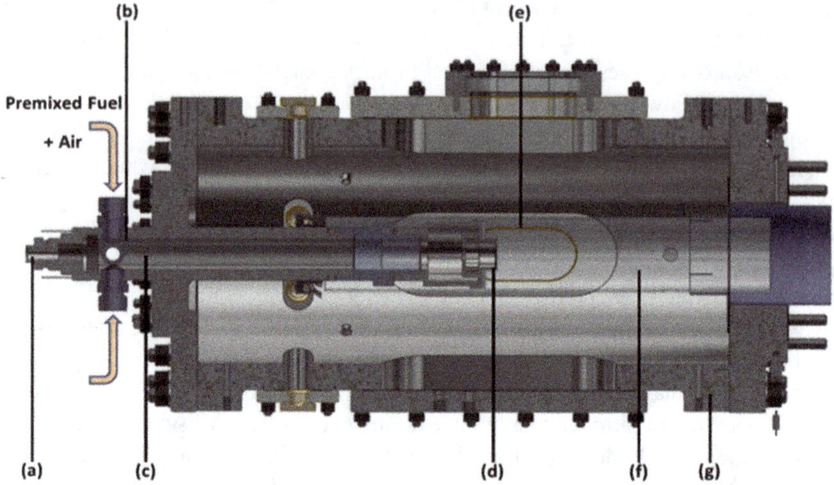

Figure 6.33 Generic tangential swirl burner at the Gas Turbine Research Centre, Cardiff University

(Signal 4000VM). Unburned NH_3 measurements were obtained by redirecting the sample through a NO converter (Signal 410) to measure unreacted concentrations (80% conversion efficiency). All NH_3 and NO concentrations were measured hot/wet and normalised to equivalent dry conditions (ISO-11042). Dry O_2 concentrations were quantified using a paramagnetic analyser (Signal 9000MGA) and used to subsequently normalise NO to equivalent 15% O_2 (ISO-11042). After changing experimental conditions, burner temperatures, pressures, flows and emissions were monitored and, once stable, held for a minimum of 60 samples to be taken. Systematic uncertainties comprising analyser specification, linearisation and span gas certification were combined with any fluctuations in measurement to give the total uncertainty represented within the data of $\sim$5%. Further tests were also obtained using an Emerson QCL system capable of measuring NO, NO_2, N_2O, NH_3, H_2O, O_2, CO_2 and CO at a frequency of 1 Hz and errors estimated at 0.5%.

The analyses depicted a great variety of complex characteristics of ammonia flames when exposed to high turbulent, swirling flows. Starting with the atmospheric analyses, results have been conducted to evaluate the potential of using stratified hydrogen injection. The results denoted promising patterns in the reduction of NOx formation. Novel measuring techniques (i.e. PLIF NO) denoted that stratification could also reduce NO production. It was observed that as hydrogen increased its stratification, there was an acute reduction of NO formation across the flame. This was evidenced by the change in radicals (NH2* and NH*) across the flame front. It was recommended that further works on the front of non-intrusive optical analyses were needed to fully characterise these types of stratified flames.

In parallel, chemiluminescence studies using OH* (Figure 6.34) denoted the impact of pressure and equivalence ratio on the formation of OH* across ammonia/

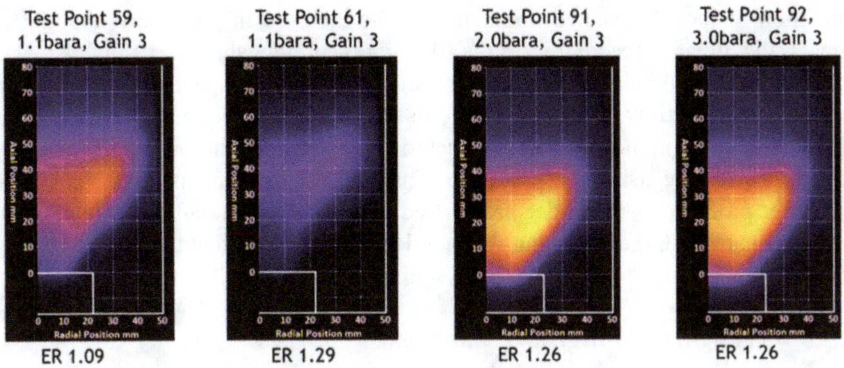

Figure 6.34 OH measurements using a 60/40% methane/ammonia blend at various conditions*

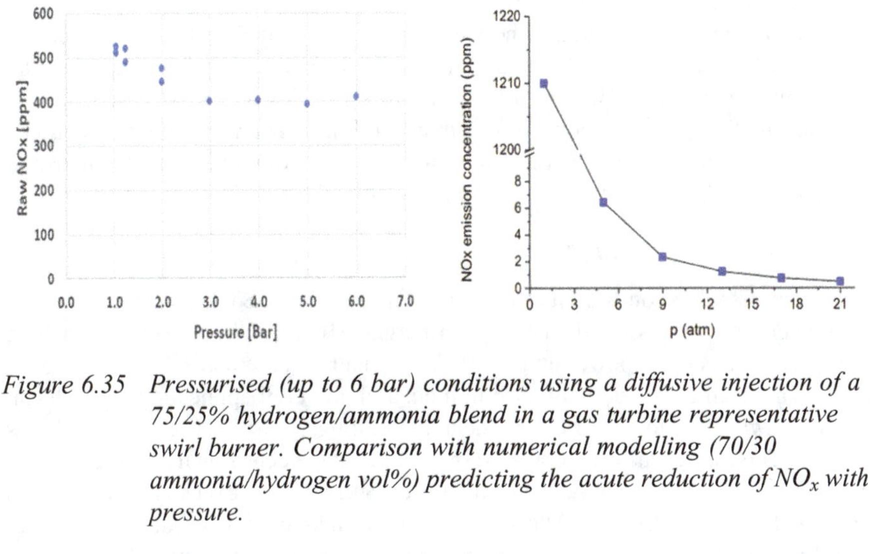

Figure 6.35 Pressurised (up to 6 bar) conditions using a diffusive injection of a 75/25% hydrogen/ammonia blend in a gas turbine representative swirl burner. Comparison with numerical modelling (70/30 ammonia/hydrogen vol%) predicting the acute reduction of NO_x with pressure.

methane flames. As expected, pressure tends to retract the flame, with a more intense front, evidenced by the increase in reactivity that can be observed in these tests.

The works enabled the selection of specific conditions for further visualization and emissions characterization. The injection of air/fuel was changed according to industrial requirements put forward by Baker Hughes. A central lance which included a central orifice was installed in the burner to try various injection strategies (i.e. partially premixed or diffusive), following up some of the results obtained in previous campaigns. Power outputs were scaled at 12.5kW/1.1 bar due to ammonia flow rates, and a blend of 25/75% ammonia/hydrogen was used as requested.

Using diffusive injection had a clear impact on the reduction of NOx emissions (Figure 6.35). Further, it was also observed that pressure has a direct impact on the formation of these unwanted pollutants, as the increase in pressure is thought to impact the greater reactivity of the flame (as previously raised), whilst also enhancing the recombination of species. Another potential aspect is the increase of reactions with tertiary bodies like NH_3 itself. These assertions are under current evaluation future publication. The subject to address will be the change in reactivity at greater pressure using these blends. Finally, it is also important to notice the flattening profile of NOx under pressurised conditions. Although the reduction of NOx was expected, the flattening was not depicted in numerical models previously studied. Another point for further analysis.

6.5.8 GTI/Florida high-pressure tests

GTI Energy and the Electric Power Research Institute (EPRI) have embarked on fundamental and applied investigations of ammonia as an alternative energy carrier under a joint programme known as Low-Carbon Resources Initiative (LCRI). The focus is on applications under turbine operating conditions for both ammonia and ammonia-hydrogen mixtures.

The project, which includes the University of Central Florida (UCF), Georgia Institute of Technology and CRAFT Tech has been looking into experimental campaigns, conceptual designs of combustors and computational analyses to demonstrate the use of ammonia power in gas turbines. The works are also coupled with efforts taken through the 'Investigation of Ammonia for Combustion Turbines' (IACT) sponsored by the Department of Energy (DOE) [141].

6.5.9 KAUST microGT

The Clean Combustion Research Center (CCRC) in KAUST has commissioned a commercial micro gas turbine (Ansaldo Energia T100, 330 kWt–100 kWe) which can generate electrical power up to 100 kW with a rated thermal efficiency of 30% [142]. This facility is generally used in industrial and civil applications, which can supply power/heat to factories, residential buildings, hospitals, etc. It works as follows: The filtered air is compressed by a single-stage centrifugal compressor to 5 bar at full load, then it is introduced to a co-current heat exchanger where it is preheated to 500 °C by the exhaust gas. The preheated air is distributed to a single-barrel burner, and the fuel is ejected and burned in an integral rarefied state at 4.5 bar. The high-temperature gas expands through a single-stage centripetal turbine and transfers heat to a co-current heat exchanger before leaving the turbine.

The schematic of the micro-gas turbine (mGT) system is shown in Figure 6.36, including the fuel delivery system, ammonia evaporator, mixing panel, combustion, exhaust emissions analysis, data acquisition, etc. An NH_3 plant was also employed, which consists of four NH_3 drums and controlling instruments to allow continuous fuel supply. Each drum contains 500 kg of liquid ammonia at 10 bar (absolute). During operation, the liquid ammonia is fully converted to a gaseous state after flowing through the evaporator and its temperature can be increased up to 60 °C. The ammonia flow is controlled by an MFC, which is then mixed with $CH_4/H_2/N_2$ gases in a mixing panel (Figure 6.37)

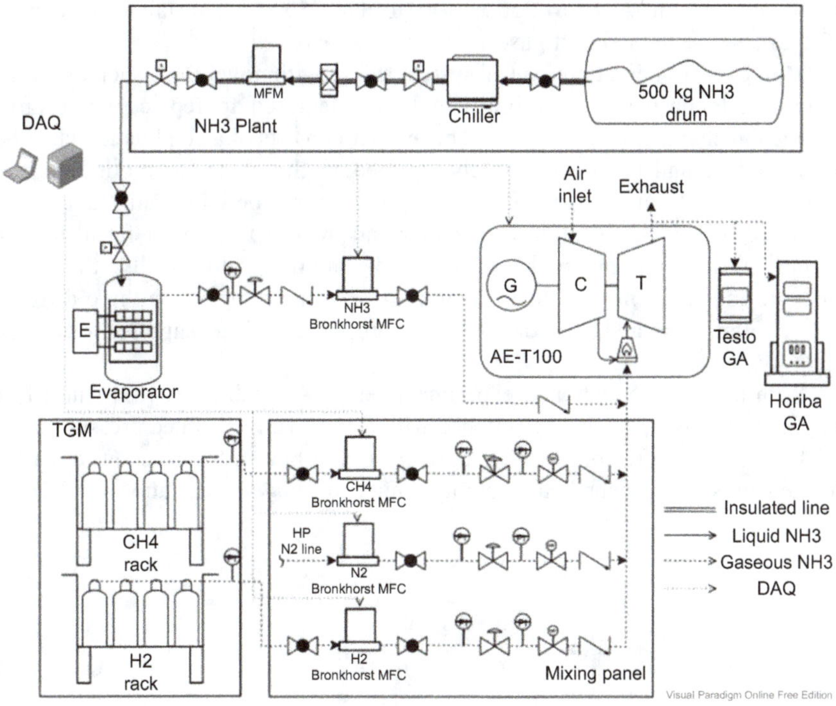

Figure 6.36 Schematic of the mGT system. Courtesy of KAUST [153].

Figure 6.37 Fuel evaporator (left) and mixing panel (right). Courtesy of KAUST [153].

before combustion. A Testo 350 and a Horiba V5000 are installed next to the mGT to analyse the exhaust gases.

Moreover, to operate the mGT system safely with novel fuels such as NH_3, a reduced-scale burner was designed and manufactured to replace the original combustion chamber (Figure 6.38). This burner consists of a combustion chamber in which fuel and air are separately injected in the centre to produce a non-premixed swirl flame; a main swirl mixing chamber, where the main fuel and air are mixed before combustion; a secondary air swirler to prevent flashback and enhance air/fuel mixing in the combustion chamber. Additionally, the burner allows to independently change the proportion of air supplied to three different intakes (pilot, main and secondary) and the proportion of fuel supplied to two fuel inlets.

With the above-mentioned experiment setup, Khateeb *et al.* [83] studied the premixed swirl NH_3/air flames enriched with H_2 or CH_4 at elevated pressures of up to 5 bar, representative of commercial micro gas turbines. Results showed that the elevated pressure widened the stability limits of equivalence ratios for NH_3/air

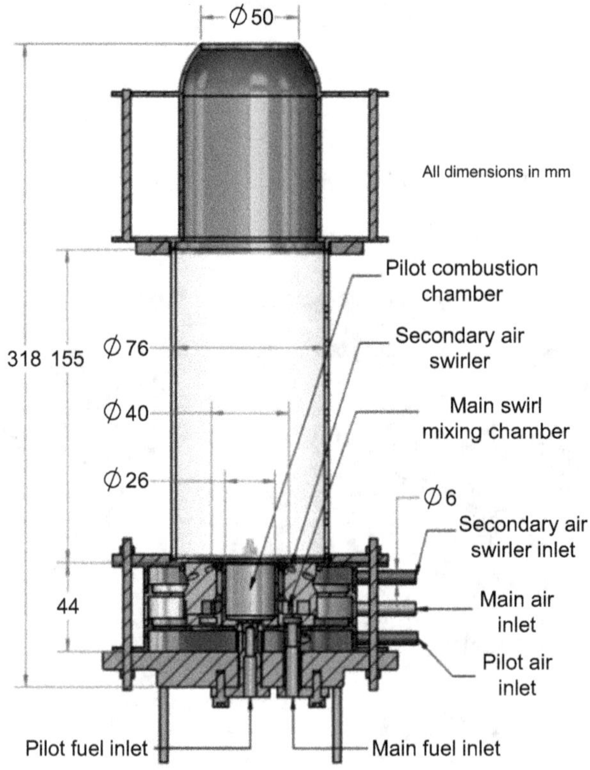

Figure 6.38 Schematic of the reduced-scale burner. Reprinted, with permission, from [154].

flames. Moreover, the ultra-lean swirl NH_3/H_2/air flames were promising for micro gas turbines since they generated low NO emissions (<200 ppm), however, the approach may produce high N_2O emissions due to heat loss or short residence time. Ávila *et al.* [142] conducted experiments at a similar mGT relevant pressure (4.5 bar) to explore the lean stability limits and emissions of NH_3/CH_4/air swirl flames. They reported the lean blowout limit of the equivalence ratio going up whilst the CO_2 was monotonically dropped when increasing the ammonia fraction. Only at rich conditions the NH_3/CH_4/air swirl flames exhibited good NO and N_2O performance.

Recently, Ávila *et al.* [143] tested the NH_3/CH_4 flames in the Ansaldo Energia T100 mGT. The electrical output power and ammonia volumetric fraction for the stable operation could reach 60 kWe and 0.63, respectively. Whilst CO_2 emissions were reduced at ~41% with ammonia addition, the increased NO_x and N_2O emissions may offset the benefits of partial decarbonization.

6.6 Propulsion

6.6.1 Aerospace satellites

Space propulsion using ammonia has also received some attention, especially in small space vehicles and satellites. Blinov *et al.* [144] presented some work in terms of the design features and performance of ammonia electrothermal micro-thrusters, showing that they can become a competitive, cost-effective option due to their specific impulse increase of ~20%. Fatuev *et al.* [145] have also presented work on the development of ACETAM, a rocket fuel based on the fluidization of gaseous acetylene by a highly concentrated dilution in liquefied ammonia. The characteristics of the fuel show improvement and higher stability when compared to other blends at various operating pressures, nozzle expansion degrees and efficiency through various stages during space launching.

Chinese groups have also studied the thermodynamic properties and potential of propellants using acetylene and ammonia [146]. Similarly, Chinese groups have assessed the potential of acetylene-ammonia with liquid oxygen (LOX) and other fuels such as rocket kerosene and methane finding out that a 25% acetylene content was the best in terms of specific impulse and security.

Some published work can also be found for the development of ammonia micro-resistor-jets (MRJ) for micro-satellites. Robin *et al.* [147] evaluated the use of ammonia MRJs to cope with weight and complexity constraints whilst creating a high thrust/power system. Moreover, they showed that at ambient temperature and vapour pressure of 0.85 MPa (123 psi), ammonia can also serve as a propellant for altitude control. Due to the small nature of the system and its innovative design using platinum tubes, the total device weight is about 15 grams. A successful 100-hour Engineering Model (EM) was carried out, leading to further development plans. Similarly, an ammonia self-managed vaporization propulsion (ASVP) system for micro-nano satellites was presented by Sun *et al.* [148]. Ground tests indicate that the ASVP system weighs 1.8 kg (with 0.34 kg liquefied ammonia propellant) and reaches a specific impulse of more than 100 s, whilst the power consumption is less than 10 W, hence meeting multiple requirements for micro-nano satellite applications. The use of ammonia in these and related applications (i.e. Cubesat satellites) has been supported by the industry due to ammonia's self-pressurising, relatively high energy density and its well-suited cold gas propellant features for propulsive attitude and control system [149].

Another use of ammonia in satellites is currently under evaluation for the use of novel blends. Officially called LMP-103S, this new fuel is a blend of ammonium dinitramide (ADN) with water, methanol and ammonia. The European Space Agency (ESA) and the Swedish Space Corporation have been researching storable ADN-based liquid mono-propellants through a series of contracts since 1997. It appears that ADN has a 30% better performance than hydrazine and is much less toxic. A project called 'RHEFORM' has looked at the compound to launch 13 propulsion systems [139].

6.6.2 Aerospace propulsion - Viking project

The onset of ammonia technology in propulsion systems was seen in other aerospace applications as well. The NACA Langley Aeronautical Laboratory (now NASA Langley RC) designed the X-15 aircraft powered by the throttleable XLR-99 engine, Figure 6.39, which could produce up to 250,500 N of thrust. The aircraft was powered by a combination of liquid ammonia and liquid oxygen (LOX), which provided the necessary energy to the system. The X-15 was launched from a B-52 research aircraft flying at high speeds (800 km/h) [117]. However, due to a lack of understanding of the need for an ammonia system and the requirement for higher gravimetric energy density fuels, the programme was discontinued [118]. Despite

Figure 6.39 The X-15 hypersonic aircraft powered by the ammonia-fuelled XLR99 engine. Courtesy of NASA [152].

this, after 199 missions, the speed of the X-15, which reached up to 6.7 Ma, was not surpassed until the launch of the space shuttle [119].

6.7 Current R&D developments

As previously raised, there are many advantages to the use of ammonia as a fuel for gas turbines. Not only the chemical can provide a large energy density compared to other net-zero carbon fuels, but also the potential of using it whilst integrating different parts of the cycle is an option. Various works have demonstrated that the use of ammonia can deliver efficiencies between 50% and 60% [150], whilst recent works have also demonstrated that the use of cooling systems and partial blending (up to 40% vol hydrogen) can ensure high flame stability with efficiencies >61%, as previously mentioned. These works, still under publication, are a precedent to the potential that this chemical can have for the replacement of fossil fuels in large or medium-scale power gas turbines. As previously mentioned, although fossils can reach efficiency figures above 60% (in some cases up to 90%), the net zero feature of ammonia makes it a good contender for the replacement of hydrocarbon fuels.

However, a big question raised by developers and funding bodies is 'how close are we?'. To answer the question, the technological stage of ammonia gas turbines can be divided into three different phases, Table 6.5.

To the knowledge of the authors, the block diagram (Figure 6.40) shows the current development stage that many institutions and consortia have reached by the time of writing this chapter. As observed, there are still many challenges before the technology reaches full maturity and can be deployed at a commercial scale worldwide. However, works are progressing faster than in other fields, and it is foreseen that before the end of the decade, the technology will be mature enough to have commercial units on most continents.

6.8 Conclusions

Gas turbines are complex systems that utilise thermodynamics, fluid mechanics, material science and combustion principles to generate power and propulsion, offering high efficiency, fuel flexibility, low emissions and low maintenance costs. In response to the trend towards decarbonization in the power and aerospace industries, ammonia has emerged as a potential fuel for gas turbines. To address undesirable emissions from ammonia-powered gas turbine engines, several ongoing initiatives aim to stabilise and mitigate them. Despite the ever-changing energy demands, gas turbines are poised to remain a crucial technology, with the utilization of ammonia as a fuel source offering a promising path forward.

Three primary thermodynamic cycles can be used to describe gas turbine engines: the ideal Brayton cycle of gas power turbines, the combined cycle and the humidified cycle. The Brayton cycle is used to model gas turbine engines, offering higher

Table 6.5 Technological stage of ammonia gas turbines

Phase 1 – fully cracked NH_3	**Phase 2 – partially cracked NH_3 (for H2 delivery) or other ammonia blends (usually with methane/natural gas)**	**Phase 3 – pure NH_3**
Pros	*Pros*	*Pros*
• Technology based on the use of hydrogen/nitrogen gas from cracked ammonia; • The technology can benefit from the use of humidified conditions to reduce the flame speed of the hydrogen flame; • Nitrogen is used as a diluent to reduce further the flame speed; • Several companies are currently working on the use of hydrogen blends for power purposes in gas turbines (i.e. Siemens, Kawasaki, MHI, GE); • Various methodologies of injection are currently under evaluation for commercial purposes and global deployment.	• The use of ammonia blends (especially with hydrogen) can increase the flammability of the mixture, getting closer to methane flames; • Greater efficiencies since less ammonia is cracked; • The stability of ammonia blends ensures wider operability limits than using pure ammonia or pure hydrogen, as the blend can be controlled to deliver specific flame speeds; • The temperatures, power outputs and ignition energies are closer to fossil-fuel-based systems, hence delivering better thermodynamics and efficiencies.	• The use of pure ammonia enables higher efficiencies since there is no need to sacrifice energy in cracking ammonia; • Ammonia can be stabilised under certain conditions in the combustor to ensure appropriate combustion efficiency (measured at ~95–96%); • The system and its auxiliaries will be less complicated than using cracking devices; • Liquid injection could be accomplished, avoiding evaporation of the NH_3.
Cons	*Cons*	*Cons*
• Cracking of ammonia requires high energy ~14% of the total energy in the fuel; • Cracking is not 100% efficient, and ammonia traces remain in the fuel; • Controlling a hydrogen flame is more challenging than other types of flames; • Embrittlement (H) and nitration in these systems have been barely analysed.	• Partial cracking is still a topic of research; • There is still debate on which blend is the most suitable for these systems; • Lack of control in the delivery of hydrogen could trigger unwanted instabilities; • Similar to Phase 1, embrittlement and nitration need to be properly studied; • Inadequate blending can deliver higher emissions than fossil fuels.	• Ammonia is very slow compared to the rest of the molecules used in gas turbines, hence having a very narrow operability limit; • In the event of blowoff (highly likely with pure ammonia), the concentration of ammonia will be extremely high, above any H&S limit; • There is limited data on the impact of pure ammonia on materials at high pressure/temp; • Liquid injection is still less efficient during combustion and delivers higher NO_x than stage 2.

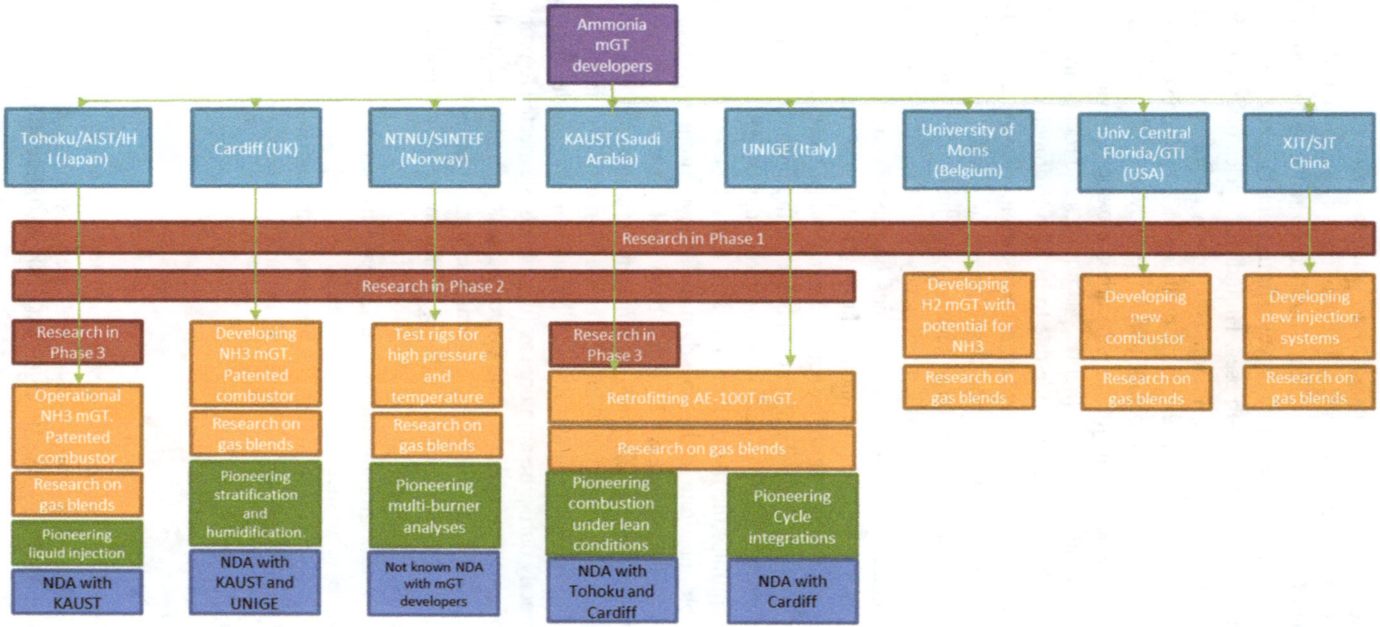

Figure 6.40 Currently demonstrated R&D developments for the use of ammonia in micro gas turbines (end of 2022). Each institution/consortium has been approaching the subject from different angles to tackle different questions about the use of ammonia and its blends in gas turbine systems. Mitsubishi's programme (still not demonstrated), and the Viking, Solar and US Military programmes (ended years ago) are not included.

efficiency and benefits compared to other cycles. Combined cycles merge two or more thermodynamic cycles and offer an efficient way to generate more heat and power using the same fuel source. The cogeneration cycle is one of the most widespread forms of combined cycles, recovering waste heat, significantly reducing fuel costs and achieving high-efficiency levels. Trigeneration cycles further enhance cogeneration by utilising the produced heat for cooling purposes. Humidified cycles offer significant advantages in power plant efficiency and emission reduction by adding water vapour to the working fluid, achieving higher efficiency and reducing NO_x emissions. They are commonly used in gas turbine power plants and have shown great potential for cogeneration and power production integrated with carbon dioxide removal.

The use of ammonia as an alternative fuel for gas turbines and rocket propulsion systems has been gaining attention worldwide due to its high hydrogen content, free carbon footprint and potential for low-emission combustion. However, ammonia combustion presents challenges such as low flame speed, low reactivity and NO_x emissions. Swirling flows and other injection technologies have been proposed to address these issues and improve combustion efficiency whilst reducing emissions. The use of liquid ammonia injection in gas turbines has also been explored, and recent studies have shown promise in enhancing flame stability and reducing emissions. The detonation process offers another solution to ammonia's poor combustion characteristics and NO_x emissions, with recent studies showing promising results in using ammonia-hydrogen mixtures as fuel in detonation engines. However, more research is needed to optimise the use of ammonia as a fuel for power generation.

Ammonia is a viable substitute for hydrocarbon fuels in gas turbines, and studies have shown that it is possible to achieve high combustion efficiencies with a lower environmental impact. Japan is actively pursuing ammonia as a decarbonised fuel for power generation, with targets set for hydrogen or ammonia to provide 10% of Japan's electricity by 2050 and at least 1% by 2030. Mitsubishi Power is developing a gas turbine that can directly burn pure ammonia as fuel, which is expected to promote the decarbonization of small to medium-scale power stations for industrial applications and remote areas. The FLEXnCONFU programme aims to develop and test a gas turbine combustion system capable of burning hydrogen and/or ammonia, and other programmes are already ongoing to achieve this aim.

In conclusion, the use of ammonia as a fuel for gas turbines presents several advantages compared to other net-zero carbon fuels. Although fossil fuels can reach higher efficiency figures, the net-zero feature of ammonia makes it a promising candidate for replacing hydrocarbon fuels in large, medium or small scale gas turbine systems.

References

[1] Vincent P. *Natural gas cleanliness*. Siemens energy [Internet]. 2022. [cited 2023 Mar 24]; Available from: https://assets.siemens-energy.com/siemens/assets/api/uuid:11e04dd1-5b2b-4e25-a80a-4bc5305b37e0/natural-gas-cleanliness.pdf

[2] Siemens-Energy. *Gas turbines or gas engines?* [Internet]. 2022. [cited 2023 Mar 24]; Available from: https://www.siemens-energy.com/global/en/offerings/power-generation/rice.html

[3] Chen L, Zhang W, and Sun F. Power, efficiency, entropy-generation rate and ecological optimization for a class of generalized irreversible universal heat-engine cycles. *Appl Energy* 2007;84:512–25.

[4] Proctor II CL. Internal combustion engines. In: *Encyclopedia of Physical Science and Technology (Third Edition)*. 2003 pp. 33–44.

[5] Cengel YA, and Boles MA. *Thermodynamics: An Engineering Approach*. 5th edn. Boston, MA: McGraw-Hill Higher Education; 2006.

[6] Mason LS, Shaltens RK, Dolce JL, and Cataldo RL. Status of Brayton cycle power conversion development at NASA GRC. *In: AIP Conference Proceedings*. College Park, MD: American Institute of Physics; 2002. pp. 865–71.

[7] Tanuma T. *Advances in Steam Turbines for Modern Power Plants*. London: Woodhead Publishing; 2017.

[8] Jansohn P. *Modern Gas Turbine Systems: High Efficiency, Low Emission, Fuel Flexible Power Generation*. Cambridge: Woodhead Publishing Limited; 2013.

[9] Radcenco V, Vargas JVC, and Bejan A. Thermodynamic optimization of a gas turbine power plant with pressure drop irreversibilities. *Journal of Energy Resources Technology, Transactions of the ASME*. 1998;120(3): 233–240.

[10] Wang WH, Chen LG, Sun FR, and Wu C. Performance optimisation of open cycle intercooled gas turbine power plant with pressure drop irreversibilities. *J Energy Inst* 2008;81:31–7.

[11] Chen L, Li Y, Sun F, and Wu C. Power optimization of open-cycle regenerator gas-turbine power-plants. *Appl Energy* 2004;78:199–218.

[12] Bejan A. *Entropy Generation Minimization: The Method of Thermodynamic Optimization of Finite-Size Systems and Finite-Time Processes*. Boca Raton, FL: CRC Press; 2013.

[13] Chen L, Bi Y, and Wu C. The influence of nonlinear flow resistance relations on the power and efficiency from fluid flow. *J Phys D Appl Phys.* 1999;32:1346.

[14] Angelino G, and Invernizzi C. Real gas Brayton cycles for organic working fluids. *Proc Inst Mech Eng A: J Power Energy* 2001;215:27–38.

[15] Chen J, and Engeda A. A review of the prospect and challenges for developing and marketing a Brayton-cycle based power genset gas-turbine using supercritical CO_2: part I—the general system challenges. In: *Turbo Expo: Power for Land, Sea, and Air.* 2020. 84157:V006T08A002.

[16] Invernizzi CM, and van der Stelt T. Supercritical and real gas Brayton cycles operating with mixtures of carbon dioxide and hydrocarbons. *Proc Inst Mech Eng A: J Power Energy* 2012;226:682–93.

[17] Siddiqui ME, and Almitani KH. Energy analysis of the S-CO_2 Brayton cycle with improved heat regeneration. *Processes* 2018;7:3.

[18] Chen Y, Lundqvist P, Johansson A, and Platell P. A comparative study of the carbon dioxide transcritical power cycle compared with an organic

Rankine cycle with R123 as working fluid in waste heat recovery. *Appl Therm Eng* 2006;26:2142–7.

[19] Bhargava RK, Bianchi M, De Pascale A, Negri di Montenegro G, and Peretto A. Gas turbine based power cycles—a state-of-the-art review. In: Cen K, Chi Y, Wang F (eds). *Challenges of Power Engineering and Environment.* Berlin: Springer; 2007. pp. 309–19.

[20] Lucas K. On the thermodynamics of cogeneration. *Int J Therm Sci [Internet]* 2000;39:1039–46. Available from: https://www.sciencedirect.com/science/article/pii/S1290072900011716

[21] Guteša Božo M, Mashruk S, Alnajideen M, and Valera-Medina A. A novel trigeneration gas turbine cycle powered by humidified ammonia/hydrogen with ammonia refrigeration. *1st Symposium on Ammonia Energy.* 2022. https://orca.cardiff.ac.uk/id/eprint/152462/

[22] Jaskólski M, Reński A, and Minkiewicz T. Thermodynamic and economic analysis of nuclear power unit operating in partial cogeneration mode to produce electricity and district heat. *Energy [Internet]* 2017;141:2470–83. Available from: https://www.sciencedirect.com/science/article/pii/S0360544217307120

[23] Smith CB, and Parmenter KE. *Energy, Management, Principles: Applications, Benefits, Savings*. Amsterdam: Elsevier; 2013.

[24] Hu SD. *Cogeneration*. Reston, VA: Reston Publishing; 1985.

[25] Pereira JS, Ribeiro JB, Mendes R, and André JC. Analysis of a hybrid (topping/bottoming) ORC based CHP configuration integrating a new evaporator design concept for residential applications. *Appl Therm Eng* 2019; 160:113984.

[26] Yağlı H, Koç Y, and Kalay H. Optimisation and exergy analysis of an organic Rankine cycle (ORC) used as a bottoming cycle in a cogeneration system producing steam and power. *Sustain Energy Technol Assess* 2021; 44:100985.

[27] Alam S, and Chenna S. Greener process for syngas cogeneration: bi-reforming of methane coupled with chemical looping combustion. *Int J Energy Res* 2022;46:3476–89.

[28] UK-Government. *Combined heat and power* [Internet]. 2023. Available from: https://www.gov.uk/guidance/combined-heat-and-power

[29] Bhatia SC. *Advanced Renewable Energy Systems, (Part 1 and 2)*. Boca Raton, FL: CRC Press; 2014.

[30] Breeze P. *Combined Heat and Power.* New York: Academic Press; 2017.

[31] Sokołowski MM. *European Law on Combined Heat and Power.* Milton Park: Routledge; 2020.

[32] Jimenez-Navarro JP, Kavvadias K, Filippidou F, Pavičević M, and Quoilin S. Coupling the heating and power sectors: the role of centralised combined heat and power plants and district heat in a European decarbonised power system. *Appl Energy [Internet]* 2020;270:115134. Available from: https://www.sciencedirect.com/science/article/pii/S0306261920306462

[33] Krawinkler R, and Simader G. Meeting cooling demands in SUMMER by applying HEAT from cogeneration, technology report for Austria, deliverable from the project SUMMERHEAT. 2007. http://systemlab.dk/smart-varme/summerheat-dk.pdf

[34] Sala-Lizarraga JM, and Perez AP. *Exergy Analysis and Thermoeconomics of Buildings: Design and Analysis for Sustainable Energy Systems*. Oxford: Butterworth-Heinemann; 2019.

[35] Gluesenkamp K, and Radermacher R. Heat-activated cooling technologies for small and micro combined heat and power (CHP) applications [Internet]. In: Beith R. (ed.). *Small and Micro Combined Heat and Power (CHP) Systems*. London: Woodhead Publishing; 2011. pp. 262–306. Available from: https://www.sciencedirect.com/science/article/pii/B9781845697952500118

[36] Dincer I, and Rosen MA. Chapter 15 – exergy analyses of integrated trigeneration and multigeneration systems [Internet]. In: Dincer I, and Rosen MA. (eds). *Exergy. 3rd edn*. Amsterdam: Elsevier; 2021. pp. 411–38. Available from: https://www.sciencedirect.com/science/article/pii/B9780128243725000142

[37] Dincer I, and Rosen MA. Chapter 14 – exergy analysis of integrated trigeneration and multigeneration systems [Internet]. In: Dincer I, and Rosen MA. (eds). *Exergy. 2nd edn*. Amsterdam: Elsevier; 2013. pp. 303–17. Available from: https://www.sciencedirect.com/science/article/pii/B9780080970899000140

[38] Sonar D. Chapter 4 – renewable energy based trigeneration systems—technologies, challenges and opportunities [Internet]. In: Ren J. (ed.). *Renewable-Energy-Driven Future*. New York: Academic Press; 2021. pp. 125–68. Available from: https://www.sciencedirect.com/science/article/pii/B9780128205396000042

[39] Guteša Božo M, Mashruk S, Zitouni S, and Valera-Medina A. Humidified ammonia/hydrogen RQL combustion in a trigeneration gas turbine cycle. *Energy Convers Manag [Internet]* 2021;227:113625. Available from: https://www.sciencedirect.com/science/article/pii/S0196890420311523

[40] Elmer T, Worall M, Wu S, and Riffat S. Assessment of a novel solid oxide fuel cell tri-generation system for building applications. *Energy Convers Manag* 2016;124:29–41.

[41] Jradi M, and Riffat S. Tri-generation systems: energy policies, prime movers, cooling technologies, configurations and operation strategies. *Renew Sustain Energy Rev* 2014;32:396–415.

[42] Dunham MT, and Lipiński W. Thermodynamic analyses of single Brayton and combined Brayton–Rankine cycles for distributed solar thermal power generation. *J Sol Energy Eng* 2013;135:031008.

[43] Mossi Idrissa AK, and Goni Boulama K. Advanced exergy analysis of a combined Brayton/Brayton power cycle. *Energy [Internet]* 2019;166: 724–37. Available from: https://www.sciencedirect.com/science/article/pii/S0360544218321078

[44] Rogalev A, Rogalev N, Kindra V, Komarov I, and Zlyvko O. Research and development of the combined cycle power plants working on supercritical carbon dioxide. *Inventions* 2022;7:76.

[45] Aliyu M, AlQudaihi AB, Said SAM, and Habib MA. Energy, exergy and parametric analysis of a combined cycle power plant. *Therm Sci Eng Prog* 2020;15:100450.
[46] Mohamed AMO, Paleologos E, and Howari F. *Pollution Assessment for Sustainable Practices in Applied Sciences and Engineering*. Oxford: Butterworth-Heinemann; 2020.
[47] Jonsson M, and Yan J. Humidified gas turbines—a review of proposed and implemented cycles. *Energy [Internet]* 2005;30:1013–78. Available from: https://www.sciencedirect.com/science/article/pii/S0360544204003573
[48] Silde A, and Lindholm I. *On Detonation Dynamics in Hydrogen-Air-Steam Mixtures: Theory and Application to Olkiluoto Reactor Building*. Roskilde: Nordic Nuclear Safety Research NKS; 2000.
[49] Gao Y, and Law CK. Rankine–Hugoniot relations in relativistic combustion waves. *Astrophys J* 2012;760:122.
[50] Chapman DL. VI. On the rate of explosion in gases. *London Edinburgh Dublin Philos Mag J Sci.* 1899;47:90–104.
[51] Zeldovich YB. On the theory of the propagation of detonation in gaseous systems. 1950. https://ntrs.nasa.gov/citations/19930093969
[52] Von Neuman J. *Theory of Detonation Waves*. Princeton, NJ: Institute for Advanced Study; 1942.
[53] Döring W. On detonation processes in gases. *Ann Phys.* 1943;43:9.
[54] Al-Dulaimi Z. Non-aqueous shale gas recovery system. PhD Thesis, Cardiff University; 2017. https://orca.cardiff.ac.uk/id/eprint/104172/
[55] Ogle RA. *Dust Explosion Dynamics*. London: Butterworth-Heinemann; 2016.
[56] Morris CI. *Pulse Detonation Rocket Engine Research at NASA Marshall.* 2004. Available from: https://ntrs.nasa.gov/api/citations/20030062180/downloads/20030062180.pdf (accessed 25 August 2023).
[57] Vutthivithayarak R, Braun EM, and Lu FK. On thermodynamic cycles for detonation engines. In: Kontis K (ed). *28th International Symposium on Shock Waves*. Berlin: Springer; 2012. pp. 287–92.
[58] Heiser WH, and Pratt DT. Thermodynamic cycle analysis of pulse detonation engines. *J Propuls Power* 2002;18:68–76.
[59] Nalim MR. Thermodynamic limits of work and pressure gain in combustion and evaporation processes. *J Propuls Power* 2002;18:1176–82.
[60] Stathopoulos P. Comprehensive thermodynamic analysis of the Humphrey cycle for gas turbines with pressure gain combustion. *Energies (Basel)* 2018;11:3521.
[61] Bellini R. *Ideal Cycle Analysis of a Regenerative Pulse Detonation Engine for Power Production*. Arlington, TX: The University of Texas at Arlington; 2010.
[62] Naked Science. The rotating detonation engine was tested in space. 2021. https://vpk.name/en/533940_the-rotating-detonation-engine-was-tested-in-space.html

[63] Aerodynamic Research Centre. *Pulsed detonation engines* [Internet]. 2014. [cited 2023 Apr 3]; Available from: https://arc.uta.edu/research/pde.htm

[64] Okafor EC, Somarathne KDKA, Hayakawa A, *et al.* Towards the development of an efficient low-NOx ammonia combustor for a micro gas turbine. *Proc Combust Inst* 2019;37:4597–606.

[65] Reddy VM, Katoch A, Roberts WL, and Kumar S. Experimental and numerical analysis for high intensity swirl based ultra-low emission flameless combustor operating with liquid fuels. *Proc Combust Inst* 2015;35:3581–9.

[66] Hayakawa A, Arakawa Y, Mimoto R, Somarathne KDKA, Kudo T, and Kobayashi H. Experimental investigation of stabilization and emission characteristics of ammonia/air premixed flames in a swirl combustor. *Int J Hydrogen Energy* 2017;42:14010–8.

[67] Syred N. A review of oscillation mechanisms and the role of the precessing vortex core (PVC) in swirl combustion systems. *Prog Energy Combust Sci.* 2006;32:93–161.

[68] Kurata O, Iki N, Matsunuma T, *et al.* Performances and emission characteristics of NH3–air and NH3CH4–air combustion gas-turbine power generations. *Proc Combust Inst* 2017;36:3351–9.

[69] Somarathne KDKA, Hatakeyama S, Hayakawa A, and Kobayashi H. Numerical study of a low emission gas turbine like combustor for turbulent ammonia/air premixed swirl flames with a secondary air injection at high pressure. *Int J Hydrogen Energy* 2017;42:27388–99.

[70] Somarathne KDKA, Colson S, Hayakawa A, and Kobayashi H. Modelling of ammonia/air non-premixed turbulent swirling flames in a gas turbine-like combustor at various pressures. *Combust Theory Model* 2018;22:973–97.

[71] Mashruk S, Xiao H, and Valera-Medina A. Rich-Quench-Lean model comparison for the clean use of humidified ammonia/hydrogen combustion systems. *Int J Hydrogen Energy 2020*;46:4472–84.

[72] Valera-Medina A, Marsh R, Runyon J, *et al.* Ammonia–methane combustion in tangential swirl burners for gas turbine power generation. *Appl Energy* 2017;185:1362–71.

[73] Yang Y, Huang Q, Sun J, Ma P, and Li S. Reducing NOx emission of swirl-stabilized ammonia/methane tubular flames through a fuel-oxidizer mixing strategy. *Energy Fuels* 2022;36:2277–87.

[74] Xiao H, Valera-Medina A, Marsh R, and Bowen PJ. Numerical study assessing various ammonia/methane reaction models for use under gas turbine conditions. *Fuel* 2017;196:344–51.

[75] Chai WS, Bao Y, Jin P, Tang G, and Zhou L. A review on ammonia, ammonia-hydrogen and ammonia-methane fuels. *Renew Sustain Energy Rev* 2021;147:111254.

[76] Ilbas M, Kumuk O, and Karyeyen S. Numerical study of a swirl gas turbine combustor for turbulent air and oxy-combustion of ammonia/kerosene fuels. *Fuel* 2021;304:121359.

[77] Zhang M, Wei X, Wang J, Huang Z, and Tan H. The blow-off and transient characteristics of co-firing ammonia/methane fuels in a swirl combustor. *Proc Combust Inst* 2021;38:5181–90.

[78] Valera-Medina A, Gutesa M, Xiao H, *et al.* Premixed ammonia/hydrogen swirl combustion under rich fuel conditions for gas turbines operation. *Int J Hydrogen Energy* 2019;44:8615–26.

[79] Khateeb AA, Guiberti TF, Zhu X, Younes M, Jamal A, and Roberts WL. Stability limits and exhaust NO performances of ammonia-methane-air swirl flames. *Exp Therm Fluid Sci.* 2020;114:110058.

[80] Valera-Medina A, Pugh DG, Marsh P, Bulat G, and Bowen P. Preliminary study on lean premixed combustion of ammonia-hydrogen for swirling gas turbine combustors. *Int J Hydrogen Energy* 2017;42:24495–503.

[81] Mong GR, Chiong MC, Chong CT, *et al.* Fuel-lean ammonia/biogas combustion characteristics under the reacting swirl flow conditions. *Fuel* 2023; 331:125983.

[82] Khateeb AA, Guiberti TF, Zhu X, Younes M, Jamal A, and Roberts WL. Stability limits and NO emissions of technically-premixed ammonia-hydrogen-nitrogen-air swirl flames. *Int J Hydrogen Energy* 2020;45:22008–18.

[83] Khateeb AA, Guiberti TF, Wang G, *et al.* Stability limits and NO emissions of premixed swirl ammonia-air flames enriched with hydrogen or methane at elevated pressures. *Int J Hydrogen Energy* 2021;46:11969–81.

[84] Zhu X, Khateeb AA, Guiberti TF, and Roberts WL. NO and OH* emission characteristics of very-lean to stoichiometric ammonia–hydrogen–air swirl flames. *In: Proceedings of the Combustion Institute*. Amsterdam: Elsevier; 2021. pp. 5155–62.

[85] Mashruk S, Okafor EC, Kovaleva M, *et al.* Evolution of N2O production at lean combustion condition in NH3/H2/air premixed swirling flames. *Combust Flame [Internet]* 2022 [cited 2022 Aug 17];244:112299. Available from: https://linkinghub.elsevier.com/retrieve/pii/S0010218022003145

[86] *Global Warming Potential Values* [Internet]. [cited 2021 Jul 30]. Available from: https://www.ghgprotocol.org/sites/default/files/ghgp/Global-Warming-Potential-Values%20%28Feb%2016%202016%29_1.pdf

[87] Mashruk S, Kovaleva M, Chong CT, Hayakawa A, Okafor E, and Valera-Medina A. Nitrogen oxides as a by-product of ammonia/hydrogen combustion regimes. *Chem Eng Trans [Internet]* 2021 [cited 2022 Jan 24]; 89:613–8. Available from: https://www.cetjournal.it/index.php/cet/article/view/CET2189103

[88] Mashruk S, Zitouni SE, Brequigny P, Munaim-Rousselle C, and Valera-Medina A. Combustion performances of premixed ammonia/hydrogen/air laminar and swirling flames for a wide range of equivalence ratios. *Int J Hydrogen Energy* 2022;47:41170–82.

[89] Mashruk S, Kovaleva M, Alnasif A, *et al.* Nitrogen oxide emissions analyses in ammonia/hydrogen/air premixed swirling flames. *Energy [Internet]* 2022 [cited 2022 Sep 7];260:125183. Available from: https://linkinghub.elsevier.com/retrieve/pii/S0360544222020746

[90] Mashruk S, Zhu X, Roberts WL, Guiberti TF, and Valera-Medina A. Chemiluminescent footprint of premixed ammonia-methane-air swirling flames. *Proceedings of the Combustion Institute*. 2022;39(1):1415–1423.

[91] Kailos NC. *Utilization of Ammonia as an Alternate Fuel in Army Aircraft Engines* [Internet]. Fort Eustis, VA: US Army Aviation Materiel Laboratories; 1966 [cited 2023 Mar 24]. Available from: https://apps.dtic.mil/sti/citations/AD0638360

[92] Verkamp FJ, Hardin MC, and Williams JR. Ammonia combustion properties and performance in gas-turbine burners. *In: Symposium (International) on Combustion*. Amsterdam: Elsevier; 1967. pp. 985–92.

[93] Bull MGSOLAR TURBINES INTERNATIONAL SAN DIEGO CA. *Development of an Ammonia-Burning Gas Turbine Engine* [Internet]. San Diego, CA: Solar Turbines International; 1968 [cited 2023 Mar 24]. Available from: https://apps.dtic.mil/sti/citations/AD0671667

[94] Okafor EC, Yamashita H, Hayakawa A, *et al.* Flame stability and emissions characteristics of liquid ammonia spray co-fired with methane in a single stage swirl combustor. *Fuel* 2021;287:119433.

[95] Okafor EC, Kurata O, Yamashita H, *et al.* Liquid ammonia spray combustion in two-stage micro gas turbine combustors at 0.25 MPa; relevance of combustion enhancement to flame stability and NOx control. *Appl Energy Combust Sci.* 2021;7:100038.

[96] Thome JR, Cheng L, Ribatski G, and Vales LF. Flow boiling of ammonia and hydrocarbons: a state-of-the-art review. *Int J Refrig* 2008;31:603–20.

[97] Chang M, Park JH, Kim HI, and Park S. Flash boiling macroscopic spray characteristics of multi-hole direct injection injectors with different hole arrangement. *Appl Therm Eng* 2020;170:114969.

[98] Pelé R, Mounaïm-Rousselle C, Bréquigny P, Hespel C, and Bellettre J. First study on ammonia spray characteristics with a current GDI engine injector. *Fuels* 2021;2:253–71.

[99] Li S, Li T, Wang N, Zhou X, Chen R, and Yi P. An investigation on near-field and far-field characteristics of superheated ammonia spray. *Fuel* 2022; 324:124683.

[100] Cheng Q, Ojanen K, Diao Y, Kaario O, and Larmi M. Dynamics of the ammonia spray using high-speed schlieren imaging. *SAE Int J Adv Curr Pract Mobil* 2022;4:1138–53.

[101] Sullivan N, Jensen A, Glarborg P, *et al.* Ammonia conversion and NOx formation in laminar coflowing nonpremixed methane-air flames. *Combust Flame* 2002;131:285–98.

[102] Woo M, Choi BC, and Ghoniem AF. Experimental and numerical studies on NOx emission characteristics in laminar non-premixed jet flames of ammonia-containing methane fuel with oxygen/nitrogen oxidizer. *Energy* 2016;114:961–72.

[103] Colson S, Kuhni M, Galizzi C, Escudié D, and Kobayashi H. Study of the combined effect of ammonia addition and air coflow velocity on a non-premixed methane jet flame stabilization. *Combus Sci Technol* 2022;194:1747–67.

[104] Colson S, Kuhni M, Hayakawa A, Kobayashi H, Galizzi C, and Escudie D. Stabilization mechanisms of an ammonia/methane non-premixed jet flame up to liftoff. *Combust Flame* 2021;234:111657.

[105] Jing Q, Wang D, Liu Q, *et al.* Inhibition effect and mechanism of ultra-fine water mist on CH4/air detonation: quantitative research based on CFD technology. *Process Saf Environ Prot* 2021;148:75–92.

[106] Xiao H, Valera-Medina A, and Bowen PJ Modeling combustion of ammonia/hydrogen fuel blends under gas turbine conditions. *Energy Fuels* 2017;31:8631–42.

[107] Sun Z, Huang Y, Luan Z, Gao S, and You Y. Three-dimensional simulation of a rotating detonation engine in ammonia/hydrogen mixtures and oxygen-enriched air. *Int J Hydrogen Energy*. 2023;48(12):4891–4905.

[108] Jing Q, Wang D, Liu Q, *et al.* Transient reaction process and mechanism of cornstarch/air and CH4/cornstarch/air in a closed container: quantitative research based on experiments and simulations. *J Hazard Mater* 2021;409:124475.

[109] Huang SY, Zhou J, Liu SJ, Peng HY, and Yuan XQ. Continuous rotating detonation engine fueled by ammonia. *Energy* 2022;252:123911.

[110] Aakko-Saksa PT, Cook C, Kiviaho J, and Repo T. Liquid organic hydrogen carriers for transportation and storing of renewable energy–review and discussion. *J Power Sources* 2018;396:803–23.

[111] Huang SY, Zhou J, Liu SJ, Peng HY, and Yuan XQ. Ammonia/oxygen-enriched air continuous rotating detonation in the hollow chamber. *Fuel* 2022;311:122166.

[112] Yu Z, and Zhang H. End-gas autoignition and knocking combustion of ammonia/hydrogen/air mixtures in a confined reactor. *Int J Hydrogen Energy* 2022;47:8585–602.

[113] Zhu R, Fang X, Xu C, Zhao M, Zhang H, and Davy M. Pulsating one-dimensional detonation in ammonia-hydrogen–air mixtures. *Int J Hydrogen Energy* 2022;47(50):21517–21536.

[114] Zhu R, Zhao M, and Zhang H. Numerical simulation of flame acceleration and deflagration-to-detonation transition in ammonia-hydrogen–oxygen mixtures. *Int J Hydrogen Energy* 2021;46:1273–87.

[115] Jing Q, Huang J, Liu Q, Chen X, Wang Z, and Liu C. The flame propagation characteristics and detonation parameters of ammonia/oxygen in a large-scale horizontal tube: as a carbon-free fuel and hydrogen-energy carrier. *Int J Hydrogen Energy* 2021;46:19158–70.

[116] De Vries N, Okafor E, Gutesa-Bozo M, Xiao H, and Valera-Medina A. Use of ammonia for heat, power and propulsion. In: Valera-Medina A, and Banares-Alcantara R (eds). *Techno-Economic Challenges of Green Ammonia as an Energy Vector*. Amsterdam: Elsevier; 2020. pp. 105–54.

[117] Jenkins DR. *Hypersonics Before the Shuttle: A Concise History of the X-15 Research Airplane*. Washington, DC: National Aeronautics and Space Administration. 2000.

[118] Seaman RW, and Huson G. The choice of NH3 to fuel the X-15 rocket plane. *In: 8th NH3 Fuel Association Conference*. 2011.

[119] Thompson MO. *At the Edge of Space: The X-15 Flight Program.* Washington, DC: Smithsonian Institution; 2003.

[120] Iki N, Kurata O, Matsunuma T, *et al.* Micro gas turbine operation with kerosene and ammonia. *In: Proceedings of the 11th Annual NH3 Fuel Conference*, Iowa, IA, USA. 2014. pp. 21–4.

[121] Iki N. Demonstration of ammonia fuel with small gas turbines. In: Aika K, and Kobayashi H (eds). *CO_2 Free Ammonia as an Energy Carrier: Japan's Insights*. Singapore: Springer Nature; 2023. pp. 487–99.

[122] Shiozawa B. A Deep Dive into SIP "Energy Carriers" Ammonia Combustion Research. 2020. https://ammoniaenergy.org/articles/sip-energy-carriers-research-on-ammonia-combustion-technologies/

[123] Arai K, Kawano M, Kudomi N, Sugiura H, Yasui Y, and Ishiwaki F. Development of 300 kW ammonia combustion micro gas turbine. In: Aika K, and Kobayashi H (eds). *CO_2 Free Ammonia as an Energy Carrier: Japan's Insights*. Singapore: Springer Nature; 2023. pp. 501–13.

[124] Ito S, Uchida M, Suda T, and Fujimori T. Demonstration tests of ammonia/natural gas co-firing power generation with a 2-MW-Class gas turbine. In: Aika K, and Kobayashi H (eds), *CO_2 Free Ammonia as an Energy Carrier: Japan's Insights*. Singapore: Springer Nature; 2023. pp. 515–22. Available from: https://doi.org/10.1007/978-981-19-4767-4_35

[125] Co IHIPS. *IM270 Series. Powerful and clean original IHI gas turbines* [Internet]. 2019;2023. Available from: https://www.ihi.co.jp/powersystems/en/lineup/IM270/index.html

[126] Ito S, Uchida M, Kato S, Fujimori T, and Kobayashi H. Emission characteristics of a lean-premixed ammonia/natural-gas gas-turbine combustor and effect of secondary ammonia injection. *Mech Eng J* 2019;6:19-00266-19–00266.

[127] Ito S, Uchida M, Fujimori T, and Kobayashi H. A study of low NOx combustion method for ammonia/natural gas co-fired gas turbine using detailed chemistry. 日本燃焼学会誌 2019;61:368–75.

[128] *Electricity Business Act* [Internet]. 1995;2023. Available from: https://ca.practicallaw.thomsonreuters.com/5-630-3729?transitionType=Default&contextData=(sc.Default)&firstPage=true#:~:text=The%20Electricity%20Business%20Act%20requires,obtain%20a%20licence%20from%20METI

[129] Crolius SH. Ammonia's role in the hydrogen society. In: Aika K, and Kobayashi H (eds). *CO_2 Free Ammonia as an Energy Carrier: Japan's Insights*. Singapore: Springer Nature; 2023. pp. 49–61.

[130] Patel S. *Ammonia gas turbine combustion has economic potential, GE-IHI study suggests* [Internet]. 2023;2023. Available from: https://www.powermag.com/ammonia-gas-turbine-combustion-has-economic-potential-ge-ihi-study-suggests/

[131] Ditaranto M, Saanum I, and Larfeldt J. Experimental study on high pressure combustion of decomposed ammonia: how can ammonia be best used in a gas turbine? In: *Turbo Expo: Power for Land, Sea, and Air*. New York: American Society of Mechanical Engineers; 2021. p. V03BT04A027.

[132] Božo MG, Mashruk S, Zitouni S, and Valera-Medina A. Humidified ammonia/hydrogen RQL combustion in a trigeneration gas turbine cycle. *Energy Convers Manag* 2021;227:113625.

[133] Guteša M. *The Numerical Simulation Model of Gas Turbine Facility for Biomass Gasification Gas Application*. Novi Sad: Novi Sad University; 2017.

[134] Power M. *Mitsubishi Power commences development of world's first ammonia-fired 40MW class gas turbine system–targets to expand lineup of carbon-free power generation options, with commercialization around 2025* [Internet]. March 2021 [cited 2023 Mar 24];1:2021. Available from: https://power.mhi.com/news/20210301.html#:~:text=YOKOHAMA%2C%20JAPAN%20(March%201%2C,increasing%20global%20focus%20on%20decarbonization

[135] Agustin V-M, Syed M, Daniel P, and Phil B. Ammonia. In: O'Connor J, Noble B, and Lieuwen T (eds). *Renewable Fuels: Sources, Conversion, and Utilization*. 2022. pp. 245–74.

[136] Nose M, Kawakami T, Nakamura S, Kuroki H, Kataoka M, and Yuri M. Development of hydrogen/ammonia firing gas turbine for decarbonized society. *Mitsubishi Heavy Ind Tech Rev* 2021;58:1.

[137] I MH. *MHI and Institut Teknologi Bandung launch joint R&D for ammonia-fired power generation using gas turbines in Indonesia* [Internet]. *2022*; 2023. Available from: https://www.mhi.com/news/220928.html

[138] Patel S*Mitsubishi Power developing 100% ammonia-capable gas turbine* [Internet]. 2021;2023. Available from: https://www.powermag.com/mitsubishi-power-developing-100-ammonia-capable-gas-turbine/

[139] Bilby E. *Greener Fuels to Propel Rocket Launches into Space* [Internet]. New York: Horizon; 2019. [cited 2023 Mar 24]; Available from: https://ec.europa.eu/research-and-innovation/en/horizon-magazine/greener-fuels-propel-rocket-launches-space

[140] British Standard ISO. 11042-1: 1996. *Gas Turbines. Exhaust Gas Emission Measurement and Evaluation.* London: British Standards Institution, 1996.

[141] Steele R, and Subbaraman G. *Program on combustion of ammonia and ammonia-hydrogen mixtures for power generation* [Internet]. Ammonia Energy2022 [cited 2023 Mar 24];Available from: https://www.ammoniaenergy.org/paper/program-on-combustion-of-ammonia-and-ammonia-hydrogen-mixtures-for-power-generation/

[142] Avila C, Wang G, Zhu X, *et al.* Lean stability limits and exhaust emissions of ammonia-methane-air swirl flames at micro gas turbine relevant pressure. In: *Turbo Expo: Power for Land, Sea, and Air*. New York: American Society of Mechanical Engineers; 2022. p. V03AT04A004.

[143] Avila C, Cardona S, Abdullah M, *et al.* Experimental assessment of the performance of a commercial micro gas turbine fueled by ammonia-methane blends. *Appl Energy Combust Sci* 2022;13:100104.

[144] Blinov VN, Shalay VV, Kositsin VV, Vavilov IS, Lukyanchik AI, and Ruban VI. Experimental testing of electrothermal microthrusters with

autonomous heating elements for orbital maneuvering of small space vehicles. *Indian J Sci Technol* 2016;9:10.

[145] Fatuev IY, Likhvantsev AA, and Khazov VN. Acetam-new high-effective fuel. *In: Proceedings of the International Astronautical Congress*. IAC: 2012. pp. 8072–8.

[146] Wei H, Shiqun S, Zonggang D, *et al.* Calculation of thermodynamic properties for a new propellant acetylene-ammonia. *Chin J Energ Mater* 2014; 22:161–4.

[147] Robin M, Brogan T, and Cardiff E. An ammonia microresistojet (MRJ) for micro satellites. *In: 44th AIAA/ASME/SAE/ASEE Joint Propulsion Conference & Exhibit*. 2008. pp. 5288.

[148] Sun SJ, Meng T, and Jin ZH. Design, development, and performance of an ammonia self-managed vaporization propulsion system for micro-nano satellites. *Front Inf Technol Electron Eng* 2019;20:1516–29.

[149] SBIR-STTR. Ammonia Micro-Resistojet Propulsion System for CubeSats [Internet]. 2010 [cited 2023 Mar 24]; Available from: https://www.sbir.gov/sbirsearch/detail/2637

[150] Cesaro Z, Ives M, Nayak-Luke R, Mason M, and Bañares-Alcántara R. Ammonia to power: forecasting the levelized cost of electricity from green ammonia in large-scale power plants. *Appl Energy* 2021;282:116009.

[151] LaRock K. *Reaction Motors XLR99 Rocket Engine* [Internet]. *National Museum of the United States Air Force; 2017.* [cited 2023 Mar 24]; Available from: https://www.nationalmuseum.af.mil/Upcoming/Photos/igphoto/2001842719/

[152] Kurt's MOCs *North American X-15* [Internet]. Flickr Pro2021 [cited 2023 Mar 24]; Available from: https://www.flickr.com/photos/kurtsmocs/45280786951

[153] Roberts WL. *Micro gas turbine test facility* [Internet]. [cited 2023 Mar 24]; Available from: https://robertsgroup.kaust.edu.sa/facilities/detail/facilities/micro-gas-turbine-test-facility

[154] Avila Jimenez CD, Cardona S, Juaied MA, *et al.* Influence of the pilot flame on the morphology and exhaust emissions of NH3-CH4-Air swirl flames using a reduced-scale burner at atmospheric pressure. *Energies (Basel)* 2022;16:231.

Chapter 7
Economic considerations for ammonia as fuel

C. Monacchini[1], C. Anfosso[1] and D. Bellotti[1]

7.1 Introduction

This chapter explores the economic aspects of ammonia as a fuel in GT and ICE. In the first part, production costs are evaluated, focusing on the Haber–Bosch process used to produce ammonia, which is energy-intensive and generally relies on natural gas. Then, the introduction of carbon capture and storage (CCS) (blue ammonia) to achieve a low-carbon production system is studied and the resulting additional costs and impact on the overall cost of production are assessed. Moreover, ammonia production exploiting renewable energy sources (green ammonia) is analysed. To conclude, the production costs of grey, blue and green ammonia are compared.

Next, the chapter examines the infrastructure needs for ammonia, which poses unique challenges for storage and transport due to its physical properties, ammonia is a liquid at relatively low pressures and moderate temperatures, making its storage and transport more challenging compared to conventional fuels. This also implies establishing or retrofitting existing facilities for ammonia production, storage tanks, transportation infrastructure (such as pipelines or tankers) and modifications to end-use equipment (such as engines or fuel cells). The initial capital investment and ongoing operational costs of this infrastructure need to be considered. Then, an overall comparison of traditional and alternative fuels costs is presented. Finally, examples of real-world applications are given, and the economic outcomes of these projects are analysed.

- **Overview of ammonia as a fuel**

Today, the reference markets of ammonia are the agriculture sector for the production of fertiliser and the chemical industry as a building block for the production of chemicals, plastics and pharmaceuticals. Ammonia is gaining more and more interest as a futural energy vector both as a great hydrogen carrier and as a fuel itself to be used mostly for power generation and shipping. The global ammonia production for 2023 was estimated at around 150 million metric tons [1]. It has been estimated that the ammonia market will grow by around 300% by 2050,

[1]Thermochemical Power Group (TPG), Department of Mechanical, Energetic, Management and Transport Engineering, University of Genoa, Italy

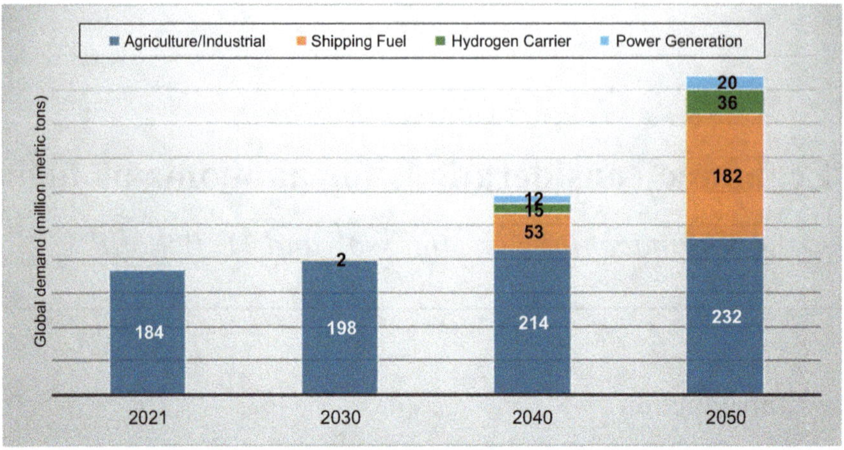

Figure 7.1 Forecast demand for ammonia worldwide from 2021 to 2050, by application (in million metric tons) (based on data from [2])

increasing from approximately 70 billion U.S. dollars in 2021 to around 224 billion U.S. dollars. This is due not only to an expected growth of the traditional agriculture and industrial sector due to an increase in demand but also to the ammonia market expansion to the energy and transport sectors. In 2050, the global ammonia demand is expected to be around 470 million metric tons (+200% compared with 2023), of which around 50% for the agriculture and industrial sectors, around 40% for shipping fuel, and the remaining 10% for power generation, as shown in Figure 7.1.

This underlines the key role of ammonia in the current context as the world moves towards a carbon-neutral future, with ammonia offering several key advantages that make it crucial to decarbonisation efforts. For example, it has a relatively high energy density compared with other hydrogen carriers, which makes it a feasible alternative to fossil fuels [3]. It can be stored and transported as a liquid at moderate pressures and temperatures, which facilitates its uses. Furthermore, burning ammonia does not increase greenhouse gas emissions, making it a carbon-neutral fuel. This is a significant advantage over traditional fossil fuels, which are major contributors to GHG emissions, indeed, ammonia is emerging as a promising alternative fuel, particularly for internal combustion engines (ICE) and gas turbines (GT). Moreover, ammonia has a higher octane number than gasoline and diesel, which allows it to withstand higher compression ratios in engines without causing knock [4–7]. This has the potential to improve engine performance and efficiency. On the other hand, there are still challenges to overcome, such as the low flame speed and low ignition temperature of ammonia. The main properties are reported in Table 7.1. To address these limitations, research is focused on advanced combustion technologies and dual-fuel strategies to improve performance and emissions.

The use of ammonia as an alternative fuel looks promising not only for ICEs but also for GTs. One of the main issues is the reduction of nitrogen oxide (NO_x)

Table 7.1 Fuel properties [8]

Species	Ammonia	Methanol	Hydrogen	Methane	Gasoline
Formula	NH_3	CH_3OH	H_2	CH_4	–
Storage	Liquid	Liquid	Compressed	Compressed	Liquid
Storage temperature (K)	300	300	300	300	300
Storage pressure (MPa)	1.1	0.1	70	25	0.1
Density @ storage conditions (kg.m^{-3})	600	784.6	39.1	187	~740
FL in air (vol.%)	15–28	6.7–36	4.7–75	5–15	0.6–8
LBV @ stoichiometry (m.s^{-1})	0.07	0.36	3.51	0.38	0.58
Auto-ignition t (k)	930	712	773–850	859	503
Research Octane Number	130	119	>100	120	90–98
LHV (MJ/kg)	18.8	19.9	120	50	44.5

emissions that can occur during ammonia combustion and overcoming low combustion efficiency due to ammonia's slow flame speed and long quenching distance. Recent progress has shown its potential as a clean energy source, a few ammonia-fired GTs are currently in operation worldwide.

7.2 Production costs of ammonia

- **Conventional production (grey ammonia)**

Ammonia production is a well-established industrial process, with several technologies available to produce ammonia at large scales. The only process available at the industrial scale is the Haber–Bosch loop (Figure 7.2). However, since this process is characterised by significant energy consumption, the research aims to identify innovative solutions with a competitive ammonia yield, low costs and environmental impacts.

The reaction that takes place in the processes can be defined as follows:

$$N_2 + 3H_2 \rightarrow 2NH_3 \quad \Delta H^0 = -46.35\ \mathrm{kJ/mol} \tag{7.1}$$

Most of the ammonia (70%) is produced using the conventional Haber–Bosch process fed by natural gas, which, at a large-scale industrial level, requires combining nitrogen and hydrogen at high temperatures (350–500 °C) and pressures (100–250 bar). Although ammonia is composed of nitrogen and hydrogen, the reaction between them does not occur at standard temperature and pressure. Therefore, the reaction is conducted under high temperatures and pressure, with the aid of a catalyst. The ammonia formation reaction is exothermic, releasing approximately 2.7 GJ of energy per tonne produced [8]. In the reactor, only a

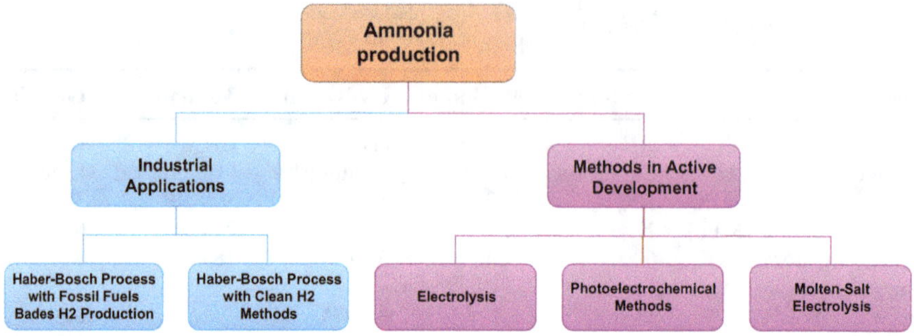

Figure 7.2 Ammonia production methods

portion of the hydrogen and nitrogen react at any given time, therefore ammonia is separated from the hydrogen and nitrogen at the reactor outlet and the unreacted gases are recirculated back into the reactor. The reaction rate of hydrogen in a single cycle ranges between 25% and 34% [9].

Nowadays, most of the ammonia (70%) is produced using the conventional Haber–Bosch process fed by natural gas [10], this is called 'grey ammonia'. Natural gas is used to produce hydrogen, while nitrogen is separated from air. The most common method for producing hydrogen on an industrial scale is steam methane reforming (SMR) (7.2), which involves reacting natural gas with steam to produce hydrogen and carbon monoxide, then this further reacts with water in the water-gas shift reaction (7.3) to produce additional hydrogen and carbon dioxide.

$$CH_4 + H_2O \rightarrow CO + 3H_2 \ \Delta H^0 = 206 \text{ kJ/mol} \tag{7.2}$$

$$CO + H_2O \rightarrow CO_2 + H_2 \tag{7.3}$$

Water gas shift reactions are exothermic while the steam reforming reaction is endothermic, so natural gas is used both as a reagent and as a fuel to heat the process. Therefore, the heavy reliance on natural gas for hydrogen production has significant environmental implications, In SMR is a carbon-intensive process, as it generates large amounts of CO_2 (2–2.5 tons of CO_2e per ton of NH_3) [11].

The high dependence of this process on natural gas also has consequences from an economic point of view as the cost of producing ammonia via the Haber–Bosch process is heavily influenced by natural gas prices. These can fluctuate due to various factors, such as supply and demand, availability of sources and geopolitical equilibrium, considering that fossil fuel reserves are not globally homogeneously distributed but are under the control of a few countries and environmental policies. Figure 7.3 shows ammonia prices, comparing prices recorded by EU customs flows with those collected through a panel of experts. The sharp increase in 2022 is a direct consequence of the rise in gas prices related to the Russian–Ukrainian crisis.

Similarly, in Figure 7.4 the trend of ammonia costs in America is depicted, showing a clear correlation with natural gas prices. As observed, there is a

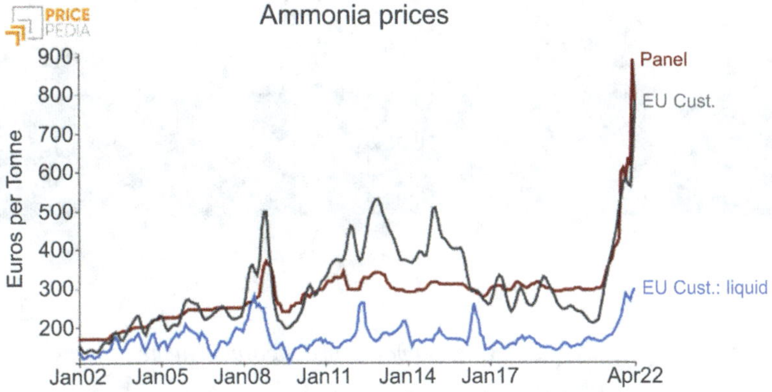

Figure 7.3 Ammonia prices in Europe. Reprinted, with permission, from [12].

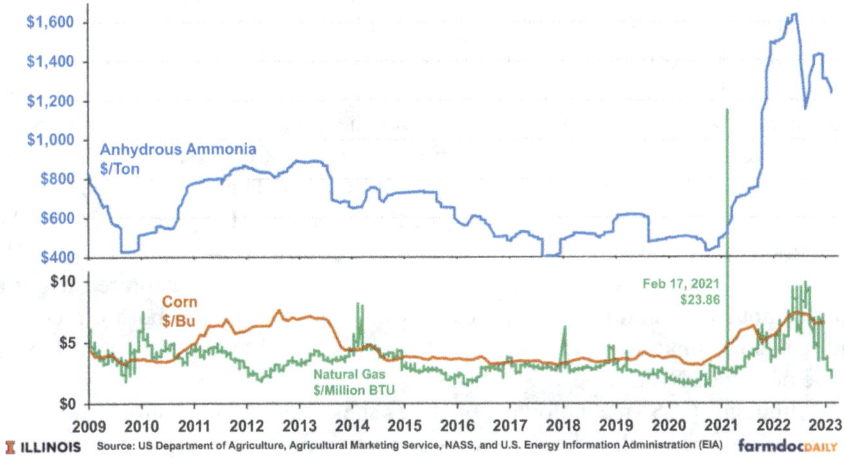

Figure 7.4 Ammonia prices in USA. Reprinted, with permission, from [13].

significant rise in both ammonia and natural gas prices in 2022. The data underscores the strong connection between the two commodities, where fluctuations in natural gas prices directly influence the cost of ammonia. As ammonia is essential for fertilisers used in agriculture, higher costs led to increased expenses for farmers, resulting in higher corn prices.

- **Low-carbon production (blue ammonia)**

To reduce carbon dioxide emissions, a CCS unit can be introduced, in this case, we define it as blue ammonia (Figure 7.5). It is produced using traditional methods, such as the Haber–Bosch process. It is essential to utilise post-combustion CO_2 capture technologies to reduce the CO_2 emissions associated with the SMR process [14].

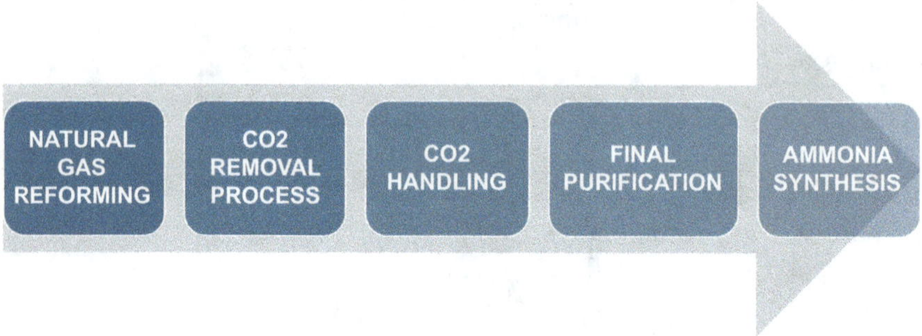

Figure 7.5 Blue ammonia production scheme

In literature, examples demonstrate how CCS technology can be successfully implemented in SMR plants to significantly reduce CO_2 emissions, contributing to global decarbonisation goals. Tarun *et al.* [15] employed both MEA absorption and membrane technologies to capture and separate CO_2 produced during the SMR process. Their findings showed that CO_2 capture technologies can achieve lower energy penalties when the feed stream entering the reformer reactor is at a higher temperature, whereas the inlet streams to the water-gas shift reactors are at a lower temperature. Shi *et al.* [16] utilised vacuum swing adsorption (VSA) to capture CO_2 from an SMR plant. The results demonstrated that the CO_2 capture system they designed was capable of capturing 90% of the CO_2 in the tail gas with a purity of 95%. Wang *et al.* [11] estimated that if 90% of the GHG emissions from H_2 feedstock production can be captured using a post-combustion capture unit, the carbon intensity (including both direct and indirect emissions) of the entire process can be reduced to approximately 0.5 kg CO_2e per kg of NH_3.

Although CCS technologies provide substantial opportunities for decarbonisation, the production of low-carbon ammonia still results in significant emissions. Considering SMR with carbon capture only from process gases, the carbon intensity of blue ammonia could be from 1.3 tCO_2/tNH_3, while the emissions associated with the hydrogen feedstock would be 6.4 tCO_2/tH_2 [17]. According to the EU taxonomy for sustainable finance, for anhydrous ammonia manufacturing to be considered sustainable and contribute significantly to climate change mitigation, the hydrogen used must have a life cycle assessment (LCA) carbon footprint of no more than 3.0 tCO_2e/tH_2 [17]. This implies that ammonia produced through SMR with CCS would not meet the EU's sustainability threshold unless carbon capture is also applied to heat generation. Alternatively, to achieve the required carbon footprint, the natural gas feedstock would need to be replaced with a low-carbon alternative, such as biomethane or low-carbon hydrogen.

While CCS significantly reduces the carbon footprint of ammonia production, it also adds to the overall cost due to the additional infrastructure and energy required for capturing, compressing, transporting and storing CO_2. According to Collodi *et al.* [18], reaching a 90% rate of capture raises the cost of CO_2 avoidance from 51.8\$/ton to 76.8\$/ton, accounting for both the capital (CAPEX) and operational (OPEX) expenses involved in capturing and compressing CO_2. Therefore, assuming a 76.8\$/t CO_2 cost for CO_2 avoidance, a 500,000 t/y ammonia plant would see around 63.7 M\$ additional costs every year or 128.3\$/t NH_3 [19].

Asgharian *et al.* [19] conducted a comprehensive techno-economic analysis of blue ammonia production, assessing the effectiveness of the cryogenic CO_2 capture (CCC) process in mitigating CO_2 emissions during hydrogen production using the SMR. The results showed that the energy penalty of the CCC process for this application ranges between 0.724 and 0.838 MJe/kg CO_2 when capturing 90–99% of CO_2 emissions in the liquid phase. On the other hand, the amine-based process energy penalty is 1.2 MJe/kg CO_2 due to the high specific heat requirements in the stripper column. A straightforward comparison with conventional amine-based processes indicated a 40% reduction in CO_2 capture costs by using the CCC method.

- **Renewable production (green ammonia)**

Green ammonia is produced using the Haber–Bosch process, like grey and blue ammonia, but hydrogen is obtained through water electrolysis powered by renewable energy such as wind, solar, hydropower and geothermal energy, while nitrogen comes from an air separation unit [20]. When both hydrogen and energy are sourced renewably, the process can achieve zero carbon emissions. To produce green ammonia, green hydrogen must first be obtained through a process of water electrolysis (Figure 7.6).

The process of producing ammonia using electrolysis involves four key units: a renewable energy source (such as solar or wind), an electrolyser, an air separation unit and the Haber–Bosch process. Hydrogen and oxygen are produced through the

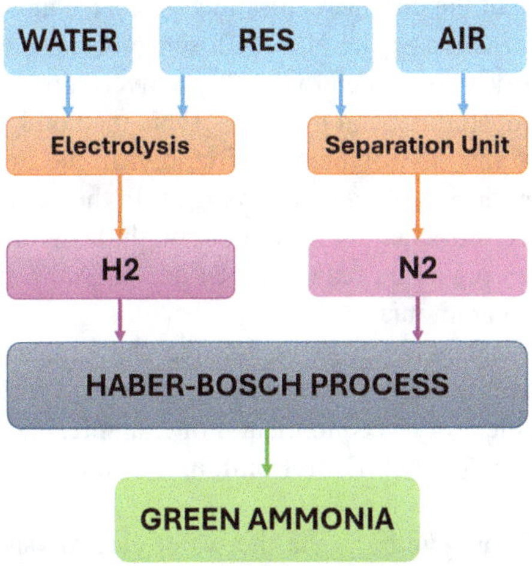

Figure 7.6 Green ammonia production

electrolysis of water [21]. This water-splitting process occurs in electrochemical cells, which are classified based on their configuration into polymer electrolyte membrane electrolysers, alkaline electrolysers and solid oxide electrolysers. Each cell consists of an anode, cathode, electrolyte, power source (typically renewable) and sometimes a membrane. When an electric current is applied to the electrodes, the covalent bonds in the water molecules break, resulting in the production of hydrogen and oxygen gases. In alkaline electrolysers, potassium hydroxide is commonly used as the electrolyte to prevent acid corrosion, while nickel is the most widely used material for electrodes due to its anodic and cathodic stability. Conventional AEC achieves maximum utilisation at operating temperatures of 40–90 °C and a pressure approximately equal to the ambient one even if some applications are available at operating pressures up to 25 bars [21–23]. Alkaline electrolysis is a mature technology, suited for stationary applications, with a significant operating record in industrial applications that allows remote operation. The electrolyte used in proton exchange membrane (PEM) electrolysers is a solid polymer membrane, such as Nafion, which acts as a proton conductor. PEM electrolysers can operate at pressures up to 200 bar or more, making them suitable for both stationary and mobile applications. Compared to alkaline electrolysers, PEM electrolysers offer several key advantages [23]: they have a faster start-up time, enhanced safety due to the absence of corrosive potassium hydroxide (KOH) electrolytes, a more compact design due to higher energy densities and the ability to operate at higher pressures. Solid oxide electrolysis (SOE) has garnered attention due to its ability to produce ultra-pure hydrogen with higher efficiency, all without the need for noble catalysts. This process utilises an ionic conductor ceramic electrolyte, which distinguishes it from other electrolysis methods. A key characteristic of SOE technology is its high operating temperature, ranging from 500 to 1000 °C, significantly higher than that of proton exchange membrane electrolysers (PEMEC) and alkaline electrolysers (AEC). However, the elevated operating temperature also accelerates the degradation rate, presenting a challenge to the commercialisation of this technology [24].

An emerging alternative to the Haber–Bosch process, electrochemical synthesis [22] uses an electrochemical cell to directly convert nitrogen and hydrogen into ammonia at lower temperatures and pressures. This method has the potential to be more energy-efficient and environmentally friendly. In the first step, water undergoes electrolysis to produce hydrogen and oxygen. In the second step, nitrogen gas is introduced into the electrochemical cell along with the hydrogen produced in the first step. The nitrogen reduction reaction then occurs at the cathode, where nitrogen is reduced to form ammonia.

From an economical point of view, beyond the expenses associated with a conventional Haber–Bosch plant, the overall capital cost is further increased by the introduction of water electrolysis equipment. For example, CRU [25] estimated that the total capital cost would be USD 51.1 billion, in addition to an operating cost of USD 570 million per year (Figure 7.7).

The cost of ammonia synthesis via the water electrolysis route is considered high primarily due to the significant capital costs of electrolysers. Additionally,

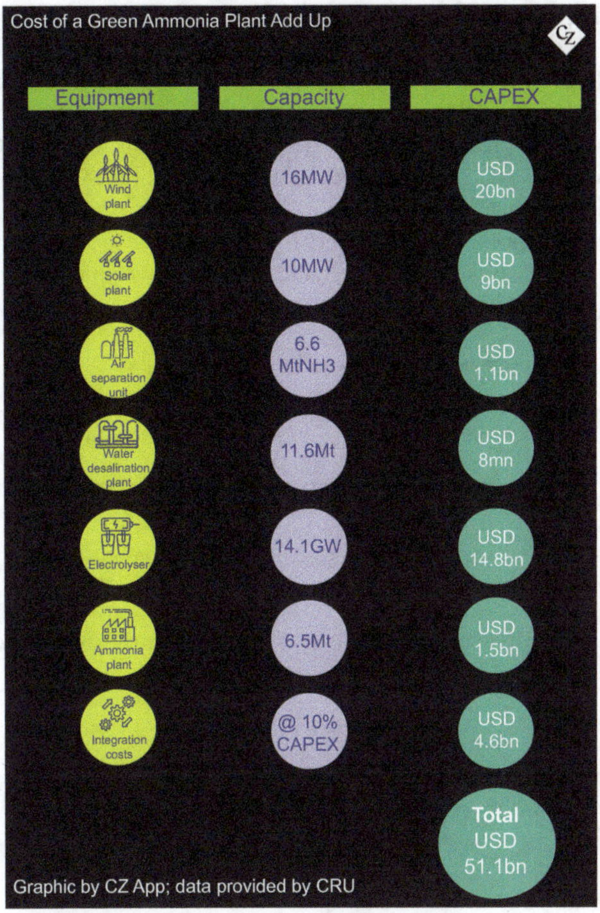

Figure 7.7 Costs of a green ammonia plant. Reprinted, with permission, from [29].

electrolysers consume a large amount of energy to produce green hydrogen, which serves as an intermediate feedstock.

Different values of the capital and operational costs are reported in the literature, depending on the plant size. Proost J. performed a comparative analysis between PEMEC and AEC CAPEX showing that for a 2 MW electrolyser the PEMEC CAPEX is around 1320 \$/kW, whereas the AEC one is about 770 \$/kW [26]. The R&D IEA department reports that, nowadays, CAPEX requirements are currently in the range of 500–1400 \$/kW for alkaline electrolysers and 1100–1800 \$/kW for PEM electrolysers, while estimates for SOEC electrolysers a range across 2800–5600 \$/kW [27]. Sousa *et al.* compared a 30 MW PEM electrolyser-based system for ammonia production with a traditional SMR using an electrolyser investment cost equal to 1100 \$/kW and operational cost equal to 2.5%

of the CAPEX [28]. Schmidt *et al.* analysed the three major electrolysers investment capital costs taking data from a pool of experts in the sector resulting in a range of 1100–2150 \$/kW for PEMEC, 880–1450 \$/kW for AEC and 3250–5500 \$/kW for SOEC [29]. Bellotti *et al.* investigated two power-to-fuel solutions both of which consider a pressurised PEM electrolysis section, the used CAPEX was 1100 \$/kW, whereas the OPEX has been assumed 4.5% of the CAPEX based on the literature review (Table 7.2) [30].

Continued advances in electrolyser technology are expected to increase efficiency and reduce costs, as shown in Figure 7.8. SOE is supposed to compete with

Table 7.2 Electrolysers technologies [23,30–34]

Type of the electrolyser	Efficiency [%]	Cost [\$/kW]	Stack life time [h]	Maturity	Pressure [bar]
AEL 'Alkaline electrolyser'	50–60	770–1400	60,000–90,000	Mature	1–30
PEM 'Polymer electrolyte membrane electrolyser'	46–60	1100–1800	20,000–600,000	Commercial	20–50
SOE 'Solid oxide electrolyser'	Up to 65	> 2500	<10,000	Demonstration	1–15

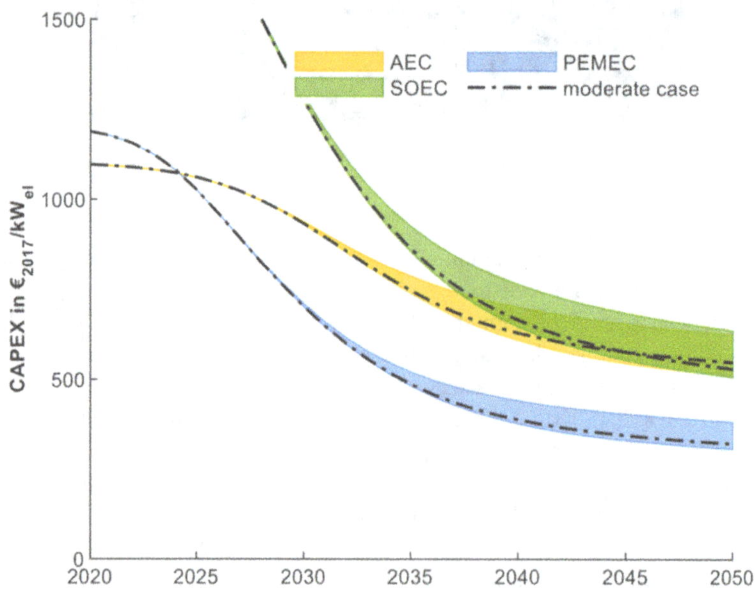

Figure 7.8 Estimated electrolysers cost. Reprinted, with permission, from [31].

alkaline cells in terms of capital costs over the long term, and although alkaline electrolysers are still the cheapest solution, PEM electrolysers cost has recently decreased in a considerable way and it is reasonable to assume a further reduction in the coming years [27].

Another factor to consider is the cost of renewable energy. Over the past decade, the cost of renewable electricity has decreased significantly, driven by advancements in technology, economies of scale, more competitive supply chains and increased experience among developers. Since 2010 [32], utility-scale solar PV has experienced the most drastic cost reduction, with an 82% decline. Concentrating solar power (CSP) has seen a 47% decrease, onshore wind has dropped by 39% and offshore wind has reduced by 29% (Figure 7.9). As of now, solar and wind power are among the most cost-competitive energy sources.

- **Comparison and cost projections**

The cost of ammonia production varies significantly depending on the method used, with grey, blue and green ammonia, what characterises the different methods of ammonia production is how the hydrogen is obtained, i.e. from fossil or renewable sources. Figure 7.10 shows that in the short term, in most cases green hydrogen is the most expensive one (3.2–7.3 \$/kg H_2) [33], but at the lower end of this range it is cost-competitive with blue hydrogen. Focusing on the long-term,

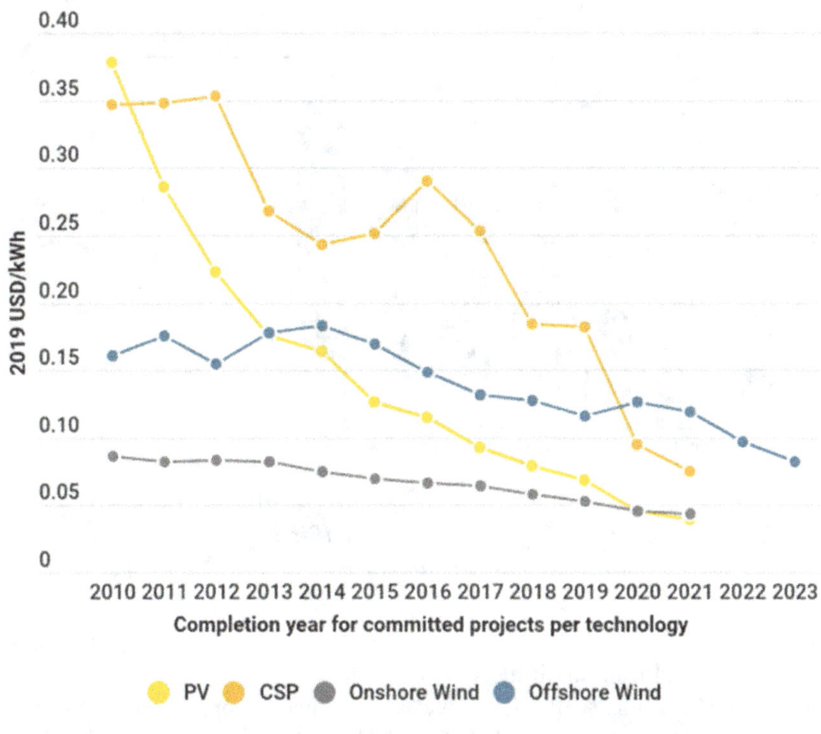

Figure 7.9 Renewable energy cost. Reprinted, with permission, from [28].

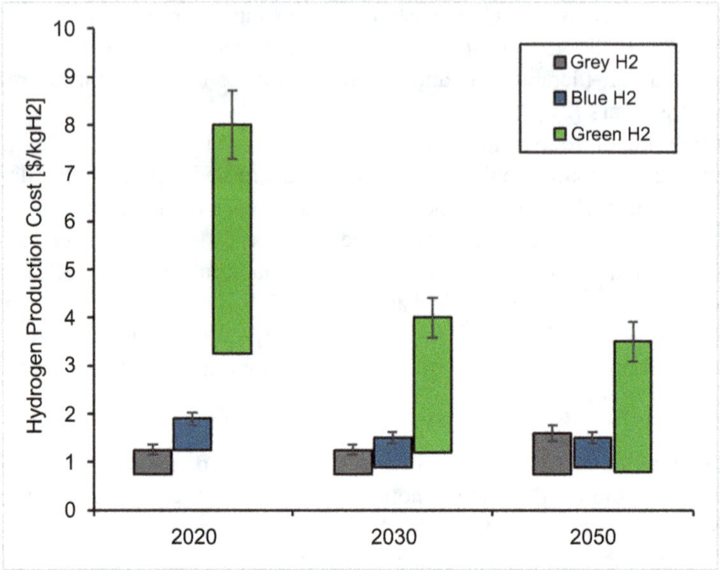

Figure 7.10 *Cost of hydrogen production in three different scenarios. Reprinted, with permission, from [20].*

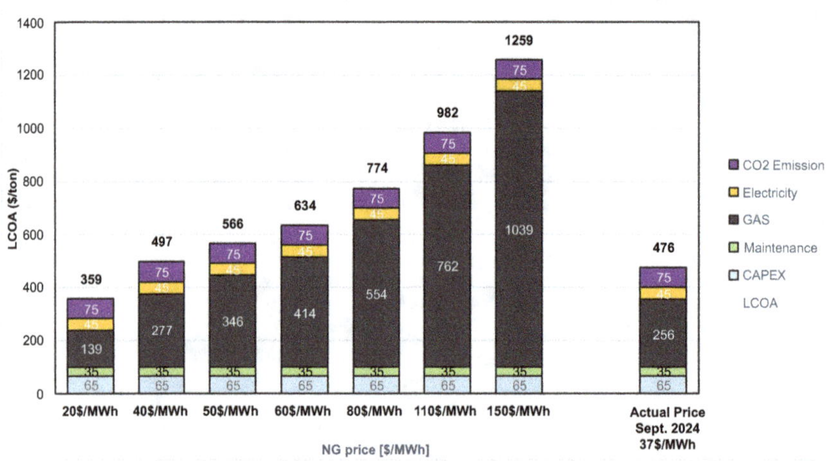

Figure 7.11 *Levelised cost of ammonia in an SMR-based process: natural gas price sensitivity (based on data from [17])*

scaling up and technological development could help achieve cost parity with hydrogen produced from fossil fuels.

Consequently, this will also have a significant impact on the cost of ammonia production. As shown in Figure 7.11, grey ammonia, produced via the

Haber–Bosch process using hydrogen derived from natural gas through SMR, is the most cost-effective option, however, this method has a high carbon footprint. Since the natural gas consumption for SMR is the most energy-intensive step in the ammonia manufacturing process, a recent analysis conducted by Hydrogen Europe [34], with the increase in natural gas prices (November 2022 – attributable to a transitional situation linked to the Russia-Ukraine conflict) of around 110 \$/MWh, the total cost of grey ammonia production reached around 982\$/ton of NH_3. In this latter case, the natural gas cost is responsible for almost 75% of the total cost. Another important cost element is the one related to CO_2 emissions, adding around 12% of the cost [34]. The CO_2 associated with SMR can be considered equal to 1.8 tonCO_2/tonNH_3 with a corresponding taxation of 75\$/tonCO_2 (established by the European Commission) [34]. Considering, instead, the actual natural gas price of 37\$/MWh, CO_2 taxation has an even higher impact: 16% of the overall cost. Figure 7.11 shows, in fact, the traditional plant LCOA (levelised cost of ammonia) at different gas prices, starting from 20\$/MWh up to 150\$/MWh [34].

Blue ammonia, on the other hand, also uses the Haber–Bosch process with hydrogen from natural gas, but it incorporates CCS technologies to reduce emissions. This leads to an increase in production costs depending on the efficiency of the CCS technology and the cost of natural gas. While more expensive than grey ammonia, blue ammonia offers a lower environmental impact. In this case, CO_2 emissions are significantly reduced thanks to the high efficiency of the carbon capture system (90%), however, it is necessary to foresee a higher cost for the storage and transport of the CO_2. Regarding the specific cost for CCS installation, it doesn't heavily affect the overall CAPEX cost as it is coupled with the ATR system which, differently from SMR, has a CO_2-concentrated production which makes this system more amenable to CCS application (Figure 7.12).

Green ammonia is produced using hydrogen obtained from water electrolysis powered by renewable energy sources like wind or solar. This method, which

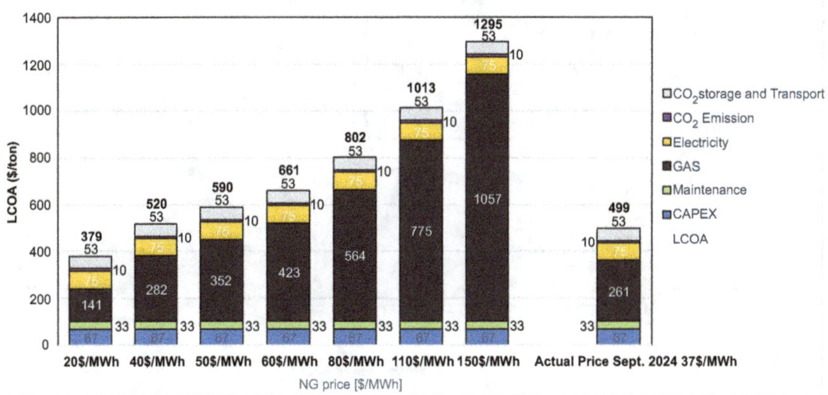

Figure 7.12 Levelised cost of ammonia in a ATR+CCS-based process: natural gas price sensitivity (based on data from [17])

results in near-zero carbon emissions, is the most environmentally friendly but also the costliest one. The high costs are primarily due to the expensive electrolysers and the large amount of renewable energy required. Despite its higher cost, green ammonia is seen as a crucial component in the transition to sustainable industrial practices, as it offers the potential for completely carbon-free ammonia production. As renewable energy becomes more affordable and technology advances, the cost of green ammonia is expected to decrease, making it a more competitive option in the future. Figure 7.13 reports an analysis carried out by IEA where, for different Countries, the LCOA produced via electrolyser taken the electricity from the national grid (left side) is compared with the 2030 scenario in which the electrolyser is totally fed by renewable intermittent energy sources (right side) [35].

Figure 7.14 reports an ammonia production feasibility map developed by Bellotti *et al.* [36]. This map plot offers an overview of the ammonia production

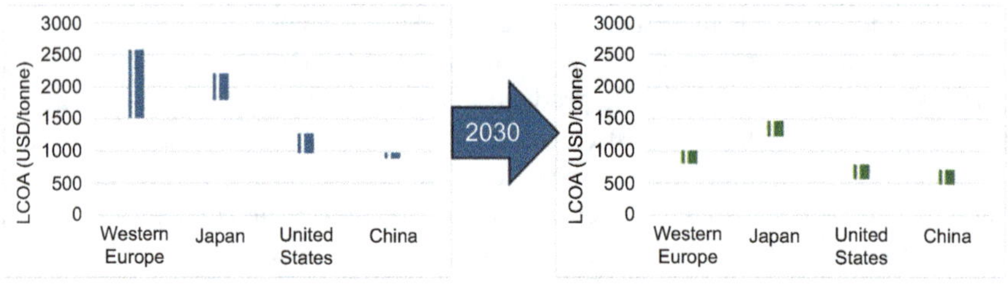

Figure 7.13 Indicative production costs for ammonia via electrolysis in selected regions (based on data from [35])

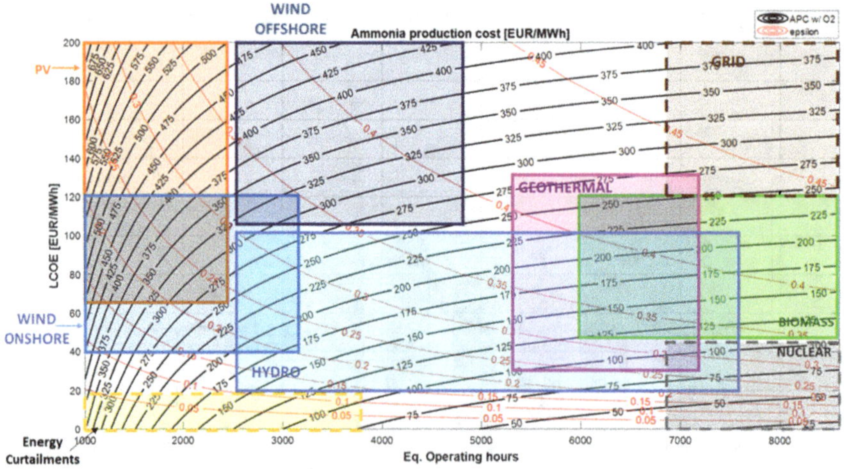

Figure 7.14 Map-plot of ammonia production cost in [€/MWh] as a function of the equivalent operating hours and the LCOE (the O_2 sale option at 100€/ton is considered). Reprinted, with permission, from [36].

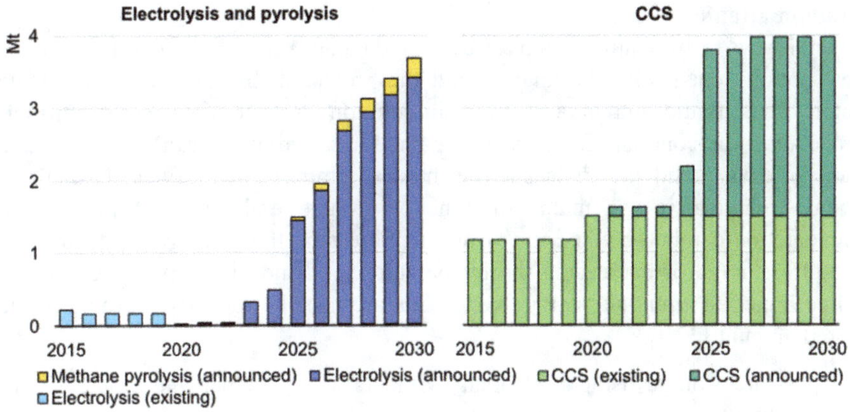

Figure 7.15 Current and announced projects for near-zero-emission ammonia production. Reprinted, with permission, from [35].

cost (€/MWh) as a function of the LCOE range of different energy sources (both renewable and fossil) and the equivalent operating hours, EOH (a parameter strongly related to the type of energy sources). The cost of electricity impacts over 50% of the overall fuel production cost, while the sensitivity analysis over the EOH showed that its influence on the LCOE is not linear and a variation of +/−50% could lead to a APC variation from −20% to +60%, respectively [36].

In the next year, in addition to more widespread application of CCS in already existing ammonia production plants, also electrolysis-based plants are gaining interest in meeting the target of near-zero CO_2 emission. Therefore, multiple projects are scheduled to come online in the coming years, bringing total electrolytic ammonia production for conventional uses to more than 3 Mt of capacity by 2030, considering announced projects as of June 2021. Moreover, it must be underlined that a large proportion of planned capacity will use variable renewable electricity as opposed to dispatchable large-scale hydropower. It is also worth it to mention that some applications are also foreseen for methane pyrolysis-based plants even if they will cover just a small amount of the overall capacity (Figure 7.15).

7.3 Infrastructure and capital expenditure

- **Storage and transportation**

Another crucial element of the economic analysis involves the infrastructure needed for the production, storage, distribution and use of ammonia as a fuel. This includes building or retrofitting facilities for ammonia production, storage tanks, transportation infrastructure (like pipelines or tankers) and modifications to end-use equipment (like engines or fuel cells).

Ammonia tanks

Several ways to store ammonia have been studied and tested in recent decades: the best option depends on the amount and the projected final use. For large-scale applications, liquid ammonia is the main solution for both storage and transport procedures. In contrast, for small-scale facilities, ammonia can be stored and transported in solid materials such as metal ammines and zeolites. Nowadays, liquid ammonia is the main solution for storage and transport procedures. Liquefaction can be obtained in two ways: increasing the pressure for room temperature storage or decreasing the temperature for atmospheric pressure storage. Three liquefied ammonia storage conditions are usually applied according to the stored amount [37]:

- pressure storage using horizontal cylinders at 16–18 bar and ambient temperature for capacities up to 2000 tons;
- semi-refrigerated storage using insulated reduced-pressure spherical vessels at 0 °C with capacities up to 2500 tons;
- low-temperature storage using vertical tanks at 1.1–1.3 bar and −34 °C (ammonia boiling temperature) with capacities of up to 50,000 tons.

The low-temperature storage is useful for higher capacities. Tanks used for storing anhydrous ammonia at or near atmospheric pressure and −33 °C are typically designed following a suitable design code, such as API 620R [38]. The standard material for these tanks is low-temperature certified carbon manganese steel, which is impact-tested at or near −40 °C. The risk of stress corrosion cracking increases as the yield strength of the steel rises. Therefore, materials with a minimum yield strength between 290 and 360 MPa are commonly used. For new tanks, selecting materials with a minimum yield strength at the lower end of this range is advisable.

The main types of atmospheric tanks operating at −33 °C are steel tanks with single or double external insulation. In single-wall tanks, the insulation consists mainly of panels of rock wool, foam glass, Styrofoam and polyurethane foam. In double wall tanks, the walls are insulating materials with dried gases in the interstitial space, such as air, nitrogen or another inert. Indeed, water cannot be used to avoid moisture freezing. However, tanks never have perfect insulation. The stored liquid can absorb heat because of the temperature difference between the tank and the surroundings. Here the NH_3 fraction can evaporate and create boil-off gases at the thermodynamics equilibrium with the stored liquid.

It is recommended that each atmospheric storage tank be equipped with at least two pressure relief valves and two vacuum relief valves to protect against potential overpressure or vacuum conditions [39]. An isolation valve should be placed between the tank and each relief valve, and a mechanical linkage system must be incorporated to ensure that only one valve can be isolated at a time. The sizing of relief valves and safety devices should adhere to the specifications outlined in API 2000.

The preferred method to store ammonia at a small scale (up to 2000 tons of capacity) is using a pressurised storage cylinder have an installed CAPEX of around 3 $/kg$_{NH3}$ [40]. Refrigerated tanks are preferred for large-scale storage (up to 50,000 tons of capacity) because of the lower capital expenditure. Particularly, the CAPEX for double-walled tanks is greater than for single-walled tanks, but maintenance and operating costs are lower. Estimates for the CAPEX of installed refrigerated storage have been reported in the literature, ranging from 0.561 to 1.062 $/kg$_{NH3}$.

Based on literature data cost functions for both the pressurised and cryogenic storage types have been estimated [41]:

$$CAPEX_{Press.storage} = 28,000 * (Storagecapacity\,[\text{tons}])^{0.7} \quad [\$_{2024}]$$

$$CAPEX_{Cryo.storage} = 63,000 * (Storagecapacity\,[\text{tons}])^{0.7} \quad [\$_{2024}]$$

Pipelines

Pipeline use has several advantages such as higher safety, a reduced carbon footprint and good efficiency due to minimum leaks. Transporting ammonia through pipelines requires following several key conditions to ensure safety and integrity. The ammonia must be heated to at least 2 °C to prevent brittle fracture of the pipeline, which can occur at lower temperatures [42]. Therefore, the ammonia is warmed at the supply terminal and then cooled back down to −33 °C at the receiving terminal. Additionally, the ammonia must be transported at a minimum pressure of 20 bar to prevent gas formation within the pipeline. Vapourisation of the ammonia could reduce transport capacity, complicate flow measurement and control, and cause cavitation damage to the pipeline's inner walls. The Interstate Commerce Commission also mandates that anhydrous ammonia transported across state lines in the USA must contain at least 0.2 wt.% of water. This small amount of water acts as a corrosion inhibitor, reducing the risk of stress corrosion cracking.

Regarding materials, steels are used in ammonia transfer lines pipelines with a diameter ranging from 0.15 to 0.25 m. Although stainless steel is highly resistant to corrosion, it is quite expensive. Materials like copper, zinc and aluminium are not utilised because they are easily corroded by ammonia.

Capital costs vary depending on the required flow rate, distance, temperature and elevation changes of the pipeline, and they include the cost of the pipe, booster stations and construction. In addition, an urban pipeline costs more than a rural one (857,000 $/km and 1,469,000 $/km) [43] and to maintain required flow rates and temperatures pumping boosters are necessary for long-distance pipelines. Operational costs are estimated at 500$ per km per year and the operating cost of each booster station would be 408,000$ per year, assuming each station is operational for 8500 hours.

A recent study, performed by Vu Dinh *et al.* shows the levelised cost of transmission (LCOT) of hydrogen stored in two different forms: liquid and under the form of ammonia [43]. Vu Dinh, compared the obtained results with those proposed by IEA (Figure 7.16) [44]. and it is underlined the economic advantage of using NH3 as a hydrogen carrier: the ammonia tanker cost is not affected by transmission distance and the fact that ammonia is liquid under favourable ambient

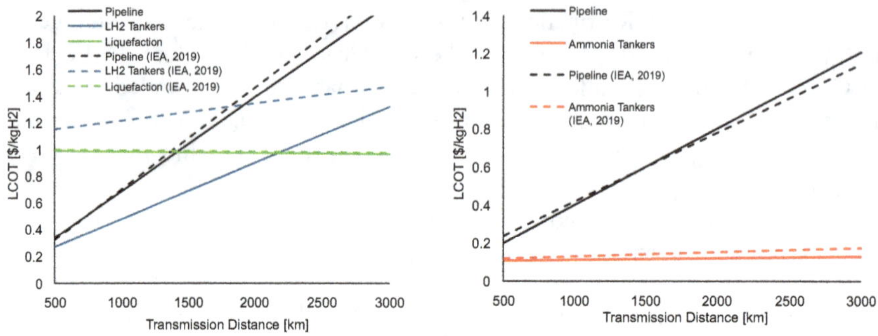

Figure 7.16 Comparison of LCOT of hydrogen (right) and LCOT of hydrogen in form of ammonia (left). Reprinted, with permission, from [43].

conditions means that the cost of ammonia liquefaction can be neglected. In this way, considering the Vu Dinh approach and a distance of 500 km the LCOT of hydrogen stored under the form of ammonia is 0.31 \$/kg H_2 whereas if liquid H_2 is taken into account, the LCOT for the same distance is 1.6 \$/kg H_2.

Sea transport

Anhydrous ammonia is today typically transported in gas carriers designed for LPG as NH3 shares similar properties to propane (i.e. saturated vapour pressure). Usually, ships within the range of 20,000–60,000 m^3 are used for ammonia transportation, whereas larger ships are mostly employed for LPG transportation. In Russia, however, ships larger than 600,000 m^3 are also used to transport ammonia [45]. The ships are typically in the range of 15,000–85,000 m^3, with three common commercial sizes for LPG/Ammonia trades of 30,000 m^3, 52,000 m^3 and 80,000 m^3 [46].

The cost of transport will depend on the financial arrangements of exporters and importers. The charter rates of energy carriers are highly volatile based on market conditions and energy demands (i.e. the COVID-19 pandemic caused the charter rate of LNG ships to fall from 87,500\$ to 30,000\$ per day). Even under less extreme economic conditions, the charter cost can be highly variable [47]. To establish the transport price of ammonia by ship, Salmon compared the transport prices available in the literature as a function of the distance (km) to apply a linear regression (Figure 7.17).

Ammonia bunkering

Ammonia's characteristics, such as its toxicity and corrosiveness, will significantly impact bunkering procedures [48]. To mitigate risks related to corrosion and potential harm to humans, aquatic life and the environment these risks, it is recommended to insert nitrogen gas into hoses and pipelines to eliminate moisture and oxygen. Then, the system should be purged with ammonia vapour to remove any residual nitrogen.

Moreover, given the fact that anhydrous ammonia remains in its liquid state through refrigeration or pressure, a BOG management system, ammonia vapour return line and ventilation system are necessary to manage the boil-off gas

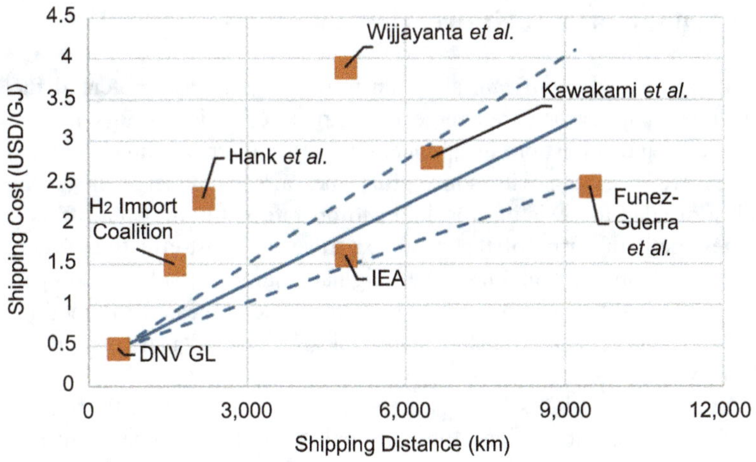

Figure 7.17 Ammonia shipping cost. Reprinted, with permission, from [49].

Table 7.3 Economic parameters for MFO and ammonia bunkering [50]

Parameters	MFO bunkering	Ammonia bunkering	Source
Bunkering licence fee	150 SGD/year/vessel	150 SGD/year/vessel	MPA (2021b)
Port dues	100 SGD/year/vessel/10 GT	100 SGD/year/vessel/10 GT	MPA (2021b)
Charter fee	350 USD/vessel/hour (6000 MT bunker supply vessel)	450 USD/vessel/hour (Handy size ammonia carrier)	Interviews of bunker suppliers, HELLENIC SHIPPING NEWS (2021)
Bunker price	451 USD/tonne	300 USD/tonne	Ship & Bunker (2021), MacFarlane *et al.* (2020)
Disputed bunker volume	15 tonnes	33 tonnes	Interviews of bunker suppliers

generated from temperature increases or pressure drops. Additionally, ammonia fuel tanks can only be filled to 94% of their total capacity, with a 4% heel left in the tank to prevent overpressure, as the supply port's maximum filling limit is 98%.

Ammonia-fuelled vessels need more volume than MFO-fuelled ones to provide the same energy, as ammonia has a lower heating value of 18.6 MJ/kg compared to 40.4 MJ/kg for MFO. Therefore, transitioning to ammonia requires adjusting the flow rate to maintain the bunkering schedule. Furthermore, an after-treatment system to reduce NO_x emissions is needed as a compression-ignition fuel.

Yang *et al.* [50] found out that number of ammonia bunker supply vessels is the most sensitive parameter for annual operational cost. When it is changed by ±50%, annual operational cost variability is as high as 15.2% (Table 7.3).

7.4 Ammonia as a fuel: overview and costs

Ammonia has gained significant attention in recent years as a potential fuel, particularly due to its carbon-free molecular structure, which results in zero CO_2 emissions during combustion. Its application is currently being investigated in both the GT and ICE sectors, with some prototypes already undergoing testing. The possibility of adapting existing machines to operate with ammonia offers several advantages, both in terms of technical expertise and cost-effectiveness. From a technical standpoint, maintaining the original equipment allows companies to leverage their existing knowledge and expertise. Economically, the modifications required to convert existing machines are typically less expensive than developing entirely new systems.

Ammonia-fuelled ICEs have garnered significant attention from companies worldwide. For instance, Bigas International, an Italian company, has adapted engines to use ammonia, as demonstrated in the Toyota GT86-R Marangoni sports car [50]. Additional research and tests have been conducted by organisations such as the Korea Institute of Energy Research (KIER), Xiamen University (China), the University of Orléans (France), Sturman Industries (USA), Iowa State University, Hydrofuel Inc. (Canada), Caterpillar, Toyota and the Hydrogen Fuel Center (USA). In the maritime sector, MAN Energy Solutions is pioneering ammonia as a fuel. Notably, Toyota has filed a patent using plasma jet igniters and a spark plug to facilitate ammonia ignition, eliminating the need for hydrogen as a fuel promoter.

Stakeholders in this sector are generally categorised into three groups [51]:

- Spark Ignition, carbon-based dual fuels: Companies like Biogas International (Italy), Savia Coalition (Italy) and Green Transportation Technology Group (South Korea) are working in this category.
- Compression Ignition, carbon-based dual fuels: Sturman Industries (USA), Hydrofuel Inc. (Canada) and Iowa State University (USA) are notable examples.
- Hydrogen-doped ammonia engines: Toyota (Japan) and Siemens (UK) are leading this approach.

Additionally, researchers have investigated ammonia blends with conventional fuels to address ammonia combustion challenges. For example, KIER's AmVeh project developed a spark ignition engine using a hybrid fuel consisting of 70% ammonia and 30% gasoline, which resulted in reduced CO emissions [51].

Several ammonia-fired GTs are also operational worldwide [3]:

- Turboden's 15 MW ammonia-fired GT was installed in a cogeneration plant in Germany in 2018. With a capacity of 15 MW, this turbine is engineered to run on either pure ammonia or a blend of ammonia and natural gas.
- MAN Energy Solutions is developing an ammonia-fired GT with an expected capacity of approximately 20 MW. This turbine is being designed to run on either pure ammonia or a blend of ammonia and natural gas.

- Kawasaki Heavy Industries has developed a 1 MW ammonia-fired GT, which is currently undergoing testing in Japan. This turbine is capable of running on pure ammonia or a mixture of ammonia and natural gas.
- Siemens has developed a 2 MW ammonia-fired GT, which is currently undergoing testing in Sweden and is designed to run on either pure ammonia or a blend of ammonia and natural gas.

7.4.1 Cost considerations about the use of ammonia as fuel in power generation systems

Changes to the combustion system are necessary to accommodate ammonia's unique combustion characteristics, such as its lower flame temperature and higher ignition energy. The fuel injectors and combustion chamber may need redesigning to ensure stable and efficient burning of ammonia.

In addition, even if most common metals are not affected by dry ammonia when combined with water vapour, ammonia will attack copper, zinc or alloys containing copper as a major alloying element. Therefore, these materials must not be used in contact with ammonia. Liquid ammonia can also attack rubber and certain plastics. While gaseous ammonia can react violently with nitrogen oxides and strong acids. In Table 7.4, the major chemical compatibilities and incompatibilities of ammonia are listed.

Therefore, the stainless steel AISI 304 and AISI 316 are suitable for use with ammonia, its cost is around 1.8–2$/kg.

To reduce the criticality of burning direct ammonia in GTs, several fuel blends composed of both hydrogen and nitrogen are currently under investigation and obtained through an ammonia cracking reactor. The ammonia decomposition reaction is an equilibrium endothermic reaction, in which two moles of NH_3 decompose into one mole of nitrogen and three moles of hydrogen with a reaction enthalpy of around 46 kJ per mole of NH_3 [52]:

To obtain pure hydrogen from thermocatalytic ammonia decomposition, a separation step is required. Based on the technology used, two integrated processes can be identified:

$$2NH_3 \rightarrow N_2 + 3H_2 \tag{7.4}$$

- **Traditional fixed bed reactor-based system**: In this case, the integrated system is mainly composed of: (i) ammonia storage, (ii) ammonia heat exchanger, (iii) ammonia evaporator, (iv) catalytic fixed bed decomposition reactor, (v) purification system to separate H_2 from N_2 and unreacted ammonia, usually using a PSA unit, (vi) H_2 compression unit: After the purification system, the H_2 pressure is relatively low (<10 bar), and for most applications, an H_2 blower or compressor is required to meet the requirements of the integrated power unit/end user (e.g. GT, ICE, fuel cell).
- **Membrane reactor-based system:** This system differs from the previous one in the type of reactor and subsequent steps. A membrane reactor decomposes ammonia into H_2 and N_2 while simultaneously separating pure H_2 from the

Table 7.4 Ammonia chemical incompatibilities [53]

Material	Compatibility	Material	Compatibility
ABS plastic	D – Poor	Kalrez	A – Excellent
Acetal (Delrin®)	D – Poor	Kel-F®	A – Excellent
Aluminium	A[1] – Excellent	LDPE	B[2] – Good
Brass	D – Poor	Natural rubber	D – Poor
Bronze	D – Poor	Neoprene	A – Excellent
Buna N (Nitrile)	B – Good	NORYL®	B[1] – Good
Carbon graphite	A – Excellent	Nylon	A[1] – Excellent
Carbon steel	B – Good	Polycarbonate	D – Poor
Carpenter 20	A – Excellent	Polyether ether ketone (PEEK)	A – Excellent
Cast iron	A – Excellent	Polypropylene	A – Excellent
Ceramic AI203	N/A	Polyurethane	D – Poor
Ceramic magnet	N/A	PPS (Ryton®)	A[1] – Excellent
ChemRaz (FFKM)	B – Good	PTFE	A – Excellent
Copper	D – Poor	PVC	A[2] – Excellent
CPVC	A[1] – Excellent	PVDF (Kynar®)	A – Excellent
EPDM	A – Excellent	Silicone	C – Fair
Epoxy	A – Excellent	stainless steel – 304	A – Excellent
Fluorocarbon (FKM)	D – Poor	stainless steel – 316	A[2] – Excellent
Hastelloy-C®	B – Good	Titanium	C – Fair
Hypalon®	D – Poor	Tygon®	A[2] – Excellent
Hytrel®	D – Poor	Viton®	D – Poor

Ratings – chemical effect:
A – excellent;
B – good: minor effect, slight corrosion or discoloration;
C – fair: moderate effect, not recommended for continuous use. Softening or loss of strength, and swelling may occur;
D – severe effect: Not recommended for any use; E – Information not available. Explanation of Footnotes
1 – satisfactory to 22 °C;
2 – satisfactory to 48 °C.

other gases. In this case, a double compressor step may be required: the first before the reactor to guarantee the proper pressure ratio for the membrane, and the second after the reactor to meet the end user's requirements.

The cost functions for both the system configurations (MR and FBR+PSA) and the hydrogen compressor are reported below. The curves have been estimated from literature data [41,54,55]:

$$CAPEX_{NH_3\ FBR+PSA} = 800{,}000 * (Daily\ H_2\ production\ [\text{tons per day}])^{0.79}\ [\$_{2024}] \tag{7.5}$$

$$CAPEX_{NH_3\ MR} = 990{,}000 * (Daily\ H_2\ production\ [\text{tons per day}])^{0.82} [\$_{2024}] \tag{7.6}$$

$$CAPEX_{H_2\ compressor} = 24{,}000 * P_{inst}[\text{kW}]^{0.7} \tag{7.7}$$

In terms of cost breakdown of the different components of the FBR+PSA unit, as reported in Figure 7.18, the main share of capital costs is associated with the purification system followed by the cracking reactor. The capital cost of the ammonia decomposition plant decreases as the capacity increases. Particularly, in the case of a system based on a fixed-bed reactor (FBR) and a purification unit (PSA) the cost is in the range of 350–400 k\$/tpd at 50 tpd-H_2 to 240–260 k\$/tpd at 350 tpd-H_2, as shown in Figure 7.19.

From a capital cost standpoint, the FBR+PSA system presents a slightly lower value compared to the MR-based system, primarily due to its greater technological maturity. However, in terms of operational costs, particularly electrical energy consumption, the FBR+PSA system incurs higher costs because of the larger number of auxiliaries and increased system complexity, as noted by Richard *et al.* [55]. Additionally, when the cost of feedstock (e.g., ammonia) is considered, the impact of CAPEX drops to less than 10%, while fuel purchasing accounts for approximately 60–70% of the total annual cost (Figure 7.20) [41,55].

Richard *et al.* carried out a techno-economic analysis comparing two cracking technologies, a membrane reactor (MR) and a conventional FTR (fired tubular reactor), for the cofiring of ammonia with hydrogen in a CCGT (combined cycle gas turbine) plant. In comparison with the cost of the turbomachinery, the fuel process systems accounts around 50–65% of the CAPEX (Figure 7.21).

Also Kanaan *et al.* investigate the LCOH related to different cracking technologies in three process scales: (i) a hydrogen refuelling station (1000 kgH_2/day), (ii) an ammonia import terminal (1500 ton NH_3/day), (iii) a mobile on-board reconversion for a heavy-duty truck (22 kgH_2/day) [41]. The results show that, currently, the cheapest option to produce H_2 from NH_3 for large-size applications is the thermal decomposition with membrane technology (7.72 \$/kg H_2 where the fuel cost represents 87% of the overall cost). In the

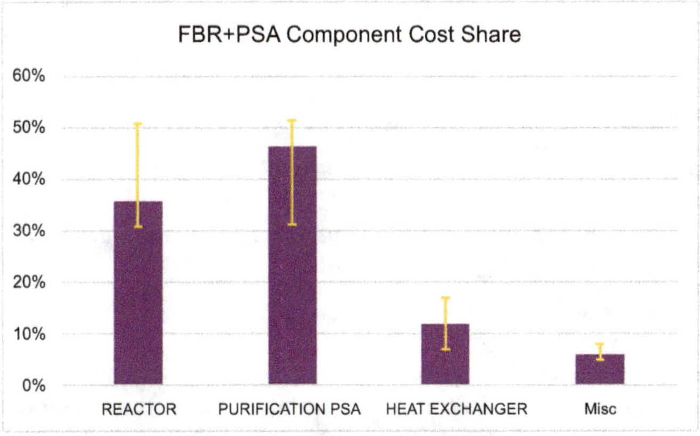

Figure 7.18 FBR-PSA integrated system capital cost breakdown (based on data from [54,55])

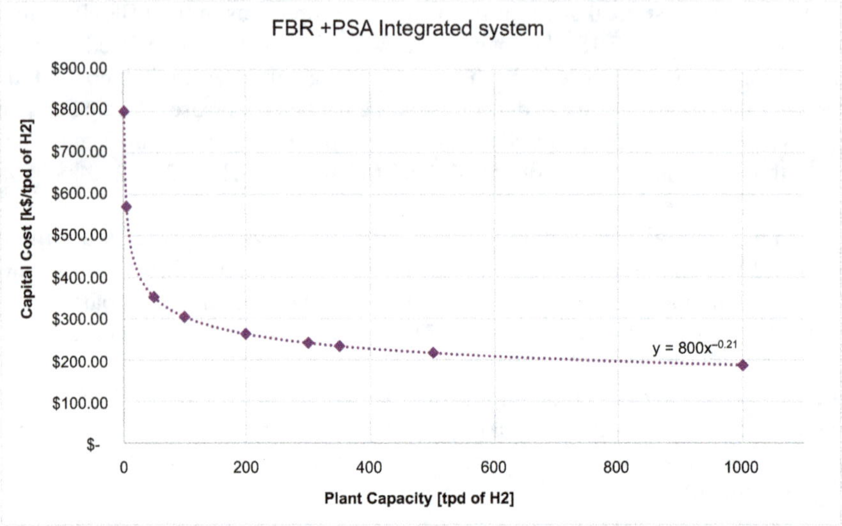

Figure 7.19 Capital cost curve of ammonia decomposition plant (FBR +PSA) (based on data from [41,54,55])

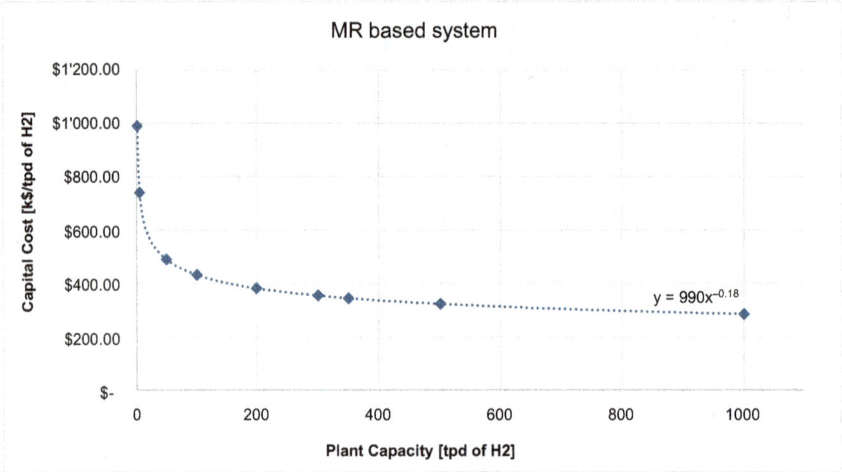

Figure 7.20 Capital cost curve of ammonia decomposition plant (MR) (based on data from [54,55])

study, a sensitivity analysis has been also performed reducing the fuel cost by 50% (from 1100/ton to 550$/ton). In this latter case, the membrane reactor remains the best choice from an economic point of view reporting an LCOH of 4.22 $/kg H_2 (Figure 7.22).

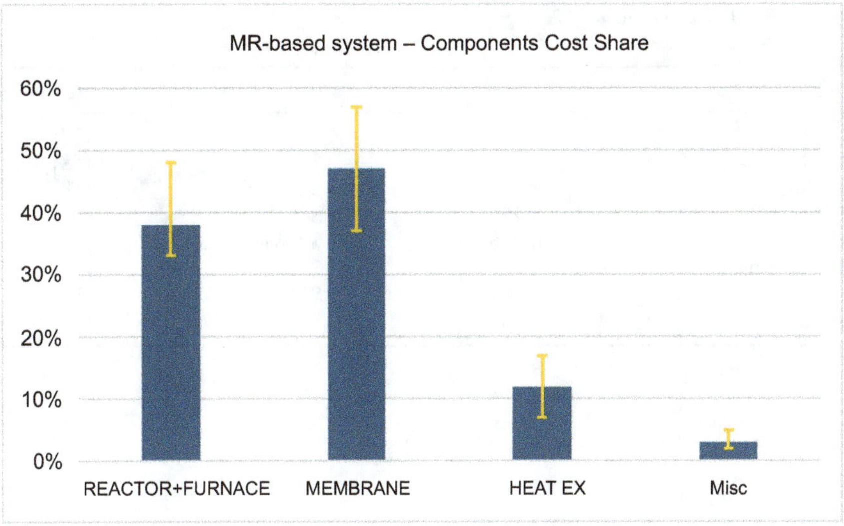

Figure 7.21 MR integrated system Capital Cost Breakdown (based on data from [54])

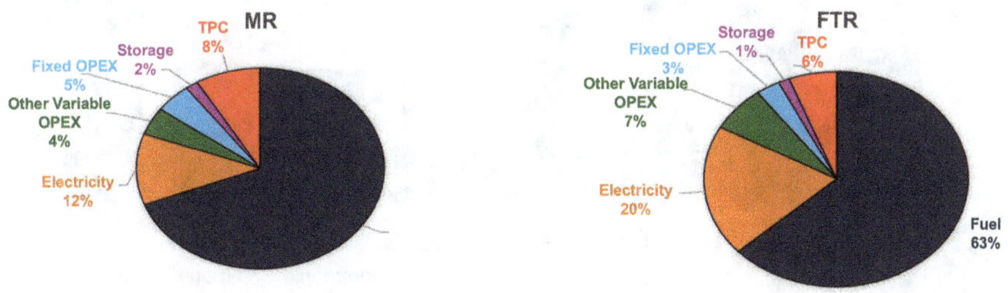

Figure 7.22 LCOH cost breakdown for an MR and a FBR-based system (based on data from [55])

7.5 Comparative analysis with other fuels

To enable a comprehensive comparison between different alternative and traditional fuels, the associated cost per megawatt-hour ($/MWh) has been calculated. By standardising the comparison, the study aims to provide a clear and consistent metric for evaluating the economic viability of different fuel options. The main economic data are listed in Table 7.5.

From Figure 7.23, it is clear that ammonia can be competitive with current fuel prices. This potential for cost competitiveness, combined with the environmental benefits of using zero-carbon, makes it a suitable candidate for future energy systems, provided that the associated infrastructure and operational issues can be effectively addressed.

Table 7.5 Fuel prices

Fuel	Cost		Reference
Methanol	588	\$/ton	[57]
Diesel	1.86	\$/l	[58]
Gasoline	1.99	\$/l	[58]
LPG	0.79	\$/l	[58]
NG	0.066	\$/kWh	[55]
Grey ammonia	476	\$/ton	[34]
Blue ammonia	499	\$/ton	[34]
Green ammonia	1000	\$/ton	[36]

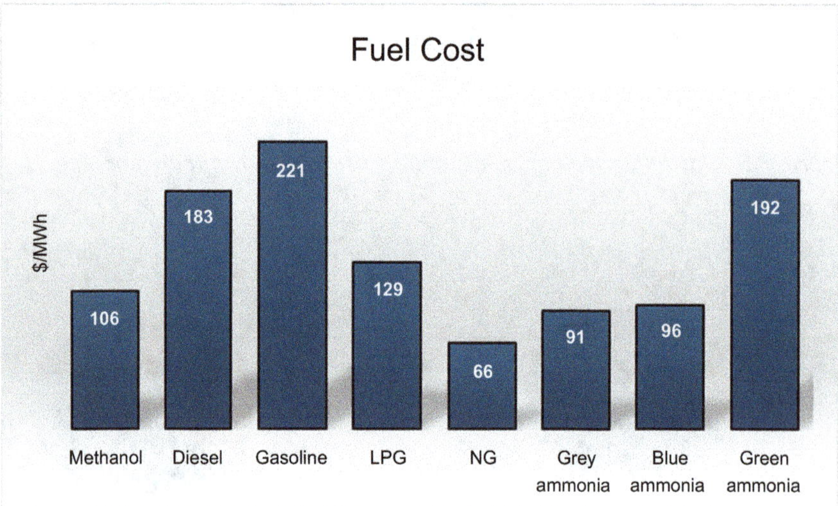

Figure 7.23 Fuel cost comparison [\$/MWh] (based on data from [56–58])

7.6 Conclusion

Ammonia is an important energy vector and is expected to play a significant role in the future energy economy. Its adoption as a fuel offers a considerable opportunity, especially in the context of transitioning to sustainable energy sources. Since ammonia does not contain carbon atoms, its combustion does not produce carbon dioxide, thereby contributing to the goal of achieving net-zero carbon emissions. However, the current ammonia production process is responsible for substantial CO_2 emissions, primarily due to the SMR process used for hydrogen production.

To manage the transition to net-zero carbon systems, 'blue ammonia' can be produced by incorporating CCS with the traditional process or 'green ammonia' can be produced through water electrolysis. The latter option is less likely to be a long-term solution in terms of economic feasibility, as the capital expenditure (CAPEX) for electrolysers significantly impacts the overall cost, bringing the LCOA to around $1500 per ton, depending on the geographical area. On the other hand, the introduction of CCS units could be considered a viable intermediate solution. Taking into account the increasing CO_2 taxation, it presents a competitive alternative to the production of ammonia via the traditional SMR process, with LCOA estimates of $476 per ton for SMR versus $499 per ton for SMR with CCS, assuming a natural gas price of $37 per MWh.

Another advantage of ammonia, despite its high toxicity, is its favourable conditions for transport and storage. Being a liquid at atmospheric pressure and −33 °C, ammonia is economically much more advantageous to transport than hydrogen. For long-distance transport (>1500 km), if pipelines are used, the levelised cost of transporting ammonia (LCOTNH3) is almost four times lower than that of liquid hydrogen (0.66$/kg H_2 for ammonia vs. 2.64$/kg H_2 for liquid hydrogen).

Using ammonia as a clean fuel for power generation, however, requires substantial modifications to existing production facilities. Changes to the combustion system are necessary to accommodate ammonia's unique combustion characteristics, such as its lower flame temperature and higher ignition energy. An alternative approach involves using a mixture of hydrogen and nitrogen derived from ammonia decomposition, which requires the installation of dedicated equipment, resulting in increased investment costs. For instance, in the case of using an FBR coupled with a pressure swing adsorption (PSA) system, the cost of the ammonia cracking unit ranges from $350–400k per ton per day (tpd) at 50 tpd-H2 to $240–260k per tpd at 350 tpd-H2. Additionally, a cost reduction of about 10% can be achieved by using a membrane reactor instead of a fixed-bed reactor.

In conclusion, ammonia is poised to play a role in the future energy mix, but there are still significant challenges to overcome before it can be commercially adopted as a primary energy vector. Ongoing research and development in synthesis, utilisation and economics are crucial to achieving a comprehensive understanding of its use and practical implementation in the coming decades

References

[1] Statista, https://www.statista.com/statistics/1266378/global-ammonia-production/ [latest access 10/02/2024].

[2] Global ammonia demand by application 2021-2050 | Statista.

[3] Pashchenko D. Ammonia fired gas turbines: recent advances and future perspectives. *Energy* 2024;290:130275. https://doi.org/10.1016/j.energy.2024.130275.

[4] Langella G, De Joannon M, Sabia P, Iodice P, and Amoresano A. Ammonia as a fuel for internal combustion engines: latest advances and future challenges. *J Phys Conf Ser* 2022;2385(1):012036.

[5] Chiong M-C, Chong CT, Ng J-H, *et al.* Advancements of combustion technologies in the ammonia-fuelled engines. *Energy Convers Manag* 2021; 244:114460. https://doi.org/10.1016/j.enconman.2021.114460.

[6] Li T, Zhou X, Wang N, *et al.* A comparison between low-and high-pressure injection dual-fuel modes of diesel-pilot-ignition ammonia combustion engines. *J Energy Inst* 2022;102:362–73.

[7] Mounaïm-Rousselle C, Brequigny P, Dumand C, and Houillé S. Operating limits for ammonia fuel spark-ignition engine. *Energies* 2021;14:4141. doi:10.3390/en14144141.

[8] Dincer I, Erdemir D, Aydin MI, Karasu H, and Vezina G. Ammonia production. In: *Ammonia Energy Technologies. Lecture Notes in Energy*, Cham: Springer; 2022. pp. 151–233. https://doi.org/10.1007/978-3-031-13532-3_3.

[9] Morgan ER, Manwell JF, and McGowan JG. Sustainable ammonia production from US offshore wind farms: a techno-economic review. *ACS Sustain Chem Eng* 2017;5:9554–9567.

[10] IEA. *Net Zero by 2050 A Roadmap for the Global Energy Sector.* 2021b. https://www.iea.org/reports/net-zero-by-2050.

[11] Wang M, Khan MA, Mohsin I, *et al.* Can sustainable ammonia synthesis pathways compete with fossil-fuel based Haber–Bosch processes? *Energy Environ Sci.* 2021;14:2535–2548. https://doi.org/10.1039/d0ee03808c.

[12] https://www.pricepedia.it/en/magazine/article/2022/05/09/prices-of-inorganic-chemistry-between-speculations-and-great-uncertainty/ (accessed 27 August 2024).

[13] https://farmdocdaily.illinois.edu/2023/02/fertilizer-prices-and-company-profits-going-into-spring-2023.html (accessed 27 August 2024).

[14] https://www.topsoe.com/blue-ammonia-technology (accessed 27 August 2024).

[15] Tarun CB, Croiset E, Douglas PL, Gupta M, and Chowdhury MHM. Techno-economic study of CO2 capture from natural gas based hydrogen plants. *Int J Greenh Gas Control* 2007;1(1):55–61. https://doi.org/10.1016/S1750-5836(07)00036-9.

[16] Shi W, Yang H, Shen Y, Fu Q, Zhang D, and Fu B. Two-stage PSA/VSA to produce H2 with CO2 capture via steam methane reforming (SMR). *Int J Hydrogen Energy* 2018;43(41):19057–74. https://doi.org/10.1016/j.ijhydene.2018.08.077.

[17] Hydrogen Europe. Clean ammonia in the future energy system. 2023. https://hydrogeneurope.eu/events/?he_event=aX178Y9mNYPgRvD (accessed 29 August 2024).

[18] Collodi G, Azzaro G, Ferrari N, and Santos S. Techno-economic evaluation of deploying CCS in SMR based merchant H2 production with NG as feedstock and fuel. *Energy Proc* 2017;114:2690–712. https://doi.org/10.1016/j.egypro.2017.03.1533.

[19] Asgharian H, Baxter L, Iov F, *et al.* Techno-economic analysis of blue ammonia synthesis using cryogenic CO2 capture process-a Danish case investigation. *Int J Hydrogen Energy* 2024;69:608–618. https://doi.org/10.1016/j.ijhydene.2024.05.060.

[20] Olabi AG, Abdelkareem MA, Al-Murisi M, *et al.* Recent progress in green ammonia: production, applications, assessment; barriers, and its role in

achieving the sustainable development goals. *Energy Convers Manag* 2023; 277:116594. https://doi.org/10.1016/j.enconman.2022.116594.

[21] https://www.yara.com.au/crop-nutrition/grow-the-future/sustainable-farming/green-ammonia/ (accessed 29 August 2024).

[22] Rashid MD, Al Mesfer MK, Naseem H, and Danish M. Hydrogen production by water electrolysis: a review of alkaline water electrolysis, PEM water electrolysis and high temperature water electrolysis. *Int J Eng Adv Technol* 2015;4(3):80–94. https://www.ijeat.org/wp-content/uploads/papers/v4i3/C3749024315.pdf.

[23] Guo Y, Li G, Zhou J, and Liu Y. Comparison between hydrogen production by alkaline water electrolysis and hydrogen production by PEM electrolysis. *IOP Conf Ser: Earth Environ Sci.* 2019;371:042022. 10.1088/1755-1315/371/4/042022.

[24] Zhang X, O'Brien JE, O'Brien RC, Hartvigsen JJ, Tao G, and Housley GK. Improved durability of SOEC stacks for high temperature electrolysis. *Int J Hydrogen Energy* 2013;38(1):20–28. https://doi.org/10.1016/j.ijhydene.2012.09.176.

[25] https://www.czapp.com/analyst-insights/ammonia-challenged-by-cost-energy-intensity/ (accessed 29 August 2024).

[26] Joris P. State-of-the art CAPEX data for water electrolysers, and their impact on renewable hydrogen price settings. *Int J Hydrogen Energy* 2019;44(9):4406–4413. https://doi.org/10.1016/j.ijhydene.2018.07.164.

[27] Bermudez JM, Evangelopoulo S, and Pavan F. Electrolysers analysis – IEA tracking report (link).

[28] Sousa J, Waiblinger W, and Friedrich KA. Techno-economic study of an electrolysis-based green ammonia production plant. *Ind Eng Chem Res* 2022; 61(39):14515–14530. DOI: 10.1021/acs.iecr.2c00383.

[29] Schmidt O, Gambhir A, Staffell I, Hawkes A, Nelson J, and Few S. Future cost and performance of water electrolysis: an expert elicitation study. *Int J Hydrogen Energy* 2017;42:30470–30492.

[30] Bellotti D, Rivarolo M, and Magistri L. Clean fuels synthesis from green hydrogen: a techno-economic comparative analysis. *E3S Web Conf.* 2021; 238:03001.

[31] Böhm H, Zauner A, Rosenfeld DC, and Tichler R. Projecting cost development for future large-scale power-to-gas implementations by scaling effects. *Appl Energy* 2020;264:114780. https://doi.org/10.1016/j.apenergy.2020.114780.

[32] https://www.irena.org/news/pressreleases/2020/Jun/Renewables-Increasingly-Beat-Even-Cheapest-Coal-Competitors-on-Cost (accessed 30 August 2024).

[33] https://kpmg.com/be/en/home/insights/2021/03/eng-the-hydrogen-trajectory.html.

[34] Publications Office of the European Union. EUR-Lex - C(2023)1086 - EN - EUR-Lex. https://eur-lex.europa.eu/legal-content/EN/TXT/?uri=PI_COM%3AC%282023%291086.

[35] IEA. Indicative production costs for ammonia via electrolysis in selected regions compared to current references. 2023. https://www.iea.org/data-and-

statistics/charts/indicative-production-costs-for-ammonia-via-electrolysis-in-selected-regions-compared-to-current-references (accessed 29 August 2024).

[36] Bellotti D, Rivarolo M, and Magistri L. A comparative techno-economic and sensitivity analysis of Power-to-X processes from different energy sources. *Energy Convers Manag* 2022;260:115565. https://doi.org/10.1016/j.enconman.2022.115565.

[37] Elishav O, Mosevitzky Lis B, Valera-Medina A, and Grader GS. Storage and distribution of ammonia. In: *Techno-Economic Challenges of Green Ammonia as an Energy Vector*. Amsterdam: Elsevier; 2021. pp. 85–103. https://doi.org/10.1016/B978-0-12-820560-0.00005-9.

[38] https://ammoniaknowhow.com/ammonia-storage-tanks/ (accessed 27 August 2024).

[39] https://www.fertilizerseurope.com/wp-content/uploads/2019/08/Guidance_for_inspection_of_atmospheric__refrigerated_ammonia_storage_tanksVJ_website.pdf (accessed 27 August 2024).

[40] Nayak-Luke RM, Forbes C, Cesaro Z, Bañares-Alcántara R, and Rouwenhorst KHR. Chapter 8 – techno-economic aspects of production, storage and distribution of ammonia. In: Valera-Medina A, and Banares-Alcantara R. (eds), *Techno-Economic Challenges of Green Ammonia as an Energy Vector*. New York: Academic Press; 2021. pp. 191–207. https://doi.org/10.1016/B978-0-12-820560-0.00008-4.

[41] Kanaan R, Nóbrega PHA, Achard P, and Beauger C. Economical assessment comparison for hydrogen reconversion from ammonia using thermal decomposition and electrolysis. *Renew Sustain Energy Rev* 2023;188:113784. https://doi.org/10.1016/j.rser.2023.113784.

[42] International PtX Hub. *Ammonia Transport and Storage.* Deutsche Gesellschaft für Internationale Zusammenarbeit (GIZ) GmbH; 2024. https://ptx-hub.org/wp-content/uploads/2024/01/International-PtX-Hub_202401_Ammonia-transport-and-storage.pdf.

[43] Dinh QV, Todesco Pereira PH, Dinh VN, Nagle AJ, and Leahy PG. Levelised cost of transmission comparison for green hydrogen and ammonia in new-build offshore energy infrastructure: pipelines, tankers, and HVDC. *Int J Hydrogen Energy* 2024;62:684–698. https://doi.org/10.1016/j.ijhydene.2024.03.066.

[44] IEA. *The Future of Hydrogen: Seizing Today's Opportunities*. Paris: OECD; 2019. https://doi.org/10.1787/1e0514c4-en.

[45] Rouwenhorst KHR, Krzywda PM, Benes NE, Mul G, and Lefferts L. Ammonia production technologies. In: *Techno-Economic Challenges of Green Ammonia as an Energy Vector*. Amsterdam: Elsevier; 2021. pp. 41–83. https://doi.org/10.1016/b978-0-12-820560-0.00004-7.

[46] Fully refrigerated tankers that carry LPG, Ammonia & Vinyl chloride. http://www.liquefiedgascarrier.com/Fully-Refrigerated-Ships.html.

[47] Shakirov A. LNG shipping cost. Inside Logistics, March 26, 2021; https://www.insidelogistics.ca.

[48] Yang M, and Lam JSL. Operational and economic evaluation of ammonia bunkering – Bunkering supply chain perspective. *Transportation Research*

Part D: Transport and Environment. 2023;117:103666. https://doi.org/10.1016/j.trd.2023.103666.
[49] Salmon N, and Bañares-Alcántara R. Green ammonia as a spatial energy vector: a review. *Sustain Energy Fuels* 2021;5:2814. doi: 10.1039/D1SE00345C.
[50] Valera-Medina A, Xiao H, Owen-Jones M, David WIF, and Bowen PJ. Ammonia for power. *Prog Energy Combust Sci.* 2018;69:63–102. https://doi.org/10.1016/j.pecs.2018.07.001.
[51] Natural gas price statistics. eurostat; https://ec.europa.eu/eurostat/statistics-explained/
[52] Asif M, Bibi SS, Ahmed S, *et al.* Recent advances in green hydrogen production, storage and commercial-scale use via catalytic ammonia cracking. *Chem Eng J* 2023;473:145381.
[53] Chemical Compatibility Database, Cole-Parmer, https://www.coleparmer.com/chemical-resistance.
[54] Papadias DD, Peng J-K, and Ahluwalia RK. *H2@Scale: Outlook of Hydrogen Carriers at Different Scales*. https://www.energy.gov/sites/prod/files/2020/03/f72/fcto-hydrogen-carriers-workshop-2019-anl.pdf.
[55] Richard S, Santos AR, Olivier P, and Gallucci F. Techno-economic analysis of ammonia cracking for large scale power generation. *Int J Hydrogen Energy* 2024;71:571–587. https://doi.org/10.1016/j.ijhydene.2024.05.308.
[56] https://www.methanex.com/about-methanol/pricing/.
[57] Leighty B. Ammonia renewable energy fuel systems at continental scale: transmission, storage, and integration for deep decarbonization of world's largest industry at lower cost than as electricity. NH3 Fuel Association, 2017. https://nh3fuelassociation.org/2017/09/26/ammonia-renewable-energy-fuel-systems-at-continental-scal e/
[58] https://it.globalpetrolprices.com/ (accessed 30 August 2024).

Chapter 8

H&S implications, regulations and international scenarios

K. Rowenhorst[1], *A. Guati-Rojo*[1] *and A. Valera-Medina*[2]

8.1 Introduction

Although ammonia is a very promising fuelling vector, there are still many concerns about its use at commercial scale. Overall, ammonia is considered a toxic and corrosive chemical, and victims exposed to its release should be removed to an uncontaminated area wearing self-contained breathing apparatus or taken to hospital immediately to avoid tissue and/or organ dead. Ammonia impacts to the environment are also a concern, especially to water bodies where ammonia concentrations can be lethal to some species, whilst its sudden release to the environment at higher concentrations can preclude the formation of particle matter. Therefore, ammonia is still treated with lot of respect, and its global deployment will require better understanding of its nature, hence enabling better health and safety practices for its employment in fuelling systems.

Favourably, good compromise is currently being reached at various levels, from storage and distribution regulations that are improving the scaling up of ammonia power systems, through certification plans that will ensure a globally standardise supply, to ambitions political agendas that have been imposed across the world, ensuring that ammonia as an energy vector is seem more real than ever.

Therefore, this chapter will address some of the challenges of using ammonia in combustion systems, moving away from those challenges related to the technical aspect of combustion and reaching other areas of high importance for the global use of this chemical – health and safety aspects, storage and distribution methods, and finally the political scenario that is currently driving local, national and regional agendas for the implementation of ammonia combustion systems.

[1]Ammonia Energy Association, Ashburn, VA, USA
[2]Centre of Excellence on Ammonia Technologies, Cardiff University, UK

8.2 Health and safety

Ammonia is a colourless gas with a sharp, penetrating odour. Its boiling point is −33.35 °C (−28.03 °F), and its freezing point is −77.7 °C (−107.8 °F), with a density of 0.73 kg/m^3 and an auto-ignition temperature of 651°C under atmospheric conditions. With an octane number of 130, it has a high heat of vaporisation (23.3 kJ per mole at its boiling point) and can be handled as a liquid in thermally insulated containers. The ammonia molecule has a trigonal pyramidal shape with the three hydrogen atoms and an unshared pair of electrons attached to the nitrogen atom. It is a polar molecule and is highly associated because of strong intermolecular hydrogen bonding. The dielectric constant of ammonia (22 at −34 °C [−29 °F]) is lower than that of water (81 at 25 °C [77 °F]), so it is a better solvent for organic materials. However, it is still high enough to allow ammonia to act as a moderately good ionising solvent. Ammonia also self-ionises, although less so than does water [1]. The combustion of pure ammonia proceeds with difficulty but yields nitrogen gas and water, with a stoichiometric air fuel ratio (AFR) of 6.06 by weight,

$$4NH_3 + 3O_2 + \text{heat} \rightarrow 2N_2 + 6H_2O$$

Liquid ammonia is used extensively as a non-aqueous solvent. The alkali metals as well as the heavier alkaline-earth metals and even some inner transition metals dissolve in liquid ammonia.

Ammonia is such a common industrial chemical that pipelines distribute it in the farm states and various locations worldwide [2]. It is transported and contained in tanks under modest pressure, similar to propane. It is potentially hazardous to inhale; a 1% concentration inhaled for 1 hour has a 1% fatality risk. However, ammonia is readily detected by its odour, and being lighter than air it rapidly dilutes in a spill. Unlike gasoline or diesel fuel, it does not catch fire in air due to its high ignition temperature.

Ammonia can be detected at ~5 ppm, and its pungent odour makes it easy to be identified. As the concentration increases (~50 ppm), the odour becomes very penetrating. However, it is known that human exposure limits can reach much higher values, with reported cases of up to 33,370 ppm incident in South Africa where after a 5 min exposure no deaths were recorded [3]. Further, methods to determine exposure are not standardised, and many approaches have been tried over the years [4]. However, regulations have been established to ensure that these levels are never reached, and safe exposure is accomplished in sites using and/or producing ammonia. Two regulations are presented in this chapter – one from the USA and the Environmental Protection Agency (EPA), and the Health and Safety Executive (HSE) from the UK (Tables 8.1 and 8.2). As observed, values range from 25 to 35 ppm depending on the exposure length. This practice is followed by many other countries. An interesting point is that the foundations behind these numbers are mainly supported by years of practice rather than biological analyses, hence opening the possibility of further research in this area.

Table 8.1 EPA AEGL guideline values [5]

	Ammonia 7664-41-7 (final) expressed in ppm				
	10 min	**30 min**	**60 min**	**4 h**	**8 h**
AEGL-1	30	30	30	30	30
AEGL-2	220	220	160	110	110
AEGL-3	2,700	1,600	1,100	550	390

AEGL-1: Airborne concentration level at which individuals could experience notable discomfort, irritation, or certain asymptomatic, non-sensory effects. However, the effects are not disabling, and are transient and reversible.
AEGL-2: Airborne concentration level at which individuals could experience irreversible or other serious, long-lasting adverse health effects or an impaired ability to escape.
AEGL-3: Airborne concentration level at which individuals could experience life-threatening health effects or death.

Table 8.2 Occupational standards, HSE (UK) [6]

	Long-term exposure limit (LTEL; 8 h)		Short-term exposure limit (STEL; 15 min)	
Workplace exposure limit (WEL)	ppm	mg/m^3	ppm	mg/m^3
	25	18	35	25

An interesting point regarding ammonia exposure is that most publications address ammonia as a toxic chemical to humans. The point was raised a few years ago [7] and it was mentioned that ammonia exposure to humans can be more corrosive than toxic. Toxicity would require humans to exceed the natural level of ammonia in the blood to a point that cells would start dying. However, it is known that before this happens, ammonia would dehydrate the cells in the respiratory system – essentially burning them – and then stopping the inhale of ammonia. Thus, this is a point of debate relating the extent of corrosivity against toxicity to humans. However, it is absolute that ammonia will be a health and safety issue if released, exposing individuals at relatively low thresholds.

Explosions with the sudden release of ammonia have been also documented [8] (Figure 8.1). Pressure starts building-up when cylinders or gas containing components are exposed to a heat source. The increase in pressure reaches a point when the structural integrity of the component is compromised, resulting in damage and sudden release of the contained gas. The problem aggravates if the heat source or a spark get in contact with the boiling, uncontained ammonia. If this happens, the probability of explosion increases with the likelihood of catastrophic scenarios. The phenomenon is known as boiling liquid evaporating vapour explosions (BLEVEs). Ammonia explosions have only been documented for large-scale facilities and large tankers. However, precaution needs to be taken to avoid similar incidents in incipient technologies.

Figure 8.1 Explosions and cylinder damage, respectively. Courtesy of Ammoniaknowhow.com.

NH_3 is also corrosive to copper, brass and zinc alloys, forming a greenish/blue colour corrosion and NH_3 should not be mixed with bromine, chlorine, iodine and hypochlorites as NH_3 is an alkaline reducing agent and reacts with acids, halogens and oxidising agents. Other materials that are also incompatible with ammonia include ABS Plastic, Hytrel, natural rubber and even Viton. Interestingly, carbon graphite, epoxy, neoprene, nylon and aluminium have excellent compatibility with NH_3, hence showing the complex nature of the chemical and the potential for its use as long as the right materials are selected [9].

Simultaneously, ammonia is known to have an indirect impact on ozone depletion through the formation of nitrous compounds in the atmosphere. These are currently considered a negligible contribution to ozone depletion [10] but will need to be considered for large-scale ammonia utilisation. Further, an interesting point raised by environmental protecting bodies is the potential of ammonia to produce particle matter, which can also be detrimental to humans, living organisms and ecosystems.

Nitrogen oxides (NO_x), which are a product of the combustion of ammonia due to the presence of nitrogen molecules, account as one of the most detrimental barriers for the use of NH_3 as a fuelling source. Although NO_x are not only a problem for ammonia systems but for most power systems (especially those operating at high temperatures as most combustion systems do), NO_x from ammonia is particularly detrimental as the high concentrations of NO can degrade into other most harmful species such as NO_2 and N_2O.

In terms of their environmental and health impact, NO_x can lead to the formation of ozone (O_3) at ground level, acid rain (via de formation of nitric acid) and particle matter (via formation of secondary particulates such as ammonium nitrate), increasing eutrophication and deteriorating health and agriculture [11]. Although NO is not considered hazardous at normal ground levels, it can lead to blood poisoning and avoid blood cells getting oxygenated, hence causing respiratory issues. Similarly, NO_2 can decrease lung function, deteriorating how the organ functions and increasing the probability of cardiovascular and respiratory diseases. It is estimated that approximately 23,000 deaths were premature to the exposure of this contaminant in the UK alone [12]. Further exposure at high concentration levels can be fatal.

Acid rain, another detrimental consequence of the production of nitrogen oxides, contributes not only to the degradation of buildings and concrete structures but also to the acidification of water bodies, with its inherent impact to biodiversity. Forest and vegetation losses, from minimal to severe, have also been reported across the world as a consequence of acidification caused by nitric acid [13].

Nitrous oxide (N_2O) is also a species that can be produced from the combustion of ammonia blends. The chemical is a colourless, sweet-tasting gas and it is also known as 'laughing gas'. It is non-combustible but it will accelerate the burning of combustible materials. It is soluble in water. Its vapours are heavier than air. When it comes to ammonia, N_2O can be produced from the combustion of the NH_3 molecule, especially when high heat losses are experienced during the process [14]. The main problem with this pollutant relies on its elevated global warming potential. Nitrous oxide has a GWP 273 times that of CO_2 for a 100-year timescale. Furthermore, N_2O emitted today remains in the atmosphere for more than 100 years, on average [15]. The implication of this impact to the environment means that just a relatively small concentration of N_2O would be enough to produce a greater climate impact than a fossil powered system, with emissions more damaging than those produced using carbon species.

Fine particle matter ($PM_{2.5}$, diameters less than 2.5 μm) can also be formed using ammonia. The main issue with these particles is that they can be small enough to pass through skin and alveoli, reaching the blood and directly impacting on health. The main species implicated in this process include ammonium nitrate (NH_3NO_3) and ammonium sulphate ($(NH_4)_2SO_4$), which are formed as nitrogen, water and sulphur oxides, naturally available in the atmosphere, get in contact with ammonia. The promoters of the reactions involved form through chemical reactions relatively strong hydrogen bonds with the reaction centre. Results obtained by Chen *et al.* [16] suggest that there are several pathways that can be facilitated by other ambient molecules. Further, the increased concentration of particle matter can also have severe impacts on cardiovascular functions and immune system problems [13], which can have direct impact on premature deaths [17].

An interesting point raised by EPA concerns the theft of ammonia [18]. Since ammonia is employed in the illegal use of methamphetamines, the chemical is often stolen from where it is stored and used. During stealing, the pressurised ammonia can be realised, causing injuries to the culprits, law enforcement personnel, emergency services and the public in general. Most problems occur during the movement of ammonia to unsuitable containers using valves and tanks fitted for other chemicals. In addition, a recurrent problem is the bad siphoning practices combined with the removal of safety features from standard containers. Cases have been reported where residents had to be evacuated as a consequence of thieves leaving open large ammonia tanks. Since very small amounts of ammonia are needed for a batch of methamphetamines (i.e. any residual ammonia in a hose could be enough for the production of drugs), the chemical has great value in the black market as a speed-up manufacturing agent. Events have occurred in locations where the chemical is held as a refrigerant, from underground pipelines or rail tankers. Often criminals abort the theft as a consequence of an injury, leaving behind tools and bottles. Several places have passed legislation making it a felony to carry steal ammonia or carry it in non-approved containers [18].

There are many other impacts that can occur when ammonia is released by accident. Further details can be found elsewhere [4]. However, one that is important to raise is the sudden release of ammonia from pressurised tanks. Different to cooled ammonia (which is cooled down below −33 °C but remains at atmospheric pressure), pressurised ammonia (which remains at room temperature, but tends to be pressurised above 0.8 MPa for liquefaction) can more challenging in case of a sudden, unexpected release. Liquefied ammonia has an expansion ratio of approximately 850 to 1 (i.e. liquid ammonia will expand 850 times), thus covering vast areas rapidly [19]. The release would suddenly become an aerosol (due to the sudden expansion, which would cool down the gas and generate small droplets). This effect will generate a denser gas than air, hence producing a 'deadly' mist as cold ammonia would start absorbing water. This mist could travel to both populated locations or places where workers are located. Such a release will also involve the aerosol phase (with small droplets). Documented cases can be found elsewhere [4].

Finally, it must be remembered that anhydrous ammonia aggressively seeks water. When it comes in contact with water in skin and eyes, it will start producing alkaline compounds leading to burns. Further, the vaporisation of liquid ammonia, which is how ammonia is usually delivered, can end up in severe freeze burns, as the ammonia vaporises, and death skill/organs (Figure 8.2).

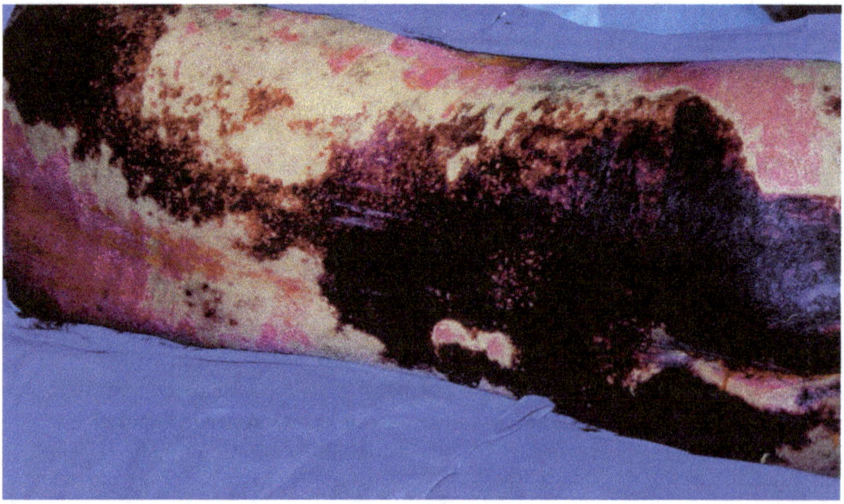

Figure 8.2 Exposure of skin to anhydrous ammonia. Reprinted, with permission, from [20].

8.3 Storage and distribution regulations

Ammonia is transported via many methods. From large-scale transportation using large pipelines to small cylinders delivered to specific sites. At the destination, ammonia usually is employed for fertilising or chemical processes, although the requirement of decarbonisation methods is leading to more companies to employ ammonia as a fuel. A vast amount of ammonia is converted into other products such as urea [21], although anhydrous ammonia with a small percentage of water content (~0.2 wt% which is added to prevent stress corrosion cracking (SCC) [22]) can also be transported in liquid form via pressurised containers or cooled systems.

When it comes to transportation of liquid ammonia, there are five methods to distribute the chemical,

- Pipelines: general used for a limited range of compounds, including ammonia, pipelines can be used to move large volumes at attractive costs. Although the pipelines suffer from flexible distribution issues (as the final distribution points would be fixed), this solution can ensure continuous delivery of ammonia. The major ammonia pipelines worldwide are located in the USA and Russia. For the US line, the reported cost of building the pipeline was estimated at US $70 million in 1969 [23], equivalent to about US $0.16 million/km in 2019 (conversion based on the Bureau of Labor Consumer Price Index). Similarly, the Russia line (of 2,471 km length) was commissioned in 1979 and used to deliver ammonia to Odessa (Ukraine), accounting for 11% of the global exports before 2022 [24]. Smaller pipelines that connect ammonia producing facilities and distribution centres can be found in Europe. In average, these pipelines range from 10 to 2 km length, although there are some that exceed 70 km (but are very specific and not as present in such a continent) [24].
- Ships: Ships equipped with insulated tanks to keep refrigerated temperatures <-33 °C are usually employed for overseas transport. The capacity of these ships varies, and due to their nature, these are capable of also transporting liquified petroleum gas (LPG). The range of these ships can vary 20,000–60,000 m^3, although there are some ships in Russia that can transport even larger quantities [25].
- Barges: Inland and intra-coastal waterways are used worldwide for passengers and freight transportation, hence opening the potential for the movement of chemicals such as ammonia. barges in the range of 500–3,000 tonnes (larger and smaller barges are fully- and semi-refrigerated, respectively) can be used. This is the case for many routes within Europe and the USA. In Europe, mostly self-propelled barges are used to transfer ammonia through narrow and shallow inland waterways. In the USA, most barge operations are along the Mississippi River and the seaside of the Gulf of Mexico [4].
- Rails: usually employed to distribute ammonia to final users. Although more expensive than pipeline or barge per tonne/km, in some instances, movement by rail can be cheaper than a combined pipeline/barge and truck transportation route. In Western Europe, more than 1,500,000 tonnes of ammonia were transported by rail in 2008 [26]. Railway tank cars (RTCs) can hold volumes in

the range of 50–110 m^3. Most cars in the USA have capacities of 65–72 tonnes, but smaller and larger cars are also available with capacities of 25–30 tonnes and 90 tonnes, respectively [22]. However, ammonia expansion with temperature limits the loading of ammonia to 57–85% of the tank's volume [27]. Tanks usually have a thickness of 1.75 cm and are covered with a 0.32 cm jacket [28].

- Trucks: similar to rail transport, trucks are also used to distribute ammonia to final users (i.e. in the USA, ammonia distribution trucks are employed to deliver the chemical to small size stations where farmers fill their tanks for ploughing activities) over distances less than 150 km. In Germany and other European countries, tank cars for liquid and aqueous ammonia have a volume of 100 and 30 m^3, respectively. These cars are usually operated at 1.0 MPa with the capability to work at a maximum pressure of 2.5–2.8 MPa [29]. Different safety measures are used to ensure safe operation and delivery, including the use of cargo tanks that can withstand impact, breakaway plumbing in case of an accident, safety valves to avoid leakage from the trailer and shut-off valves [4].

For example, the ammonia used in 2017 [30] was transported by pipelines (3%), motor vehicle (51%), rail (37%) and marine vessels (9%).

When it comes to storage of the chemical, there are several methods to do so depending on the scale of the storage site. The subject, which has required a vast amount of research and development especially for the fertilising and refrigeration industries, has reached considerable maturity worldwide. For the case of large-scale storage, there are three preferred methods [31],

1. Pressurised storage where ammonia is pressurised above 8 atmospheres (between 1.6 and 1.8 MPa) at atmospheric temperature (10–20 °C). Capacities smaller than 2,000 tonnes of ammonia are usually stored this way. The method is usually employed in trucks, trains and small barges, as this storage method is economical.
2. Low temperature storage employs insulated cylindrical tanks with capacities around 50,000 tonnes of ammonia. The chemical is cooled down below −33 °C under atmospheric pressure. In terms of health and safety, this method is preferred as it ensure that any leak would only vaporise to the atmosphere, whilst pressurised containers can produce a deadline mist as previously raised. The method can be ideal for the delivery of ammonia directly from synthesis plants or refrigerated vehicles. Also, it has a vast potential for the delivery of large ammonia quantities in combustion systems. Different to pressurised ammonia, the liquid flow can be controller better, ensuring that weather variations do not affect the state of the molecule during power producing operations.
3. Semi-refrigerated storage requires insulated tanks at low temperatures (above −33 °C) and slight pressure delivered using single-stage compressors. Capacities around 2,500 tonnes are commonly stored via this method. Similar to pressurised storage, the method is relatively economic, and it is usually employed for retailer-level distribution.

There are various methods that can be employed to maximise the use of cool liquid/high pressurised ammonia. For example, when ammonia is received pressurised, the stream can be expanded, flashing to near atmospheric pressure before being fed into the delivery tanks. When this occurs, the expansion leads to a decrease in temperature even below −33 °C, creating a liquid state. Any remaining vapour ammonia is liquified by compression and recycled to the flashing vessel. This helps in the conversion of the liquid, its cooling and easy storage, and the removal of unwanted gas traces such as nitrogen, argon and hydrogen [31].

In terms of regulations for the refrigerated tanks, they are usually constructed under the standards API 620 R or EN 14620, using carbon-manganese steel certified to withstand 233 K with minimum yield strength in the range of 290–360 MPa to limit the effect of SCC [32]. However, the welding materials have a significantly higher yield strength than the base materials, which could be a problem for SCC if water is not added to the blend. The systems can be either single or double-wall insulated, having a gas tight insulation to avoid freezing moisture. When the tank is placed on the ground or concrete platforms, although over soil it requires some heating to avoid freezing conditions.

Terminal and in-plant storage are similar. However, the size of storage terminals serving ocean vessels or barges depends on the capacity of the ship (usually 25–50% larger than the ship's capacity), and the terminals are located at an accessible location for these vessels. In addition, these facilities require some modifications to allow faster loading and unloading, decrease refrigerated ammonia vaporisation and minimise port time. Thus, the pumps required need to be able to operate at a higher capacity and a larger and well-insulated pipeline must be used. These requirements lead to higher operation and capital costs relative to in-plant storage [4].

Nurse tanks of ~3,780–5,490 litres are used on farms and at retailer filling facilities, mainly for agricultural applications (although they could also be used for fuelling). In the USA alone, 1–1.5 million ammonia filling events were recorded annually by the end of the 2010s decade [4]. In addition, dual tanks with a capacity of ~7,570 litres can be used. An additional option is using T50 ISO tank containers (Figure 8.3), which are usually constructed from stainless or carbon steel. These containers are used for ammonia transportation with a capacity range of 11,000–25,000 litre, an operating pressure of 2.1–2.2 MPa and temperatures of 233–333 K [33]. ISO tanks and Nurse tanks can be used to transport ammonia when mounted on a mobile vessel. In addition, smaller ammonia cylinders are available from a few kgs up to 700 kg, and are used in a range of applications and industrial processes [34].

8.3.1 Marine scenario – case study

Ammonia for marine utilisation is closer to become a reality for commercial purposes by the time this chapter was written (2024). However, there are many uncertainties on its use and the health and safety implication the utilisation of this chemical as a fuel will bring to the industry. For example, there are concerns about

Figure 8.3 Ammonia retailer ISO Tank. Courtesy of Eurotainer Gas Tanks for anhydrous ammonia.

refilling devices and regulations at ports, whilst leaks, maintenance and gas purges at sea present a technical and training challenge for future operators. Standards and homogeneous international approaches will be also needed to ensure equipment integrity, whilst several unresolved aspects around environmental impacts are still waiting for a demonstratable answer. All these points raise the interest from academia, industry and government bodies to move faster in the development of programs capable to enable a global ammonia fuel supply chain.

A recent analysis conducted by DNV [35] makes a strong case for the use of ammonia in future maritime applications. The current strong interesting on the use of the molecule has led to several pilot programs that seek to show the feasibility of ammonia as a fossil replacement fuel whilst also ensuring high energy density. However, the toxicity of ammonia (and also methanol) are some of the first points mentioned in the report, whilst the extreme flammability of hydrogen is also raised as a problem for the use of such a fuel. However, e-ammonia (i.e. ammonia product from renewable energy via electrolysis) appears to be the preferred option when carbon is not available or needs to be considerably reduced, leaving other fuels such as bio-gasoil (MGO) as pilot for ammonia combustion.

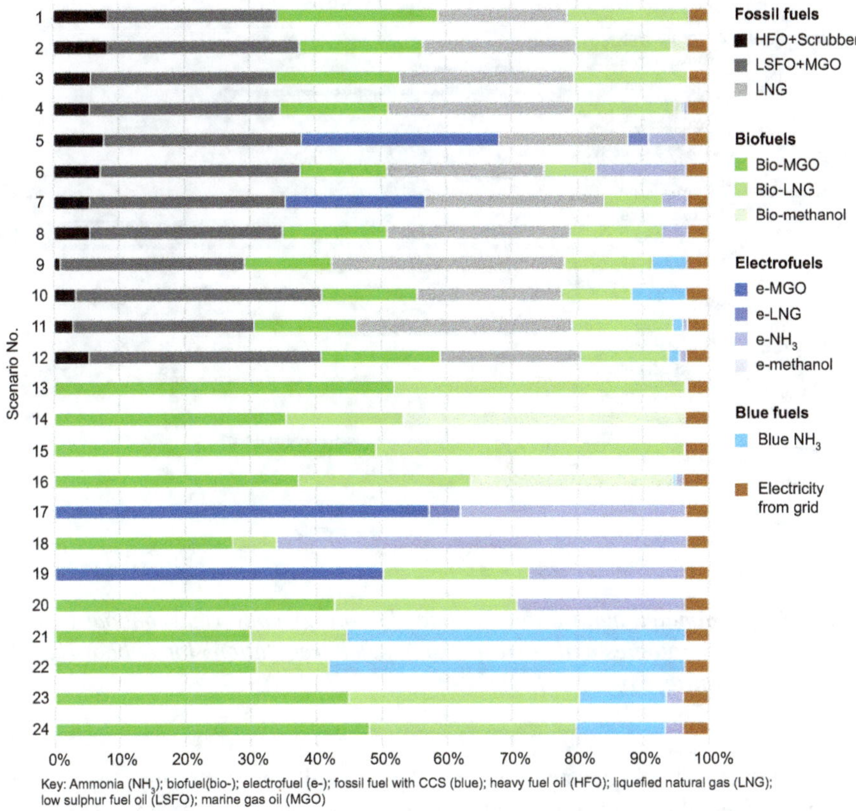

Figure 8.4 Scenarios for the maritime energy mix in 2050. Courtesy of DNV [35].

Regulatory policies and primary energy prices will be key drivers for uptake of carbon-neutral fuel and the future fuel mix. The uptake of carbon-neutral fuel needs to pick up in the mid-2030s, reaching 40% of the fuel mix in 2050 under the current IMO ambitions and 100% to decarbonise shipping fully [35]. Hence, several scenarios were assessed by the consultancy (Figure 8.4). Out of these, a few scenarios show relevant traces for the use of green e-ammonia. Special emphasis is given to scenario 18, where e-ammonia has a share close to 60% of the total energy mix. Also of interest is the use of blue ammonia (where carbon capture and storage, CCS, is used to trap CO_2 product of the formation of ammonia via steam methane reforming) in scenario 22, with a share greater than 50%. The interest in this ammonia lays on its current cost (much lower than e- or green ammonia) and its current availability. Therefore, ammonia is a good contender for decarbonisation purposes of this industry.

However, the annual investment in the area goes from 8 to 28 billion USD, with up to 150 billion USD if a full decarbonisation is sought [35]. Thus, industries and bodies are very keen on this area of development. The point has led to the use

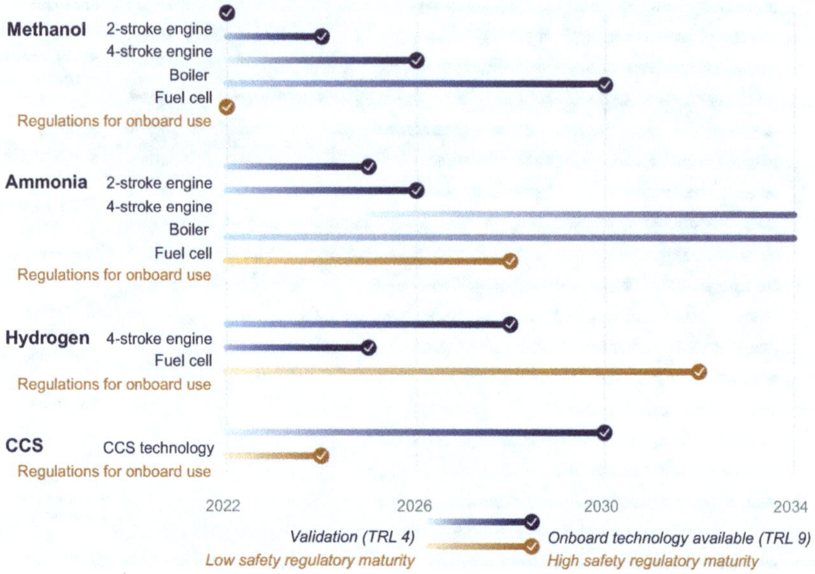

Figure 8.5 Estimated maturation timelines for energy converters, onboard CCS technologies and corresponding safety regulations for onboard use. Courtesy of DNV [35].

onset of novel technologies and concepts that are still under scrutiny. In terms of ammonia, internal combustion engines (ICEs) using two and four stroke concepts are under assessment at various levels and countries. Interestingly, technologies such as the use of boilers and fuel cells of high power output are also under review, with still not so clear time for final development (Figure 8.5). However, it is notorious that in the assessment, 'Regulations for onboard use' are set to be ready by 2028 [35].

However, and as pointed out previously, there is a critical need for the right regulations before ammonia can be used as a fuel in the maritime industry. The IMO has provided an international mandatory regulatory framework for alternative fuels through the International Code of Safety for Ships Using Gases or Other Low-Flashpoint Fuels (IGF Code). However, neither methanol, ammonia nor hydrogen are currently covered by detailed technical requirements in the Code [6]. Therefore, a shipbuilder will have to demonstrate through extensive risk evaluations that the chosen fuel system design solution meets the intent of the goal and functional requirements of the IGF Code, and that it is as safe as a conventional oil-fuelled ship [35]. It must be emphasised that ammonia use will be feasible using current technologies and marine architecture, although with some retrofitting (Figure 8.6).

Other consultancies (i.e. Lloyd's Register, Ricardo and the Environmental Defence Fund) have also set their eyes on the use of ammonia and the potential environmental implications in case of spillage when the molecule is used as a fuel

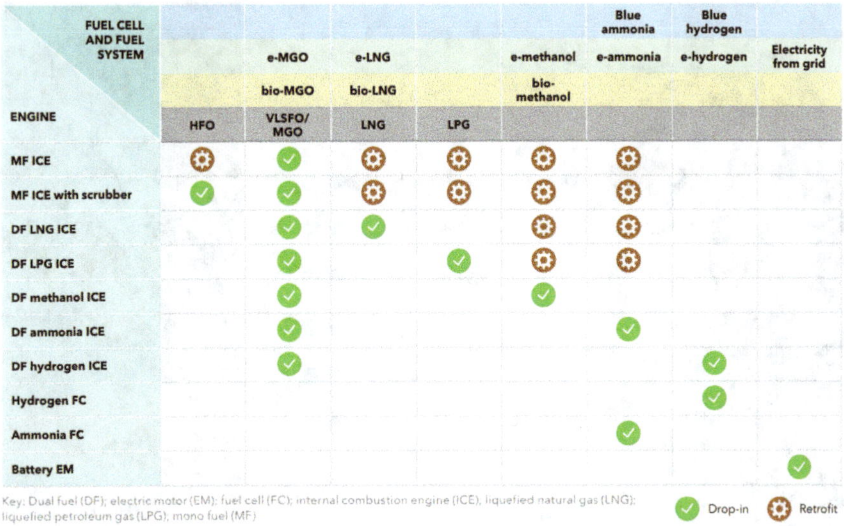

Figure 8.6 The energy converters, fuel options and transitions allowed in the GHG Pathway Mode. Courtesy of DNVl [35].

in marine systems. Their assessment, based on the collision of a ship or spillage during bunkering, shows the negative impact on aquatic life product of ammonia release without mitigation. Of special interest are estuaries, mangroves and wetlands in these scenarios where large fish concentrations co-habitat. Overall, ammonia will cause an increase in algal growth with potential eutrophication, whilst impact on the food chain can damage an entire ecosystem. Nutrient limitations could be also perceived, with greater sensitivity to drought and hypersalinity. Particularly, micro-organisms (e.g. bacteria, plankton) and invertebrates can suffer reproductive impacts whilst also increasing their mortality. Meanwhile, vertebrates might suffer physiological damage and greater mortality [36]. Overall, it seems that ammonia is more damaging to fish, whilst in comparison to oil spills, the latter has greater impact on birds and invertebrates (Figure 8.7).

The reason for this is that for the purposes of thinking about ammonia once it is in a water body, two primary realities apply. The first is that ammonia is toxic to most forms of aquatic organisms. The threshold level of harm varies by organism but is generally measured in mg per litre (part per million). The second reality is that the relationship between the gross quantity of ammonia released in a spill and the harm that is inflicted is far from straightforward. As in previous paragraphs, this is because ammonia in water exists in two forms: native, unionised ammonia (NH_3^0) and the ammonium ion (NH_4^+). The primary vector of harm is NH_3^0. This is because living cells are set up to exchange ions with their environment. They are not equipped to manage the small, neutral (non-ionic) NH_3^0 molecule, which can passively diffuse across the membranes of, for example, fish gills. In a confusing twist, after the NH_3^0 molecules appear in bodily fluids, most are converted to NH_4^+.

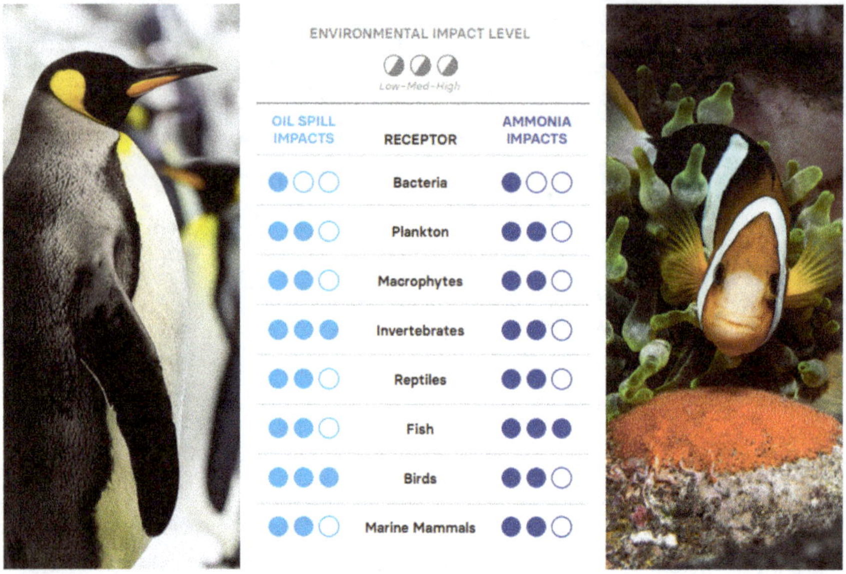

Figure 8.7 Environmental impact comparison between oil and ammonia spills. Courtesy of EdF [36].

Hence, levels of NH_3^0 above a critical threshold in a water body will in due course produce elevated quantities of NH_4^+ in the tissues of organisms that outstrip the capacity of the organisms' excretory functions [4].

Final conclusions of most of these analyses are that further research is required to fully understand the H&S implications of ammonia as a fuel, enabling the implementation of robust policies that mitigate risks whilst complying with safe procedures for humans and organisms.

8.4 Certification plans

The ammonia produced from various feedstocks is identical. Ammonia is typically traded as anhydrous ammonia (Table 8.3). However, the environmental footprint depends on the production method (Figure 8.8). If ammonia is applied as a fuel or hydrogen carrier, the exclusive use of ammonia with a reduced carbon footprint or zero-carbon footprint is highly important [37]. Otherwise, global CO_2 equivalent can be higher than for conventional fossil fuels.

Therefore, transparent frameworks for life cycle assessment are required to identify the carbon intensity of the produced ammonia in a specific location, and at a given timeframe [39]. Currently, various organisations are setting up transparent frameworks for calculating the carbon footprint of ammonia, including the Ammonia Energy Association and Fertilizers Europe. This allows for harmonisation of trade between jurisdictions.

Table 8.3 Commercial grade anhydrous ammonia [38]

Parameter	Range
Total nitrogen content	82.0 wt%
Ammonia content	99.5–99.8 wt%
Water content	0.2–0.5 wt%
Oil content	<5 ppm

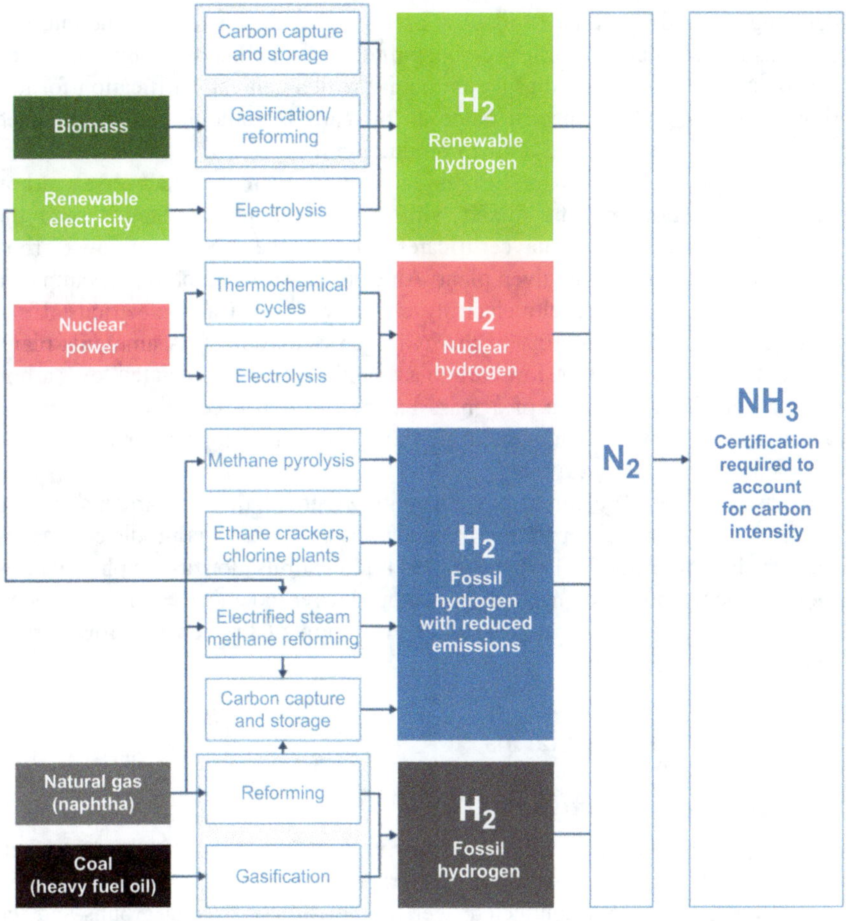

Figure 8.8 Ammonia production pathways from various feedstocks. Courtesy of IRENA [39].

Low carbon ammonia certification will be introduced in pilot schemes in 2023 and 2024, with further rollout expected thereafter. Hydrogen production typically accounts for over 90% of the energy input and carbon footprint for ammonia production, implying that ammonia certification is strongly linked with existing and future hydrogen certification schemes.

Four methods for carbon accounting can be identified, namely (1) identity preserved, (2) segregated, (3) mass balance and (4) book & claim. The implications of the chosen method have significant consequences.

Identity preserved refers to a carbon accounting method where low carbon ammonia production is produced separately from other ammonia production at a given site, and the low carbon ammonia is also stored, transported and delivered separately from the other ammonia production at a given site. A segregated carbon accounting method is similar to identity preserved, with separate production, storage, transport and delivery of low carbon ammonia and other ammonia. However, low carbon ammonia is traded as a commodity with a general certification for low carbon ammonia, and the specific batch of low carbon ammonia cannot be traced back to a specific factory, and a given production time.

Mass balance refers to a carbon accounting method where low carbon ammonia production may be mixed with other ammonia production, and the resulting low carbon ammonia certificates refer to the amount of low carbon ammonia produced from the given plant. An example of this a partial revamp of a fossil-based ammonia plant with electrolysers fed with renewable electricity. If the electrolyser produces about 10% of the hydrogen of a 100 KTPA ammonia plant, this results in about 10 KTPA low carbon ammonia with the associated low carbon ammonia certificates. In case of a mass balance carbon accounting method, the physical low carbon ammonia molecules are still traded together with the low carbon ammonia certificates.

In case of a book & claim carbon accounting method, the low carbon ammonia certificates can be traded separately from the physical ammonia produced via low carbon production methods. This can lead to trade optimisation, as the physical low carbon ammonia does not need to be shipped over potentially long distances. However, general transparency regarding the origin of the certified low carbon ammonia product is limited.

8.5 International scenarios

8.5.1 Japanese initiative

It is fair to say that Japan has been the leader in the subject, and it is thanks to their work that the subject has progressed from a mere academic topic to an industrially relevant theme. The work conducted over the years from several groups and the creation of large consortia have been key parameters to ensure that the use of ammonia as an energy vector and a combustion fuel become a reality fully accepted by the International Energy Agency (IEA) and the International Renewable Energy Association (IRENA).

Although the history of ammonia combustion in Japan can go back decades, the real breakthrough came with the setup of the 'Energy Carrier' program, a cross-ministerial Strategic Innovation Program (SIP) started initially under the framework of the Advanced Low Carbon Research and Technology and Development Program (ALCA) in 2013 [40]. The aim was to develop futuristic technologies. Interestingly, out of the entire Energy Carrier program (comprising 16 research institutes, 32 universities and 59 industrial groups), it was the 'ammonia direct combustion' subject the one with the greatest amount of participants (20), hence receiving a considerable financial support over the duration of the initiative.

The works prompted the improvement in reaction mechanisms whilst also delivering patentable technologies capable of using direct ammonia combustion concepts to avoid cracking of the molecule, hence increasing the total energy output of such systems. Of great relevance is the development of the first small gas turbine powered by ammonia. Toyota Energy Solutions in collaboration with the Advanced Institute of Science and Technology (AIST) developed devices capable of burning 50 and 300 kW pure ammonia [40]. In parallel, tests using a 50 kW micro gas turbine unit were conducted at AIST to fully replace kerosene by ammonia, hence showing the generation of stable power under these conditions [41]. However, the units still denoted the production of NOx that required further cleaning up processes.

Scaling up the technology, IHI Corporation developed a low-NOx combustor capable of co-firing methane and ammonia at 2 MW [42]. Ammonia was co-fired to reduce CO_2 emissions in the same proportion of ammonia content. The work eventually progressed to the use of liquid ammonia injection to not only remove the fossils, but also improve the injection efficiency by avoiding the vaporisation of the chemical [43]. The works expanded to the design of a Mitsubishi Power Unit rated at 40 MW. The system is expected to start operation in 2024 [44], being only fuelled with pure ammonia. This would set the precedent for the first unit rated at that power using such a chemical. Large expectations especially on flame stability and NOx emissions are awaited as the system develops. It is important to emphasise that these and further works have continued via the New Energy and Industrial Technology Development Organization (NEDO), which has secured a fund $\sim$517 million USD to continue working on ammonia energy supply chain [45,46]. However, it is also notorious that there is a gap in technological development for combined cycles, hence opening the opportunity to future developers and users of ammonia combustion systems.

Boilers and furnaces are also of great interest to the works conducted in Japan. Co-firing ammonia with pulverised coal is a promising area that can lead to a considerable reduction in CO_2 emissions. Initial analyses denoted the potential production of NO at large quantities with subsequent N_2O measured at the stag. Thus, further studies were performed and a demonstration was conducted in a large facility to fully understand the mechanism behind these emissions. Chugoku Electric Power Co. conducted a large demonstration where coal and ammonia were co-fired in a commercial power plant. The Mizushima Thermal Power Station, Unit 2 (rated at 156 MW) was employed for the program. The results demonstrated that CO_2 was considerably reduced, whilst the use of ammonia did not change significantly the NOx emissions of the system at 100% coal fuelling [47]. Although the ratio of ammonia was initially very low (up to 0.8%), further works conducted by IHI and a new burner enabled the increase of ammonia up to 20% in the mixture. The results showed that emissions were similar to full coal power, plus the heat distribution in the boiler remained almost the same (a major concern, since ammonia radiation and flame temperature are lower than the produced by fossil fuels and coal). The results were so promising, that JERA, one of Japan's largest power producers, has considered the replacement of 50% and even 100% coal by ammonia in its 1 GW power station, raising large bids in the international market to secure up to 500,000 tonnes of ammonia annually for such operations [48]. The program is supported by NEDO.

Similarly, industrial furnaces were evaluated for the replacement of fossils using ammonia. One of the main challenges of the technology is that the radiative heat of ammonia is lower than fossil flames (as a consequence of the reduction in soot). Works conducted by Osaka University and oxyfuel combustion showed that pure ammonia combustion under these conditions can overcome this barrier. The work enabled the study of various burners of small size (10–100 kW), which was later demonstrated by Taiyo Nippon Sanso Co. Ltd. And Nippon Steel Nisshin Co. Ltd. [40].

These and other works have made critical the development of a supply chain of ammonia which has been theorised to increase to up to 30 million tonnes per year by 2050 [40]. Therefore, several government initiatives led by the Ministry in Economy, Trade, and Industry have been conceived to support the introduction of ammonia. For example, the Clean Fuel Ammonia Association was established in 2019 to support private companies interested on free carbon ammonia. The association now consists of more than 100 governmental agencies, private institutions and international partners that will be part of the future green ammonia supply chains [49]. Further, the Public-Private Council on Fuel Ammonia Introduction was launched in October 2020 [50]. The body developed a report setting the ammonia targets of Japan by 2030 and 2050.

As mentioned, the progress in the topic by Japan is substantial and well ahead of the rest of the world. Private and government bodies keep signing agreements with Japan to start using ammonia in their energy mixes. It is therefore critical that continuous collaboration with consortia and partners from Japan remains substantial to keep progressing in the promotion of the use of ammonia at a global level, hence sharing experiences in the use of ammonia as a combustion fuel.

8.5.2 The Dutch plan

The Netherlands currently possess a strong fertiliser production industry, but the country also aims to play a key role in decarbonising the European energy landscape, which includes the use of ammonia as both zero-carbon fuel and hydrogen carrier.

The Netherlands have two existing fertiliser plants, namely in Sluiskil (Yara, 1.67 Mt-NH_3/y production capacity) and in Geleen (OCI, 0.996 Mt-NH_3/y production capacity). It is estimated that the CO_2 equivalent emissions are 3.4 Mt-$CO_{2,eq}$/y for

Sluiskil and 2.3 Mt-$CO_{2,eq}$/y for Geleen [51]. As for the first plant, Yara has taken various steps to decarbonise its fertiliser production, process that would lead to green ammonia not only for these applications, but also for fuelling. Firstly, about 0.022 Mt-NH_3/y have been produced based on by-product hydrogen delivered from an ethane cracker operated by Dow in Terneuzen since 2019 [52]. In parallel, starting from 2025, about 800 kt-CO_2/y will be transported from the Netherlands to Øygarden, Norway, where the CO_2 will be sequestered by Northern Lights, a joint venture between Equinor, Shell and TotalEnergies. Renewable hydrogen will also be produced based on offshore wind electricity coupled with alkaline electrolysis, totalling about 0.075 Mt-NH_3/y production [53]. The combined measures will result in a CO_2 equivalent reduction of nearly 1.0 Mt-$CO_{2,eq}$/y in Sluiskil by 2025.

Similarly, OCI aims to decarbonize part of its hydrogen supply via electrolysis-based hydrogen fed with offshore wind electricity produced in the North of the Netherlands [54]. The NorthH2 project aims to produce 1 Mt-H_2/y in the North of the Netherlands, off the coast of Eemshaven, which will be then transported to industrial clusters via pipelines. The combined carbon reduction for OCI's ammonia and methanol production will be 0.9 Mt-$CO_{2,eq}$/y by 2030.

Ammonia imports as both a zero-carbon fuel and hydrogen carrier have been gaining traction in the Netherlands recently, with industrial consortia such as the ISPT Green Ammonia Platform. An example of early work on ammonia as an energy carrier can be found in their 'Power to Ammonia' report [55]. However, recommissioning of the plant for other purposes and full coal replacement have led to the stall of the project.

Thus, greater impetus has been positioned on the marine sector. The Port of Rotterdam is the largest port in Europe, with significant infrastructure for both maritime trading and for refineries. The aim is to transition to net-zero emissions. The Port of Rotterdam estimates that up to 20 Mt-H_2/y is required by 2050, of which 18 Mt-H_2/y will be imported [56]. Ammonia is one of the best hydrogen carriers for intercontinental hydrogen trade, according to IRENA [57]. The required 18 Mt-H_2/y is equivalent to about 116 Mt-NH_3/y on an energy basis, which represents about 63% of the current global ammonia demand of about 185 Mt-NH_3. Hence, ammonia, having such a high hydrogen content, has been assessed for bunkering for ammonia-fuelled ships by various consortia. Various companies are expanding their ammonia import capacities. OCI has reached FID to increase its ammonia storage throughput from 0.4 Mt-NH_3/y to 1.2 Mt-NH_3/y by 2023, requiring a US $20 million investment [58]. Similarly, Vopak is currently converting two existing refrigerated LPG storage tanks to store ammonia in Vlissingen [59]. Various green shipping corridors are currently established to ensure availability of low carbon fuels.

On relevant notes, at least four large-scale ammonia crackers have been announced in the Netherlands, namely the ACE Terminal (Rotterdam; Gasunie, HES International & Vopak), Amplifhy (Rotterdam; VTTI, Essent & E.ON), the Green Hydrogen Import Terminal (Rotterdam; Air Products & Gunvor) and Greenpoint Valley (Vlissingen; Vesta & Uniper). Furthermore, an initiative of

18 companies, led by the Port of Rotterdam, is performing a study on a 1 Mt-H_2/y ammonia cracker to deliver hydrogen for heating and power applications [60].

8.5.3 German advancements and collaborations

Germany has also announced several projects related to ammonia for power generation as well as strategies to support the decarbonisation of its current fertilizer production. The German government has made it clear that its goal is to become a key player in the green hydrogen and ammonia production.

The Federal Government adopted the National Hydrogen Strategy in June 2020 [61]. As a consequence, several projects involving ammonia are already under development in order to support the decarbonisation of current applications. In the international sphere, the Federal Government has launched several funding measures and initiatives to promote the import of green hydrogen from outside Europe (e.g. projects in Saudi Arabia and Chile).

Yara currently operates the largest ammonia storage in Germany, located in Rostock, a city in the north coast with a strategic location for transportation into the Bay of Mecklenburg of the Baltic Sea. This deep-sea terminal currently imports 600 kt of ammonia every year. In addition, Yara operates another deep-sea terminal in Brunsbüttel. The company has announced recently that it plans to modify both ammonia terminals to reach up to 3 Mt of clean ammonia, with a potential to add additional tank capacity and expand import capacity further if needed [62]. Modifications to Brunsbüttel are planned to be completed by the end of 2023 and will be home of RWE's ammonia import terminal, which will receive shipments from Namibia in 2026. Rostock, on the other hand, is planned to be the site for an ammonia import terminal developed by Total Eren and VNG. Total Eren production projects located in Chile will be the source. VNG energy hub in Rostock is expected to become the key entry for ammonia import into Germany, with hydrogen pipeline to extend into Berlin and Leipzig [63].

BP has also announced plans for a new ammonia cracker in Wilhelmshaven, Germany. The project will provide 130 kt of low-carbon hydrogen from green ammonia per year, starting in 2028 and using the current oil and gas pipelines for hydrogen transport [64]. Also at Wilhelmshaven, Niedersachsen Ports, Uniper and Tree Energy Solutions are exploring the possibility of a brand-new infrastructure that could provide 10–20% of Germany's hydrogen demand in 2030 with the first phase to be completed by 2026 [65]. After the successful construction of the LNG terminal inaugurated in December 2022, the project is looking to import 2.6 Mt of ammonia per year, combined with an onsite electrolyser capacity of 1 GW and an industrial-scale ammonia cracker.

These projects and several more are coming up after the announcement of potential hydrogen and ammonia export locations in the Middle East, Africa and Australia, which could supply part of the European demand in the coming years.

As it has been made clear from the examples above, several ammonia projects are to be completed in the coming years; therefore, regulations are key for the development of these projects. In order to support a green hydrogen industry and expansion of an international market, H2Global Foundation was established in 2021. It is based on a 'double-auctions model' where the aim is to reduce the gap between the current high prices of hydrogen and the lower prices at which it can be sold and used at a regional level [66]. The current German Federal Ministry for Economic Affairs and Climate Action (BMWK) has approved an investment of approximately €900 million euros to support this initiative via its implementation entity Hintco, with €360 million to be used on the first H2Global green ammonia tender [67]. In more detail, the 'double-auctions model' entitles as a first step, an international action for the purchase of green hydrogen or its derivatives. The best offer wins and gets a long-term contract. This way suppliers have the certainty to plan and invest more in their hydrogen production. At the same time, they are conditioned to sell the products they produce in their partner countries to Europe.

As a second step, the hydrogen that has been delivered to EU countries in this way will go to second auction process to ensure green hydrogen available at a competitive price. This will create stronger incentives for the industrial sector and others to invest in installations that can use this hydrogen.

8.5.4 British proposals

As part of the European initiatives, the UK has also been working on several fronts to use ammonia as an energy vector.

The Department for Business, Energy and Industrial Strategy (BEIS) has recently awarded £6.7 million (~8.2 million USD) to the Tyseley Ammonia to Green Hydrogen Project. The project seeks to demonstrate the use of ammonia for cracking and production of hydrogen. The project seeks to deliver over 200 kg per day of hydrogen to a refuelling station. The cracking unit, designed by H2Site, will serve to deliver purer hydrogen for transportation purposes. The project will take place at the Tyseley Energy Park, a strategic energy and resource hub in the West Midlands, UK [68].

Similarly, Siemens Energy, Fortescue Future Industries (FFI) and GeoPura are developing a £3.5 million (~4.3 million USD) ammonia cracker prototype in Newcastle, UK. The prototype will deliver 200 kg of high-purity, PEM-suitable hydrogen fuel per day, and will be based on Fortescue's 'Metal Membrane Technology' (MMT) purification process. The MMT process developed by FFI and Australia's national science agency, the Commonwealth Scientific and Industrial Research Organisation (CSIRO) is key to producing high-purity hydrogen from the ammonia cracking process. It selectively filters the hydrogen whilst blocking other gases, allowing it to be used as a fuel and converted as needed, at the time of re-fuelling [69].

In terms of distribution, Air Products and Associated British Ports (ABP) will develop a facility at the Port of Immingham on the UK's east coast for ammonia imports and hydrogen production (i.e. cracking), providing a hydrogen supply for the industrial and heavy transport sectors. Immingham is one of the UK's largest ports and sits within Humberside: the UK's largest industrial cluster. Air Products is already planning a CCS hydrogen project in Immingham. ABP will invest in new infrastructure with a new jetty to service the import and export handling of liquid bulk products. In addition to handling green ammonia, the jetty is being designed so that it can accommodate other cargoes connected to the energy transition, including the import of liquified carbon dioxide (CO_2) from carbon capture and storage projects for sequestration in the North Sea – thereby playing a significant role in the UK's energy transition. The project will bring a wide range of benefits to Immingham and the UK, including eliminating up to 580,000 tonnes of greenhouse gas emissions each year, up to US $4.6 billion in growth and 1,400 direct jobs and approximately 1,600 indirect jobs for supply chains and local businesses [70].

Along the lines of marine applications, two major project grants sponsored by UK Research and Innovation (UKRI) also consider ammonia as a method for energy storage and delivery. The first one is called OceanREFuel. Offshore Wind in Northern Europe has to date become a tremendous success surpassing most commercial and technical expectations. This success has buoyed investor confidence and led to ambitious targets, including a detailed Offshore Wind Sector Deal committed to 30 GW of electric power with 60% local content by 2030 representing a major component of UK strategic electricity policy. The UK government extended this ambition to 40 GW by 2030 immediately following the December 2019 general election. Such a large proportion of electricity from an intermittent renewable energy resource highlights the importance of energy storage, which the Ocean-REFuel project will directly address [71]. Looking into the use of both hydrogen and ammonia, the project will evaluate and compare the methods of delivery, application and use of these two molecules, hence providing stakeholders with a quantitative measurement to define which vector is most feasible for their particular needs. Regarding ammonia, the chemical will be produced at Orkney, Scotland, after the production of hydrogen from offshore wind farms near the region. The product green ammonia will be distributed to several locations including the Teesside Free Port and Wales, where global distribution and a local network will be established, respectively. As for the local network, the evaluation will include the delivery to farms for direct use of the chemical (i.e. as a fertiliser, fuel, heating source and coolant) whilst enabling the creation of 'City Gate' terminals capable of using the ammonia for storage and then hydrogen production and injection into the gas network to cities – which are known to have greater conflicts with the use of direct ammonia [4].

On a similar path, ammonia ICEs for marine applications are under the scope of the UKRI via the program grant MariNH_3. The program, led by Nottingham University, is a 5-year research programme to develop new and disruptive engine technology that will significantly cut greenhouse gases and pollution emitted by diesel powered marine vessels. The £5.5 million (~6.7 million USD) will foresee social and regulatory aspects of the use of ammonia at port and marine offshore levels, novel technologies with strong fundamentals in the use of ammonia mixtures, and a strong approach to the demonstration of the concept at a vessel level. The decarbonisation of commercial shipping will require this ammonia approaches (i.e. which have also been taken elsewhere across Europe and Asia), the development of new engine technologies and effective policy implementation.

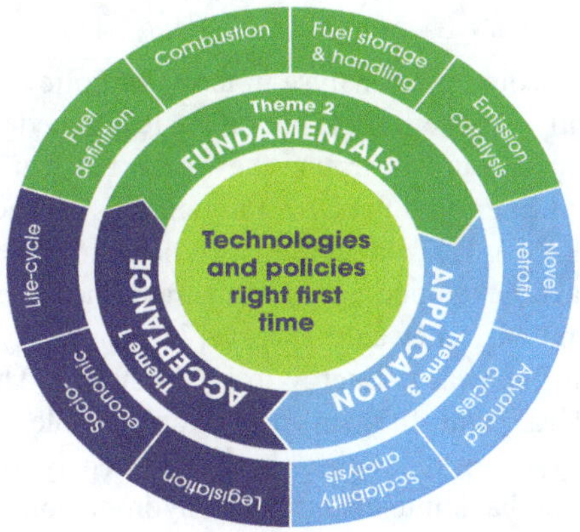

Figure 8.9 MariNH$_3$ holistic approach

The research is focused on decarbonising end use, with minimal pollutants and reduced related energy demand, whilst at the same time overcoming key technical, economic and social barriers to the adoption of green ammonia as a fuel (Figure 8.9) [72].

Along those lines, Nottingham University has joined a project led by engine developers MAHLE Powertrain and partners Clean Air Power. The trio will demonstrate decarbonisation of heavy duty engines using ammonia and hydrogen fuel (or a blend of the two), based on MAHLE's existing technology and with Clean Air Power to develop a fuel injector. The research focuses on two approaches – near-term with the retro-fit of dual-fuel arrangements, and the long-term development of MAHLE's own single cylinder engine, which uses the MAHLE Jet Ignition® (MJI) – a highly innovative form of pre-chamber ignition – replacing the traditional spark plug to ignite the ammonia, eliminating the need for a fossil-fuel ignition source. According to the manufacturer, MAHLE's Jet Ignition technology consists of a small chamber in the cylinder head that, once ignited, forces the resulting hot gas through small orifices into the main combustion chamber as a series of jets that quickly and uniformly ignite the remaining mixture, hence ensuring super-clean, high efficient combustion with no pollutants such as nitrogen oxides [73].

As this is presented, other initiatives are underway for the development of gas turbines, furnaces, boilers and other components using ammonia combustion. Some projects, already commercial, have employed ammonia via fuel cells to deliver power in far-from-the-grid locations (i.e. for telecommunication purposes) whilst also assessing the use of these cells for the transportation sector. However, ammonia combustion will see a considerable boost in a few years with the work that universities, industries, devolved governments, UKRI, BEIS and other major players are conducting for the direct utilisation of the chemical in the power, heating and cooling sectors.

8.5.5 US recent works

The USA is an interest country to analyse in terms of policy and ammonia energy developments. Currently, this country is not on the leading edge of all the ammonia projects that have been announced around the world, but that is changing.

The Department of Energy of the USA through the Advanced Research Project Agency-Energy (ARPA-E), RTI International (a non-profit research institute) and additional partners have developed and demonstrated novel processes to produce ammonia from intermittent renewable energy [74]. Similarly, in 2016, ARPA-E launched the program Renewable Energy to Fuels through Utilisation of Energy-Dense Liquids (REFUEL) with the aim to develop scalable technologies for converting electrical energy from renewable sources into energy-dense carbon-neutral liquid fuels (CNLFs) and back into electricity or hydrogen on demand [75]. Sixteen projects from this program were funded by ARPA-E involving a US $32.7 million investment, out of which 13 focused on ammonia. The aim was to accelerate the shift to domestically produced transportation fuels, improving American economic and energy security and reducing energy emissions.

Whilst Europe and East Asia committed to ambitions hydrogen targets and announced large investments through 2020–2021, the USA was the only major market without a national hydrogen strategy. Therefore, at the beginning of 2021, 11 companies and organisations across the hydrogen value chain launched a new advocacy body called ‘Hydrogen Forward’ to lobby the US government to peruse a national hydrogen strategy [76]. The scenario transformed one year later. By 2022, the US Department of Energy announced an investment of US $7 billion to boost the Clean Hydrogen Economy by funding regional hydrogen hubs [77].

This announcement was accompanied by a clear roadmap describing the pathways for clean hydrogen to decarbonize different applications by demand scenarios for 2030, 2040 and 2050. Ammonia market being one of the early opportunities for creating largescale demand for clean hydrogen. The roadmap illustrates a strategic opportunity to achieve 10 Mt of clean hydrogen annually by 2030, 20 Mt annually by 2040 and 50 Mt annually by 2050. That same year, the USA passed the Inflation Reduction Act 2022 (IRA). The act includes clean energy tax credits and other provisions to increase domestic renewable energy production. Involving incentives for clean hydrogen and fuel cell technologies, either by extending existing federal tax credits, or creating new ones.

The act accomplishes several objectives, one of the most relevant for the hydrogen market would be a new tax incentive called 45V which is the *Clean Hydrogen Production Tax Credit*, that incentivises companies to produce green hydrogen. There is a sliding scale that starts off at US $0.60 per kilogram of hydrogen produced and works its way up to US $3 per kilogram (Table 8.4). Incentivising investment in green hydrogen production and green ammonia production.

8.5.6 Southeast Asian and China approach

The role that Asia, particularly the Southeast region, is playing on the use of ammonia is increasing as the technologies mature and are demonstrated at a global

Table 8.4 The four-tier incentive breakdown

Carbon intensity (kg CO_2 per kg H_2)	Max hydrogen production tax credit (US $/kg H_2)
4–2.5	0.60
2.5–1.5	0.75
1.5–0.45	1.00
0.45–0	3.00

Table retrieved from USA Office of Energy Efficiency & Renewable Energy [78].

scale. Several countries are now joining ammonia-based programs to untangle the potential of their renewable resources and supply large economies such as Japan and China with enough net zero energy for the next decades. As it is imaginable, the number of projects in the region surpasses the scope of this chapter. However, a few projects are highlighted, hence showing the interest from industries and governments in various locations across South-East Asia to switch from fossils to ammonia-based fuelling.

During the 2021 Summer, an important agreement was signed between General Electric and IHI Corporation for the development of a gas turbine business roadmap. The initiative will support the transition towards ammonia-based units, whilst decarbonising Asia. The work, based on marketing analyses and demonstration stages to show the feasible use of ammonia as gas turbine fuel, has set a worldwide precedent for both companies in the search of better methods to use the chemical at large power scales [79].

Similarly, Thailand's state-owned oil and gas company PTT and Singapore-based AET Tankers agreed to jointly develop dual fuel vessels powered with ammonia. It is expected that by 2026 the vessels will be ready after retrofitting to receive ammonia [80]. Therefore, Singapore's interest on joining the ammonia race has been tangible over the last few months of 2022. The country has also worked on the development of investment funds, green corridors, new bunkering concepts and innovative safety strategies to employ ammonia in the marine sector (which as previously stated, will also require the use of combustion-based engines). GIC, Singapore's sovereign wealth fund, and UK-based Carlyle are currently investing in the development of Eneus Energy's 14 GW renewable ammonia project pipeline. Eneus Energy has recently received funds to develop a green ammonia facility in Orkney, Scotland. This is the second ammonia-related investment for GIC, who earlier this year invested in InterContinental Energy's portfolio of renewable ammonia Supergiant projects. Similarly, the Maritime and Port Authority of Singapore and the Port of Rotterdam have agreed to establish a green maritime corridor by 2027, hence moving in the direction of decarbonisation of maritime applications and ports [81]. The two Authorities will work with a larger group, including the Global Centre for Maritime Decarbonisation and the Mærsk Mc-Kinney Møller Center for Zero-Carbon Shipping, BP, Maersk, Ocean Network

Express and Shell (amongst others) as industry partners. The works will follow other consortia already working in the country.

Similarly, other economies such as Vietnam, Malaysia and India have jumped on the train of the use of ammonia. Vietnamese renewable energy project developer 'The Green Solutions' will partner with various large companies (i.e. ECONNECT, Thyssenkrupp and Black&Veatch) to develop novel production plants in Tra Vinh. As for Malaysia, the large energy producer TNB has joined forces with IHI to start working on the use of ammonia combustion boilers. The units, rated ~200 kW, will use natural gas from the local grid in combination with ammonia at different mixing ratios. The program promises to decarbonise power and processes at the same scale that ammonia is replaced in the natural gas blend [82].

India, also a large economy interested in the use of green ammonia due to its potential utilisation as a fertiliser for their harvesting activities, carbon mitigation and future market, has also signed an agreement with Japanese firms to conduct tests to co-fire ammonia with fossil fuels. The works, still under review by the time this section was written, show the interesting potential that the use of green ammonia can bring to the region and such a large economy. Furthermore, India has introduced a green hydrogen obligation. The National Green Hydrogen Mission, launched in 2021, aims to produce at least 5 Mt of green hydrogen by 2030. The objective will be to decarbonize the current 5 Mt of hydrogen utilised annually in the country for various industrial purposes, including ammonia production for fertilizers [83].

Of course, one powerhouse that cannot be disregarded is China. China has considerably speed up their research and development program for the use of ammonia at various scales. From the use of fuel cells to the development of combustion prototypes, China has actively engaged in the subject in fundamental and applied fields. As previously mentioned, the number of projects in such a country is vast and out of the scope of this section. However, the most representative are presented as follows.

Reports from the *South China Morning Post* announced that China's state-owned mining and energy investment corporation has successfully demonstrated the co-firing of 35% ammonia with coal in the Huaneng Yantai coal power plant, Shandong Province. The tests denoted that NOx emissions were lower than burning pure coal, with only 0.01% ammonia traces [84]. In support of these works, a vast number of research groups working on ammonia combustion have joined the efforts to combust the fuel whilst increasing efficiencies and reducing emissions. Xu *et al.* [85] performed a simulation of a 600 MWe coal power plant unit replacing coal at various ratios (5%, 10%, 15%, 20%) on a heating value basis. Using Ansys Plus and two different scenarios, i.e. identical adiabatic flame temperature (AFT) and identical excess air coefficient (EAC) for similar burnout of fuel, the analysis examined the reduction of CO_2 and ammonia impact on the exergy of the system. Initial results denoted the reduction of CO_2 and SO_2, whilst water content considerably increased. Carbon dioxide emission were decreased by 96.4 tonne/h when ammonia was co-fired with a ratio of 20% on the heating value basis. However, as the ammonia content increased so did the exergy losses due to the irreversible loss of the ammonia combustion process being greater than those for coal. Interestingly, the exergy loss of heat exchangers and fans was lower with higher ammonia content due to a decrease of heat transfer temperature difference and reduced air flow into the furnace, respectively. It was observed that the exergy efficiency of the boiler was slightly reduced using ammonia (from 53.82% to 53.22%) at identical AFT scenarios. Unfortunately, it was noted that the increase of H_2O content in flue gas might improve the acid dew point of flue gas, which suggests that during ammonia co-firing, the ash deposition and corrosion problems in the low-temperature heat recovery zones should be specially considered [85].

Finally, it is worth mentioning the work conducted in Australia. Similar to other countries interested on ammonia, Australia has pushed the local and international agenda, with projects that go from fundamental chemistry to large programs aimed to recover, store and distribute the vast renewable energies the country possesses. In terms of the use of ammonia for energy utilisation, one of the most notorious programs is the 'Asian Renewable Energy Hub (AREH)'. The proposal is based on the construction and operation of large hybrid systems comprised of solar and wind systems capable of delivering massive energy outputs ($\sim$26 GW), with the intention of producing net zero molecules (i.e. hydrogen and ammonia) for distribution of such an energy. The facility, located 220 km east of Port Hedland, has been conceived for the production and distribution of energy to the Southeast region. Initially, the plan was to use submarine high voltage cables that would feed the Asian grid. However, the project has progressed and, in order to provide greater flexibility, the cable has been replaced by hydrogen and ammonia for export. Energy would be also delivered into the Pilbara region to enable large scale mine electrification, hence reducing diesel use whilst supporting mining and metal producing activities. The environmental approval of the first phase (15 GW) was authorised in 2020. In that same year, an assessment report of the Environmental Protection Authority (EPA) of Western Australia recommended the implementation of the hybrid systems. Overall, up to 1,743 wind turbines of 260 m would be installed with 2 GW of solar photovoltaics. The project would include 37 substations and up to 1,514 km of site access tracks linking wind turbines with other infrastructure in an area of $\sim$670,000 ha [86]. It is estimated that at full capacity, the project will deliver $\sim$1.6 million tonnes of renewable hydrogen or 9 million tons of green ammonia annually, hence reducing around 17 million tonnes of CO_2.

8.5.7 Other parts of the world

Another nation already interested on ammonia is the UAE. In 2021, the UAE became the first country in the Middle East and North Africa to pledge to offset all its domestic carbon emissions by 2050, followed by Saudi Arabia and Bahrain. As part of the UAE project to develop the world's largest single-site solar power plant – Al Dhafra Solar PV IPP – the country also developed a Hydrogen Leadership Roadmap to support domestic low-carbon industries and contribute to the country's net-zero ambition and position the country as a competitive exporter of hydrogen. In this line, Helios Industry in Abu Dhabi announced a project in KISAD to design, develop, finance, construct, own and operate a project of US $1 billion to produce 40,000 Mt/y of green ammonia in the first phase. This project aims to target the European and American markets [87].

Similarly, the Company Abu Dhabi Chemicals Derivatives (Ta'ziz) announced plans to build a low carbon ammonia plant in Ruwais with a capacity of about a million tonnes per year. At the beginning of 2023 the company signed a shareholder agreement with Fertiglobe, South Korea's GS Energy and Japan Mitsui [88].

In parallel, the companies NEOM, ACWA Power and Air Products signed in 2020 an initial agreement for a US $5 billion project to build a green hydrogen-based ammonia plant. The NEOM Green Hydrogen Project is one of the largest utility scale, commercially-focused hydrogen facilities powered entirely by renewable energy. The project in Saudi Arabia will have the capability to produce 1.2 million tonnes of green ammonia per year, starting its production in the first quarter of 2025. Air Products is expected to invest a further US $2 billion in distribution [89].

Along ambitious targets to import clean hydrogen and ammonia, countries that were not key players in the energy sector are staring to also announce hydrogen strategies. Latin America, especially Chile, has announce ambitious goals. Chile's National Green Hydrogen Strategy has three key objectives: (1) to achieve 5 GW of electrolysis capacity by 2025, (2) produce the cheapest green hydrogen on the planet by 2030 and (3) to become one of the three global exporters by 2040 [90]. For this to be possible projects like H2 Magallanes in Chile have been announced and are expected to produce up to 13% of the world's green hydrogen due to large onshore wind potential in the county. This hydrogen will be likely distributed as green ammonia [91].

8.6 Conclusions

As presented in this chapter, the use of ammonia brings several health and safety challenges. However, these challenges are not unknown by the refrigeration or fertilising industries, which have been operating for decades using this chemical. Therefore, lessons learned from all those years of utilisation can be employed for the creation of new regulations that will ensure ammonia can be safely used as a fuelling vector.

Similarly, the work is leading to the development of certifications that will ensure the commercialisation of low or net zero ammonia. These certifications will precede the larger deployment of a greener molecule, whilst also advising consumers about the impact of replacing current fossil fuels using ammonia product of greener processes. Pink, green and blue ammonia will overtake the market in due course, hence mitigating the emission of CO_2 towards the atmosphere whilst also ensuring, via clean combustion technologies, that the chemical can deliver net zero carbon heating, power and propulsion.

Finally, these accomplishments are encouraging nations worldwide to switch to the use of ammonia as a green solution for decarbonisation purposes either as direct use of ammonia for fuelling or for cracking purposes to recuperate green hydrogen. News from Asia, America, Europe and Africa are constantly being published, with large project supporting the production, delivery and utilisation of greener ammonia that is being moved through intercontinental routes. The story of green, pink and blue ammonia is just starting; hence, the reader should expect to hear many more news about the pace that this subject is going to take in the coming years.

References

[1] Britannica, Encyclopaedia. 'Ammonia'. 2013. http://www.britannica.com/EBchecked/topic/20940/ammonia-NH3 (accessed January 17, 2023).

[2] Hattenbach R. 'Transportation & delivery of anhydrous ammonia'. In *Annual NH3 Fuel Conference*. 2012. https://nh3fuel.files.wordpress.com/2012/05/chemicalmarketingservices.pdf (accessed February 23, 2023).

[3] Michaels RA. Emergency planning and the acute toxic potency of inhaled ammonia. *Environ Health Perspect* 1999;107(8):617–627. https://doi.org/10.1289/ehp.99107617.

[4] Valera-Medina A, and Banares-Alcantara R. *Techno-Economic Challenges of Ammonia as an Energy Vector*. 1st edn. New York: Academic Press; 2020.

[5] Environmental Protection Agency. AEGL [Online]. 2016. Available: https://www.epa.gov/aegl (accessed February 16, 2023).

[6] HSE. *EH40/2005 workplace exposure limits*. 2011. http://www.hse.gov.uk/pUbns/priced/eh40.pdf (accessed October 20, 2022).

[7] Crolius S. An open letter to the International Energy Agency. *Ammon Energy* 2019. https://www.ammoniaenergy.org/an-open-letter-to-the-international-energy-agency/ (accessed July 12, 2019).

[8] Smith G, Mott J, and Garcia T. A hazard assessment of ammonia as a fuel: understanding the hazards and managing the risks/threats of ammonia as a fuel. In *NH3 Fuel Assoc.* 2013. https://nh3fuelassociation.org/wp-content/uploads/2013/10/nh3fcx-gsmith-jmott-tgarcia.pdf (accessed February 20, 2023).

[9] Cole-Parmer Instrument Company. Chemical compatibility database, ammonia, anhydrous. 2017. https://www.coleparmer.co.uk/Chemical-Resistance (accessed July 12, 2019).

[10] Ravishankara AR, Daniel JS, and Portmann RW. Nitrous oxide (N_2O): the dominant ozone-depleting substance emitted in the 21st century. *Science* 2009;326:123–5. https://doi.org/10.1126/science.1176985.

[11] Valera-Medina A, Gutesa M, Xiao H, *et al.* Premixed ammonia/hydrogen swirl combustion under rich fuel conditions for gas turbines operation. *Int J Hydrogen Energy* 2019;44:8615–26. https://doi.org/10.1016/j.ijhydene.2019.02.041.

[12] Coghlan A. Nitrogen oxides in car exhaust kill tens of thousands in UK. *New Sci.* 2015. https://www.newscientist.com/article/dn28245-nitrogen-oxide-is-not-so-harmless-and-could-damage-human-health/ (accessed July 11, 2019).

[13] Environmental Protection Agency. Ground-level ozone pollution. EPA; 2019. https://www.epa.gov/ground-level-ozone-pollution/health-effects-ozone-pollution (accessed September 3, 2019).

[14] Mashruk S, Okafor EC, Kovaleva M, *et al.* Evolution of N_2O production at lean combustion condition in NH_3/H_2/air premixed swirling flames. *Combust Flame* 2022;244:112299. https://doi.org/10.1016/j.combustflame.2022.112299.

[15] Environmental Protection Agency. Understanding global warming potentials. *Greenh Gas Emiss.* 2023. https://www.epa.gov/ghgemissions/understanding-global-warming-potentials (accessed February 1, 2023).

[16] Chen S, Zhao Y, and Zhang R. Formation mechanism of atmospheric ammonium bisulfate: hydrogen-bond-promoted nearly barrierless reactions of SO_3 with NH_3 and H_2O. *ChemPhysChem* 2018;19:967–72. https://doi.org/10.1002/cphc.201701333.

[17] Kurvits T, and Marta T. Agricultural NH_3 and NO_x emissions in Canada. *Environ Pollut.* 1998;102(1):187–194. https://doi.org/10.1016/S0269-7491(98)80032-8.

[18] Cerruto-Noya CA, Goad CL, and Mireles Dewitt CA. Antimicrobial effect of ammonium hydroxide when used as an alkaline agent in the formulation of injection brine solutions. *J Food Prot* 2011;74(3):475–479. https://doi.org/10.4315/0362-028X.JFP-10-343.

[19] Office of Industrial Relations Workplace Health and Safety Queensland. *Emergency Planning for Ammonia-Based Refrigeration Systems Guide*. Queensland: Office of Industrial Relations Europe Workplace Health and Safety Queensland; 2018.

[20] Amshel CE, Fealk MH, Phillips BJ, and Caruso DM. Anhydrous ammonia burns case report and review of the literature. *Burns* 2000;26(5):493–497. https://doi.org/10.1016/S0305-4179(99)00176-X.

[21] YARA. *YARA Fertilizer Industry Handbook*. Oslo: Yara; 2017.

[22] Palgrave DA. *Fluid Fertilizer Science and Technology*. Abington: Marcel Dekker; 1991.

[23] Hignett TP. Transportation and storage of ammonia. In: Hignett, T.P. (ed), *Fertilizer Manual.* Dordrecht: Springer; 1985. pp. 73–82. https://doi.org/10.1007/978-94-017-1538-6_7.

[24] Fertilizers Europe. *Guidance for Inspection of and Leak Detection in Liquid Ammonia Pipelines*. Brussels: Fertilizers Europe; 2013.

[25] UNIDO UN, International Fertilizer Development Center, United Nations Industrial Development Organization, International Fertilizer Development Center. *Fertilizer Manual*. 3rd edn. Norwell, MA: Kluwer Academic Publishers; 1998.

[26] European Fertilizer Manufacturers Association, EFMA. *Guidance for Transporting Ammonia by Rail*. 2nd edn. Brussels: European Fertilizer Manufacturers Association; 2007.

[27] Hassan CRC, Puvaneswaran B, Aziz AR, Zalina MN, Hung FC, and Sulaiman NM. Quantitative risk assessment for the transport of ammonia by rail. *Process Saf Prog* 2010;29(1):60–63. https://doi.org/10.1002/prs.10345.

[28] TRANSCAER®. Anhydrous ammonia training. Online Train. 2023. https://www.transcaer.com/training/online-training-courses/anhydrous-ammonia-safety-training (accessed October 1, 2023).

[29] Schlichthärle G, and Huberich T. Tank-car loading station for liquid and aqueous ammonia. A detailed description of the functions and safeguards of a modern, computer-controlled rail-car loading station for liquid and aqueous ammonia. *Plant/Oper Prog* 1983;2:165–7. https://doi.org/10.1002/prsb.720020310.

[30] The Fertilizer Institute. Fertilizer logistics and transportation. 2023. https://www.tfi.org (accessed August 18, 2023).

[31] Appl M. Ammonia, 2. Production processes. In: *Ullmann's Encyclopedia of Industrial Chemistry.* Weinheim: Wiley-VCH Verlag GmbH & Co. KGaA; 2011. https://doi.org/10.1002/14356007.o02_o11.

[32] EFMA. *Guidance for Inspection of Atmospheric, Refrigerated Ammonia Storage Tanks*. Brussels: European Fertilizer Manufacturers Association; 2008.

[33] Tullyn Trading. 20ft T50 Anhydrous Ammonia tank container. 2018.

[34] Linde. Ammonia. *Ind Gases.* 2018. https://www.linde-gas.com/en/products_and_supply/packaged_chemicals/product_range/ammonia.html (accessed January 10, 2018).

[35] DNV-GL. Marine forecast to 2050. *Energy Transit Outlook.* 2022. https://www.dnv.com/maritime/publications/maritime-forecast-2022/download-the-report.html (accessed December 5, 2022).

[36] Environmental Defence Fund. Ammonia at sea report. *Summary*. 2022:1–7. https://cdn.ricardo.com/ee/media/assets/ammonia-at-sea-report-summary.pdf (accessed July 15, 2023).

[37] Stocks M, Fazeli R, Hughes L, and Beck FJ. Global emissions implications from co-combusting ammonia in coal fired power stations: an analysis of the Japan-Australia supply chain. *J Clean Prod* 2022;336:130092. https://doi.org/10.1016/j.jclepro.2021.130092.

[38] CF Industries. Product specification sheet – anhydrous ammonia. 2013. https://www.cfindustries.com/globalassets/cf-industries/media/documents/product-specification-sheets/ammonia—na/anhydrous-ammonia-c-grade-pss.pdf (accessed February 23, 2023).

[39] IRENA. Innovation outlook renewable ammonia. 2022. https://www.irena.org/-/media/Files/IRENA/Agency/Publication/2022/May/IRENA_Innovation_Outlook_Ammonia_2022.pdf (accessed January 3, 2023).

[40] Aika K, and Kobayashi H (eds.) *CO2 Free Ammonia as an Energy Carrier.* Singapore: Springer Nature; 2023. https://doi.org/10.1007/978-981-19-4767-4.

[41] Iki N, Kurata O, Matsunuma T, *et al.* Micro gas turbine firing kerosene and ammonia. In: *ASME Turbo Expo 2015 Turbine Tech. Conf. Expo.*, vol. 8. New York: American Society of Mechanical Engineers; 2015. p. V008T23A023-V008T23A023. https://doi.org/10.1115/GT2015-43689.

[42] Ito S, Uchida M, Suda T, and Fujimori T. Development of ammonia gas turbine co-generation technology. *IHI Eng Rev* 2020;53:1–6.

[43] Okafor EC, Kurata O, Yamashita H, *et al.* Liquid ammonia spray combustion in two-stage micro gas turbine combustors at 0.25 MPa; relevance of combustion enhancement to flame stability and NOx control. *Appl Energy Combust Sci.* 2021;7:100038. https://doi.org/https://doi.org/10.1016/j.jaecs.2021.100038.

[44] INFORMATION-PRESS. Mitsubishi Power developing 100% ammonia-capable gas turbine. Mitsubishi Heavy Ind Engine Turbocharger, Ltd 2021; 2022.

[45] NEDO. Fuel ammonia supply chain establishment. 2022. https://green-innovation.nedo.go.jp/en/project/building-fuel-ammonia-supply-chain/ (accessed January 2, 2023).

[46] Global Trade Alert. Japan: NEDO allocates USD 517 million for the development of the fuel ammonia supply chain. 2022. https://www.globaltradealert.org/intervention/115232/financial-grant/japan-nedo-allocates-usd-517-million-for-the-development-of-the-fuel-ammonia-supply-chain (accessed April 20, 2023).

[47] Shiozawa B. A deep dive into SIP ‘energy carriers’ ammonia combustion research (second half). *Ammon Energy Assoc* 2020. https://www.ammoniaenergy.org/articles/a-deep-dive-into-sip-energy-carriers-ammonia-combustion-research-second-half/ (accessed October 1, 2021).

[48] Atchison J. JERA targets 50% ammonia-coal co-firing by 2030. *Ammon Energy Assoc* 2022. https://www.ammoniaenergy.org/articles/jera-targets-50-ammonia-coal-co-firing-by-2030/ (accessed January 2, 2023).

[49] CFAA. Clean fuel ammonia association 2022. https://greenammonia.org/en/ (accessed May 24, 2023).

[50] Crolius S. METI forms ammonia energy council. *Ammon Energy Assoc n.d.* https://www.ammoniaenergy.org/articles/meti-forms-ammonia-energy-council/ (accessed January 10, 2022).

[51] Batool M, and Wetzels W. Decarbonising options for the Dutch fertiliser industry. 2019:*39*. https://www.pbl.nl/sites/default/files/downloads/pbl-2019-decarbonisation-options-for-the-dutch-fertiliser-industry_3657.pdf (accessed February 10, 2023).

[52] Brown T. Ammonia plant revamp to decarbonize: Yara Sluiskil. *Ammon Ind* 2019. https://ammoniaindustry.com/ammonia-plant-revamp-to-decarbonize-yara-sluiskil/ (accessed March 1, 2023).

[53] Atchison J. The ammonia wrap: an ammonia-powered shipping network in northern Europe and more. 2021. https://www.ammoniaenergy.org/articles/the-ammonia-wrap-an-ammonia-powered-shipping-network-in-northern-europe-and-more/ (accessed March 12, 2023).

[54] Atchison J. OCI and NortH2 partner to develop green ammonia & methanol supply chains in the Netherlands. 2022. https://www.ammoniaenergy.org/articles/oci-and-north2-partner-to-develop-green-ammonia-methanol-supply-chains-in-the-netherlands/ (accessed January 20, 2023).

[55] ISPT. *Power to Ammonia*. Amersfoort: ISPT; 2017.

[56] Business Europe. European Hydrogen Hub by the Port of Rotterdam. 2020. http://climateyourbusiness.eu/project/european-hydrogen-hub-by-the-port-of-rotterdam/ (accessed April 20, 2023).

[57] IRENA. Global Hydrogen Trade to meet the 1.5 °C Climate Global: technology review of hydrogen carriers 2922. https://www.irena.org/publications/2022/Apr/Global-hydrogen-trade-Part-II (accessed March 1, 2023).

[58] OCI. OCI to expand Port of Rotterdam ammonia import-terminal. 2022. https://www.oci.nl/news/2022-oci-nv-to-expand-port-of-rotterdam-ammonia-import-terminal/ (accessed December 20, 2022).

[59] Atchison J. Vopak explores new ammonia infrastructure in Singapore and the Netherlands. 2022. https://www.ammoniaenergy.org/articles/vopak-explores-new-ammonia-infrastructure-in-singapore-the-netherlands/ (accessed January 10, 2023).

[60] Port of Rotterdam. Large scale Ammonia Cracker to enable 1 million tonnes of hydrogen imports. 2022. https://www.portofrotterdam.com/en/news-and-press-releases/large-scale-ammonia-cracker-to-enable-1-million-tonnes-of-hydrogen-imports (accessed December 15, 2022).

[61] Bunderministerium fur Wirtschaft und Klimaschutz. Report of the Federal Government on the implementation of the National Hydrogen Strategy. 2020. https://www.bmwk.de/Redaktion/EN/Publikationen/Energie/report-of-the-federal-government-on-the-implementation-of-the-national-hydrogen-strategy.pdf?__blob=publicationFile&v=2 (accessed October 1, 2022).

[62] YARA. Yara is speeding up the Hydrogen Economy in Germany. 2022. https://www.yara.com/news-and-media/news/archive/2023/yara-is-speeding-up-the-hydrogen-economy-in-germany/ (accessed December 1, 2022).

[63] Atchison J. H2Global launches First Green Ammonia Tender. 2022. https://www.ammoniaenergy.org/articles/h2global-launches-first-green-ammonia-tender/ (accessed February 10, 2023).

[64] BP. BP reveals plans to evaluate expansion of Germany green energy Port with new Hydrogen hub. 2022. https://www.bp.com/en/global/corporate/news-and-insights/press-releases/bp-reveals-plans-to-evaluate-expansion-of-germany-green-energy-port-with-new-hydrogen-hub.html (accessed March 23, 2023).

[65] UNIPER. Niedersachsen Ports together with Uniper and TES give the starting signal for the planning phase of a new jetty for Green Gases in Wilhelmshaven. 2022. https://www.uniper.energy/news/niedersachsen-ports-together-with-uniper-and-tes-give-the-starting-signal-for-the-planning-phase-of-a-new-jetty-for-green-gases-in-wilhelmshaven (accessed February 10, 2023).

[66] H2 Global. Shaping the Global Energy Transition. n.d. https://www.h2-global.de/ (accessed January 10, 2023).

[67] H2 Global. New German government gives the go-ahead for the H2Global initiative. 2021. https://www.h2-global.de/post/neue-bundesregierung-gibt-startschuss-fuer-h2global-initiative (accessed January 10, 2023).

[68] Gemserv. Tyseley Energy Park to host world-leading ammonia to hydrogen project with new Government backing. 2022. https://gemserv.com/our-thoughts/tyseley-energy-park-world-leading-ammonia-to-hydrogen-project/ (accessed January 9, 2023).
[69] Siemens Energy. Siemens Energy and partners developing game-changing ammonia prototype. 2022. https://www.siemens-energy.com/uk/en/energy/game-changing-green-ammonia-prototype.html (accessed January 9, 2023).
[70] Air Products. Large-scale green hydrogen for the UK: air products and associated British Ports partner on renewable hydrogen production. 2022. https://www.airproducts.com/news-center/2022/08/0830-air-products-and-associated-british-ports-partner-on-renewable-hydrogen-production (accessed January 10, 2023).
[71] Cardiff University Press. Harnessing the oceans for energy. 2021. https://www.cardiff.ac.uk/news/view/2536400-harnessing-the-oceans-for-energy (accessed January 11, 2023).
[72] MariNH3. Clean green ammonia engines for marine – MariNH3. 2022. https://marinh3.ac.uk/ (accessed February 10, 2023).
[73] Mahle. MAHLE Powertrain drives the development of ammonia combustion for heavy duty engines. 2022. https://www.mahle-powertrain.com/en/news-and-press/press-releases/mahle-powertrain-drives-the-development-of-ammonia-combustion-for-heavy-duty-engines-91713 (accessed January 12, 2023).
[74] University of Minnesota. Renewable energy ammonia production research continues at the WCRO. *West Cent Res Outreach Cent* 2022. https://wcroc.cfans.umn.edu/news/arpa-e-ammonia-project (accessed January 9, 2023).
[75] Ummary S. Advanced Research Projects Agency – Energy. Renewable energy to fuels through utilization of energy-dense liquids. *ARPA-E* 2016:1–16. https://arpa-e.energy.gov/technologies/programs/refuel (accessed December 29, 2020).
[76] Forward H. Moving hydrogen forward. 2021. https://www.hydrogenfwd.org/ (accessed October 1, 2022).
[77] Warner M. U.S. Department of Energy announces history $7 billion funding opportunity to jump-start America's clean hydrogen economy. 2022. https://www.warner.senate.gov/public/index.cfm/2022/9/u-s-department-of-energy-announces-history-7-billion-funding-opportunity-to-jump-start-america-s-clean-hydrogen-economy#:~:text=TheDOE National Clean Hydrogen,decarbonization and economic development goals. (accessed January 3, 2023).
[78] U.S. Department of Energy. Financial incentives for hydrogen and fuel cell projects. 2022. https://www.energy.gov/eere/fuelcells/financial-incentives-hydrogen-and-fuel-cell-projects (accessed January 10, 2023).
[79] General Electric. GE and IHI sign agreement to develop ammonia fuels roadmap across Asia. 2021. https://www.ge.com/news/press-releases/ge-and-ihi-sign-agreement-to-develop-ammonia-fuels-roadmap-across-asia (accessed October 11, 2022).

[80] Atchison J. Ammonia-powered Aframaxes. *Ammon Energy Assoc n.d.* https://www.ammoniaenergy.org/articles/ammonia-powered-aframaxes/ (accessed March 1, 2023).

[81] Atchison J. Geographic region: Singapore. *Ammon Energy Assoc* 2023. https://www.ammoniaenergy.org/regions/singapore/ (accessed January 2, 2023).

[82] Tenaga Nasional. TNB, Petronas & IHI successfully conduct Malaysia's first ammonia co-combustion experiment. 2022. https://www.tnb.com.my/announcements/tnb-petronas-ihi-successfully-conduct-malaysias-first-ammonia-co-combustion (accessed October 10, 2022).

[83] Government of India. National green hydrogen mission. 2023. https://mnre.gov.in/img/documents/uploads/file_f-1673581748609.pdf (accessed February 10, 2023).

[84] Atchison J. New benchmark for coal co-firing reported in China. *Ammon Energy Assoc* 2022. https://www.ammoniaenergy.org/articles/new-benchmark-for-coal-co-firing-reported-in-china/?mc_cid=9c6f6bb4c0&mc_eid=81cd7b883e (accessed May 11, 2023).

[85] Xu Y, Wang H, Liu X, Zhu J, Xu J, and Xu M. Mitigating CO_2 emission in pulverized coal-fired power plant via co-firing ammonia: a simulation study of flue gas streams and exergy efficiency. *Energy Convers Manag* 2022; 256:115328. https://doi.org/10.1016/j.enconman.2022.115328.

[86] CSIRO. Australian Renewable Energy Hub. 2023. https://research.csiro.au/hyresource/australian-renewable-energy-hub/ (accessed April 25, 2023).

[87] S&P Global. UAE's $1 bil green ammonia project to start in 2024; targets Europe, US markets. 2021. https://www.spglobal.com/platts/en/market-insights/latest-news/petrochemicals/071221-uaes-1-bil-green-ammonia-project-to-start-in-2024-targets-europe-us-markets (accessed October 12, 2022).

[88] The National News UAE. Ta'ziz to develop low-carbon ammonia plant in Ruwais. 2023. https://www.thenationalnews.com/business/energy/2023/01/18/taziz-to-develop-low-carbon-ammonia-plant-in-ruwais/ (accessed February 10, 2023).

[89] Air Products. Air products, ACWA Power and NEOM sign agreement. 2020. https://www.airproducts.com/news-center/2020/07/0707-air-products-agreement-for-green-ammonia-production-facility-for-export-to-hydrogen-market (accessed April 1, 2023).

[90] Energy Partnership Chile-Alemania. Chile presents national strategy to become a world leader in green hydrogen. 2020. https://www.energy-partnership.cl/newsroom/chile-hydrogen/ (accessed January 22, 2023).

[91] Ministerio de Energia. El más grande de Chile: Ministro Jobet anuncia nuevo proyecto de hidrógeno verde en Magallanes. 2021. https://www.energia.gob.cl/noticias/nacional/el-mas-grande-de-chile-ministro-jobet-anuncia-nuevo-proyecto-de-hidrogeno-verde-en-magallanes (accessed January 22, 2023).

Chapter 9
Outlook—unsolved challenges in combustion systems

Y. Li[1], X. Shi[1], Z. Liu[1], S. Chen[1] and S. Bin[1]

9.1 Introduction

With the targets of carbon neutrality and sustainable energy system development, ammonia (NH_3) as a zero-carbon energy storage agent generated from renewable electricity has received widespread attention. It holds great potential as a renewable synthetic fuel and hydrogen (H_2) carrier. As an effective way of large-scale ammonia utilization, combustion is shifting its role from the primary to the secondary energy supplier. Ammonia-fueled energy systems, heavily reliant on combustion devices, are gradually being developed. However, various challenges need to be resolved alongside these promising prospects. This chapter will discuss different aspects of the unsolved challenges related to ammonia combustion, including combustion enhancement and pollutant control, practical combustion devices, and economic viability, standards, regulations, and policies.

9.2 Technical challenges

9.2.1 Simultaneous combustion enhancement and emissions control

Current technologies for regulating ammonia combustion mainly include plasma-assisted combustion, ammonia pre-cracking, reactive fuel co-firing, and oxygen enrichment. However, most of these technologies face challenges in simultaneously controlling combustion reactivity and pollutant emissions.

9.2.1.1 Plasma-assisted combustion

Plasma-assisted combustion has received attention as a potential method to overcome the limitations of ammonia combustion. Plasma, the fourth state of matter, features rapid thermal effects, relatively high free electron energy, and long-lived

[1]School of Mechanical Engineering, Shanghai Jiao Tong University, China

excited particles. Plasma influences the flow field during discharge, altering gas transport properties, reducing ignition temperature, enhancing combustion efficiency, and extending combustion limits. Additionally, plasma-assisted combustion can achieve low-temperature combustion by reducing activation energy, which can significantly enhance ammonia combustion.

However, the complexity of plasma processes makes it challenging to separate plasma effects from thermal effects through conventional experiments. The dilution of different inert gases affects the electron energy distribution in plasma and consequently the reactivity of the mixture. Further experimental, kinetic, and numerical studies are needed to understand the physical and chemical models of plasma-assisted combustion.

9.2.1.2 Ammonia pre-cracking

Ammonia is an excellent hydrogen storage carrier due to its high hydrogen content. Ammonia can be pre-cracked to generate hydrogen and nitrogen, enhancing ammonia combustion due to lower ignition energy and higher combustion velocity of hydrogen. For partially pre-cracked ammonia, the relationship between the laminar flame propagation and pressure is more obvious than in pure ammonia.

However, the highly efficient and low-cost cracking of ammonia remains challenging. Preheating ammonia with a catalyst can facilitate cracking [1], but the cost and design of preheating parts will bring challenges. Besides, plasma catalysis is a promising method for low-temperature ammonia pre-cracking, but it requires further research before commercialization.

9.2.1.3 Reactive fuel co-firing

High-reactivity fuels, like gasoline, diesel, methane, dimethyl ether (DME), polyoxymethylene dimethyl ethers (PODE), and hydrogen, are widely used as co-firing fuels to enhance ammonia combustion. The presence of high-reactivity fuel promotes ammonia combustion, but it also enhances NOx emissions. The proportion of reactive fuel must be balanced and in the trade-off between optimizing combustion reactivity and controlling emissions. Fully premixed injection strategies are not ideal for pure ammonia combustion or ammonia/reactive fuel co-firing, as they can result in high NO_x emissions (exceeding 5000 ppm) under lean conditions and excess NH_3 emissions under rich conditions [2]. A staged combustion strategy with secondary air injection can ensure the complete burning of excess reactive fuel and ammonia, which is a viable solution to reduce emissions. Thus, precise control of the equivalence ratio in the primary combustion zone is essentially necessary due to the narrow window for low NO_x/NH_3 emissions.

9.2.1.4 Oxygen enrichment

Oxygen enrichment uses oxygen-rich air with an oxygen content exceeding 21% as the oxidant. Previous studies have shown that oxygen enrichment raises flame temperature and enhances laminar flame propagation [3]. Higher flame temperatures might significantly increase thermal NO_x, but oxygen enrichment also promotes ammonia pyrolysis, transforming excess ammonia into hydrogen and nitrogen under rich conditions. Consequently, oxygen-enriched ammonia combustion effectively extends the stable combustion and low NO_x/NH_3 emissions windows [4]. However, oxygen enrichment is suitable mainly for stationary combustion systems like land-based gas turbines and boilers, and further research is required for mobile combustion systems.

9.2.2 Modification in practical combustion devices

9.2.2.1 Boiler

The energy structure in China, characterized by "rich coal, poor oil, less gas," has led to a significant reliance on coal-fired power plants. CO_2 emissions from coal-fired power accounts for 34% of total carbon emissions in China [5]. Ammonia as a promising zero-carbon fuel is expected to replace the coal in boilers. However, due to current limitations in green ammonia production, cost, and combustion stability, ammonia cannot completely replace the coal in current coal-fired application. The application of ammonia in boilers is still in its infancy, focusing on pilot or semi-industrial research, with weak foundational knowledge in NH_3/coal co-firing, combustion technology, and engineering application processes.

Modifications in furnace and fuel injection systems

The fundamental combustion theory is critical for boiler design and optimization. Currently, there is limited understanding of the cofiring behaviors of ammonia/coal blends. In particular, there is a lack of in-depth knowledge about combustion characteristics such as pulverized coal ignition delay, ignition mode, volatile combustion, and coke combustion when co-firing with ammonia.

NH_3/coal co-firing can utilize existing power plant facilities without large-scale modifications to the main body of the boiler. However, there is scarce experimental data on NH_3/coal co-firing in large-scale power station boilers such as a tangentially fired boiler, opposed boilers, and W-shaped flame boilers. Therefore, it is necessary to conduct in-depth research to understand the co-firing characteristics of NH_3/coal in these boilers. As the proportion of ammonia co-firing in boilers gradually increases, it is essential to address its impact on boiler performance, slagging, particulate matter (PM), and hydrogen sulfide emissions, to overcome challenges including reduced combustibility, decreased flame emissivity, reduced furnace heat load, increased NOx emissions, and reduced efficiency.

Emission control technologies

NOx emissions are a major concern for ammonia/coal co-firing. The ammonia co-firing ratios and injection locations are key factors in NO_x emission control. Detailed studies on the competitive model between suppression of NO_x emissions formation by ammonia and the partial conversion of unburned NH_3 to NO_x are necessary. Effective NOx control can be achieved through combustion strategy management, such as air-staged combustion. This includes measures such as appropriately increasing the rate of primary air and controlling the main combustion zone. However, these measures may cause coke formation on burners and potential high-temperature corrosion.

Additionally, existing research has hardly explored pollutants like hydrocyanic acid (HCN), N_2O, and mineral impurities. In NH_3/coal co-firing, understanding the migration and conversion of elements such as N, S, and Cl in coal, as well as the emission of highly toxic gaseous pollutants like HCN and nitrogen-containing polycyclic aromatic hydrocarbons (PAHs) are yet to be explored. Further investigation is warranted into the mineral impurities in coal and the effects of ammonia co-firing on mineral transformation and deposition behaviors, particulate emissions, and soot formation.

Modification in pollutant after-treatment equipment

In terms of pollutant after-treatment equipment, there is still a lack of research on adjustments of pollutant control equipment at the boiler's tail, such as selective catalytic reduction (SCR), electrostatic precipitators (ESP), flue-gas desulfurization (FGD) units. The choice of low-temperature ammonia-SCR catalysts for boilers needs clarification, as catalyst type and structure affect deNO_x activity and resistance to H_2O. Further research on ammonia-SCR catalyst H_2O poisoning models needs to be carried out. Additionally, catalysts must have good high-temperature stability and broad adaptability to various coal compositions to maintain their effectiveness of emission reduction.

Engineering application processes

A thorough evaluation is essential for the compatibility and optimization of numerous systems when retrofitting coal-fired power units for ammonia integration. Potential issues may arise when the ammonia co-firing rate exceeds 20%. For example, decreasing flame temperature can reduce the heat transfer to the water-steam pipeline, changing the steam output of the boiler and potentially unburned carbon emissions. Oxidizer levels may need to be controlled to ensure complete combustion. Corrosion issues between unburned NH_3 emission and metal parts also need to be addressed in system design.

9.2.2.2 Internal combustion engine

Gasoline and diesel have been widely used as fuels for internal combustion engines (ICEs), categorized as spark ignition (SI) and compression ignition (CI) engines, respectively. Using ammonia in ICEs has garnered attention for decarbonization in global transportation. Ammonia has the advantages of high hydrogen content, high octane number, and safe storage. However, its lower minimum ignition energy and laminar burning velocity (LBV) limits its direct utilization in ICEs.

SI engines

In SI engines, the energy density of ammonia is more than 2.5 times lower than that of gasoline, and its auto-ignition temperature is about 350 °C higher, leading to ignition difficulties and reduced engine power [6]. Pure ammonia engines cannot achieve satisfactory combustion without modifications, resulting in a 20% reduction in power output compared to gasoline engines [7]. To improve combustion stability, ammonia is often blended with better combustion performance. Hydrogen is a preferable co-firing fuel for NH_3-fueled ICEs, as it can enhance cycle stability and prevent misfires. However, excessive hydrogen can lead to thermal loss and reduced thermal efficiency. Therefore, optimizing the hydrogen blend is crucial for balancing combustion efficiency and stability.

CI engines
Ammonia also has great potential in CI engines. NH_3/diesel-fueled engines show enhanced heat release rates and combustion efficiency, reducing the risk of knocking and pre-ignition. However, NH_3/diesel-fueled engines may experience decreased cylinder compression and peak pressure. The relatively low LBV of ammonia can lead to poor combustion performance, and ammonia injection can strongly influence diesel injection, resulting in larger IDTs. High-pressure direct injection (HPDI) strategies for liquid ammonia have shown promise in improving thermal efficiency and reducing greenhouse gas emissions. However, optimizing injection timing and direction is necessary to balance fuel oxidation and pollutant production.

Pollutant control and after-treatment systems
Controlling emissions of unburnt NH_3 and NOx emissions under different operating conditions, especially during the cold start phase at low ambient temperatures, poses challenges for after-treatment systems. Common after-treatment methods like SCR or ternary catalysts require further study to optimize catalytic conversion efficiency and the selection of catalysts. Developing new SCR systems with high catalytic conversion efficiency, low cost, and long usage life for ammonia engines is essential. Ensuring the effective oxidation of slip NH_3 to meet regulatory requirements is a significant challenge for after-treatment system design.

9.2.2.3 Gas turbine

Research on ammonia-fueled gas turbines was initiated in the 1960s but was halted due to high costs and technological limitations [8]. Stationary gas turbines usually burn gaseous fuels and are not limited by combustion chamber volume, making them more compatible with ammonia. However, challenges related to fuel supply, combustion stability, combustion efficiency, pollutant control, and material compatibility need to be overcome for large-scale applications.

Modification in fuel supply systems
NH_3/H_2 co-firing has attracted much attention, yet there are no mature products currently available ensuring the long-term reliability of ammonia and hydrogen supply systems. Compared to fossil fuels, ammonia and hydrogen have different storage and transportation requirements. Both fuels need to be handled and transported carefully to ensure safety and prevent leaks or accidents. The corrosive effect of ammonia on certain metals (e.g. copper, zinc, aluminum, and its alloys) and rubber materials necessitate specialized, corrosion-resistant tanks and pipelines [9]. Hydrogen supply systems face challenges like embrittlement and lubrication issues, requiring storage under high-pressure and low-temperature conditions.

Given the complexities posed by ammonia and hydrogen, especially at high temperatures where they can corrode and damage gas turbine components, the materials used in gas turbines should withstand these conditions and maintain their structural integrity over time. This requires the optimization of advanced materials, robust coatings, and surface treatments.

Low-emission combustion technologies
Innovative combustion methods, such as multi-stage rich-lean combustion, rich-quench-lean (RQL) combustion, and moderate or intense low oxygen dilution (MILD) combustion, are needed to address ammonia combustion difficulties. Combining steam injection with exhaust gas recirculation (EGR) can improve the efficiency of gas turbines, and combined cooling, heat, and power (CCHP) configurations are promising for adaptation in different gas turbine plants.

Overall, ammonia-fueled gas turbines are still a relatively new technology, with limited data on performance and efficiency. Recent developments in heavy-duty gas turbines have considered safely using 100% ammonia [10]. Future research on ammonia combustion technology centered around coal-fired power plants and gas turbines is expected to improve the industrial application of ammonia combustion.

By addressing these challenges, ammonia can become a viable and sustainable alternative to traditional hydrocarbon fuels, contributing to global efforts to reduce CO_2 emissions.

9.3 Economic and logistical challenges

9.3.1 Economic viability

Ammonia has long been used as one of the important industrial chemicals, particularly in agriculture as a key component in fertilizers. Driven by the urgent global target of CO_2 reduction, ammonia is emerging as a renewable, zero-carbon fuel for industries and transportation. It can replace traditional fossil fuels, significantly reducing CO_2 emissions. However, this new role of ammonia as a fuel source presents several challenges, including economic considerations, shifts in market demand, technological advancements, and the need for new standards and policies. Economic viability, in particular, requires ongoing attention and research.

The economic viability of fuels, whether traditional fossil fuels, or renewable energies such as hydrogen and ammonia, has always been closely linked to policy incentives, market demand, and technological progress. For example, the 2050 net-zero emissions policy in Europe and the "3060" carbon peak and neutrality policy in China have spurred the development of renewable energies through economic measures (e.g. loan interest rates and entrepreneurship support). These policies have initiated upgrades across the renewable energy industry chain. In China, the hydrogen energy chain is currently intensively developing, including diverse hydrogen production methods (gray, blue, and green hydrogen technologies), various hydrogen storage methods (gaseous, liquid, solid), and hydrogen transport (pipelines, tankers, etc.), along with application in metallurgy, chemicals, transportation, energy storage. However, there is no clear policy guidance for ammonia in most countries, leading to slow development of the ammonia industry chain. As carbon emission restriction policies take effect, market demand is gradually shifting towards hydrogen, including its direct use in fuel cells and combustion facilities. A portion of the market demand is gradually shifting to ammonia, a suitable hydrogen carrier for long-distance transport. Countries such as Japan and the UK are paying more attention to developing the ammonia energy chain, actively creating new policies, and using economic means to encourage its development, further promoting the upgrade of the existing ammonia industry.

The technological development and innovation of ammonia can create new economic growth points by improving the ammonia production efficiency, optimizing storage and transport, and expanding its application fields, forming a vast industry chain. This chapter primarily discusses the economic challenges and prospects of ammonia triggered by technological development. As a sustainable energy carrier, ammonia has significant advantages but also faces many economic challenges covering the entire industry chain, including infrastructure construction, logistical complexity, and safety regulation.

In the upstream industry chain, ammonia production costs vary greatly depending on the source of raw materials and preparation methods. Grey ammonia, synthesized through fossil fuels and traditional thermochemical methods, requires the least changes and infrastructure construction, thus having the lowest cost. Blue ammonia, synthesized by adding carbon capture and storage technology to grey ammonia, requires increased technology costs to improve the maturity of carbon capture technology. Green ammonia, due to high electricity prices and low maturity of renewable energy conversion technology, requires extensive and long-term technology cost input, infrastructure renovation, and construction costs. It is also affected by temporal and spatial instability, resulting in lower production volumes. Therefore, grey ammonia currently has the lowest production cost, followed by blue ammonia, with green ammonia featuring the highest. However, as carbon taxes increase and renewable energy technology develops, green ammonia production may become economically viable within 30 years or less.

In the midstream industry chain, storage method and duration determine storage costs. Since most ammonia is synthesized through hydrogen, its production cost is somewhat higher than that of hydrogen. However, if storage time and quantity are considered, the benefits of liquid ammonia far exceed those of liquid and gaseous hydrogen. Long-distance pipeline transport of liquid ammonia is more cost-effective compared to gaseous hydrogen, which requires increased safety regulation and inspection to avoid combustion and explosion risks. Transportation of liquid hydrogen involves high costs for insulation and pressurization equipment. By contrast, using liquid ammonia for long-distance transport greatly reduces transportation costs. However, ammonia which has safety hazards such as corrosiveness, toxicity, and flammability, requires consideration of its impact on transportation and storage equipment, personnel, and the environment, further increasing the costs. Technological developments are needed to reduce these safety costs.

In the downstream industry chain, ammonia has many application fields, including agriculture, industry, and transportation. This section specifically analyzes the economic challenges of ammonia application in combustion. Due to its lower reactivity, ammonia is generally mixed with higher-reactivity fuels for combustion, with hydrogen being the best choice. In NH_3/H_2 co-firing technology, adding an ammonia catalytic cracker to the system increases infrastructure costs, safety regulation costs, and operating costs of consumables such as catalysts. If ammonia is directly burned using emerging technologies such as oxygen enrichment, infrastructure costs of adding air separation devices also need to be considered. Additionally, since ammonia combustion produces NOx and NH_3 emissions, post-treatment costs need to be considered, further increasing infrastructure and operating costs. Overall, current technology requires increased infrastructure and safety regulation costs for ammonia combustion applications.

The ammonia industry chain needs synchronous development of the upstream, midstream, and downstream industries to promote further economic growth. As research deepens and new technologies effectively combine with renewable energy, ammonia production efficiency is expected to improve, reducing production costs and carbon emissions Concurrently, addressing safety concerns, minimizing storage and transportation costs, and enabling ultra-long-distance transportation and stable storage remain critical issues. Addressing the challenges of low combustion efficiency and high NO_x and NH_3 emissions is also paramount. The development of innovative catalysts and combustion enhancement processes can lead to improved combustion efficiency and reduced harmful emissions, significantly lowering the construction and maintenance costs of related facilities.

9.3.2 Standards, regulations, and policies

Ammonia is increasingly recognized as a promising player in the sustainable fuel market due to its zero-carbon potential. However, the application of ammonia as a global sustainable fuel involves a complex, multidimensional journey that requires coordinated efforts to address substantial issues of safety, public education, environmental impacts, and international cooperation.

One major obstacle is societal misconceptions and public apprehensions about ammonia. Despite its absence of carcinogenic substances found in traditional fuels like gasoline, its distinctive smell at trace concentrations frequently raises concerns about its toxicity. However, studies show that the risks of accidental ammonia leakage are comparable to those of liquefied petroleum gas (LPG). The pronounced smell of ammonia serves as an effective warning sign, enabling early mitigation measures to prevent catastrophic failures. This suggests that ammonia could be utilized as widely as LPG.

Public sentiment significantly influences the energy sectors, including hydrogen and ammonia. Innovative companies, such as the Asian Renewable Energy Hub, are driving the global interest in ammonia, aligning with sustainability goals. Public support is crucial for the recognition of ammonia as a future energy vector. This can be achieved through familiarization and comprehensive education, highlighting the necessity for ongoing exploration and advancement in the ammonia sector for carbon neutrality.

A comprehensive strategy is essential to overcome these challenges. This strategy should integrate strict safety measures, educational outreach initiatives, environmental management, and vigorous international cooperation. Promoting societal understanding of ammonia as an environmentally friendly energy sector through extensive educational programs and clear communication will galvanize public support for its adoption in key application areas.

However, the widespread adoption of ammonia faces challenges from regulatory standards and necessary infrastructure adjustments. Similar to gaseous counterparts like compressed natural gas (CNG), the storage and transportation logistics of ammonia require critical adjustments to the infrastructure to accommodate its pressurized characteristics. This requires a concerted effort to set new standards and retrofit existing facilities with enhanced safety features to address its relatively low required pressures and strong corrosion potential.

On the other hand, the tendency of ammonia combustion to produce harmful PM in urban environments, and its lower reactivity and flammability, underscore the need for policy measures that advocate for controlled use and emission reductions. A comprehensive legislative framework is required to develop safety and environmental standards for ammonia throughout its life cycle, from production and storage to transport and use. These efforts aim to protect public health and ecological integrity, considering the variable effects of ammonia associated with pH and temperature. For example, the strict regulatory environment in the EU provides valuable insights into managing ammonia-related risks, including flammability, explosion hazards, and the impact of toxic exposure on human health and the environment.

9.4 Work directions and future prospects

As a zero-carbon fuel, ammonia's potential in power fuel, clean electricity, and hydrogen storage has been emphasized. However, challenges such as low combustion activity, narrow flammability range, high auto-ignition temperature, slow LBV, and high NO_x emissions exist in ammonia combustion, which accelerates the emergence and optimization of ammonia combustion regulation technologies, such as plasma-assisted combustion, ammonia pre-cracking, reactive fuel co-firing, and oxygen enrichment, aiming to simultaneously enhance combustion intensity and control pollutant emission.

Before applying ammonia in a wide range of actual industrial fields, it is necessary to fully understand its combustion and emission characteristics under various operating conditions and more detailed, practical kinetic models. In the future, key parameters in experiments, such as fuel ratio, combustion speed, flame structure, and pollutant emission will be explored in large-scale applications. By integrating basic research with actual industrial applications, a more complete ammonia combustion system will be constructed. Moreover, the analysis of the economic viability of ammonia energy caused by technological developments is a key issue that requires long-term attention and research. This includes infrastructure construction, logistical complexity, and safety regulation in the upstream, midstream, and downstream industry chains. In the future, countries need to establish clear policy guidelines for synthetic ammonia, strengthen public education, and develop international cooperation, which further promotes the development and perfection of the ammonia energy industry chain, pushing ammonia to play a significant role in global energy development.

9.5 Conclusion

Ammonia as a zero-carbon fuel has been extensively explored. Key challenges include optimizing reactive fuel co-firing to enhance combustion, managing NOx emissions under high pressure, and developing adaptive reaction kinetics models. Research is also needed on co-firing ammonia with other fuels in boilers, engines, and gas turbines, focusing on catalysts for pollutant after-treatment. Combustion regulation technologies like oxygen enrichment offer solutions for improving combustion and controlling emissions. Future efforts will focus on clear policy guidelines for synthetic ammonia, suitable kinetics models for industrial use, key experimental parameters, and economic feasibility analyses. These steps aim to develop the ammonia industry and establish its role in global energy systems.

References

[1] F. Ma, L. Guo, Z. Li, *et al.* A review of current advances in ammonia combustion from the fundamentals to applications in internal combustion engines. *Energies* 16 (2023) 6304.

[2] L. Kang, W. Pan, J. Zhang, W. Wang, and C. Tang. A review on ammonia blends combustion for industrial applications. *Fuel* 332 (2023) 126150.

[3] B. Mei, X. Zhang, S. Ma, *et al.* Experimental and kinetic modeling investigation on the laminar flame propagation of ammonia under oxygen enrichment and elevated pressure conditions. *Combustion and Flame* 210 (2019) 236–246.

[4] Z. Liu, Y. Zhang, W. Li, X. Shi, and Y. Li. Self-promoted fuel pyrolysis under oxygen enrichment enables clean and efficient ammonia combustion. *The Innovation Energy* 1 (2024) 100006.

[5] R. Yong, C. Yang, M. Xue, F. Nie, and X. Zhao. Application status and prospect of ammonia energy. *Strategic Study of CAE* 25 (2023) 111–121.

[6] D. Erdemir, and I. Dincer. A perspective on the use of ammonia as a clean fuel: challenges and solutions. *International Journal of Energy Research* 45 (2020) 4827–4834.

[7] J.S. Cardoso, V. Silva, R.C. Rocha, M.J. Hall, M. Costa, and D. Eusébio. Ammonia as an energy vector: current and future prospects for low-carbon fuel applications in internal combustion engines. *Journal of Cleaner Production* 296 (2021) 126562.

[8] P. Dimitriou, and R. Javaid. A review of ammonia as a compression ignition engine fuel. *International Journal of Hydrogen Energy* 45 (2020) 7098–7118.

[9] J. Ordoñez, E.J. Gago, and A. Girard. Processes and technologies for the recycling and recovery of spent lithium-ion batteries. *Renewable and Sustainable Energy Reviews* 60 (2016) 195–205.

[10] D. Pashchenko. Ammonia fired gas turbines: recent advances and future perspectives. *Energy* 290 (2024) 130275.

Index

www.ingramcontent.com/pod-product-compliance
Lightning Source LLC
LaVergne TN
LVHW011836200926
848484LV00002B/362